Studies in Computational Intelligence

Volume 975

The series "Studies in Computational Intelligence" (SCI) publishes new developments and advances in the various areas of computational intelligence—quickly and with a high quality. The intent is to cover the theory, applications, and design methods of computational intelligence, as embedded in the fields of engineering, computer science, physics and life sciences, as well as the methodologies behind them. The series contains monographs, lecture notes and edited volumes in computational intelligence spanning the areas of neural networks, connectionist systems, genetic algorithms, evolutionary computation, artificial intelligence, cellular automata, self-organizing systems, soft computing, fuzzy systems, and hybrid intelligent systems. Of particular value to both the contributors and the readership are the short publication timeframe and the world-wide distribution, which enable both wide and rapid dissemination of research output.

Indexed by SCOPUS, DBLP, WTI Frankfurt eG, zbMATH, SCImago.

All books published in the series are submitted for consideration in Web of Science.

More information about this series at http://www.springer.com/series/7092

Yaochu Jin · Handing Wang · Chaoli Sun

Data-Driven Evolutionary Optimization

Integrating Evolutionary Computation, Machine Learning and Data Science

Yaochu Jin
Department of Computer Science
University of Surrey
Guildford, UK

Handing Wang
School of Artificial Intelligence
Xidian University
Xi'an, China

Chaoli Sun
School of Computer Science
and Technology
Taiyuan University of Science
and Technology
Taiyuan, China

ISSN 1860-949X ISSN 1860-9503 (electronic)
Studies in Computational Intelligence
ISBN 978-3-030-74642-1 ISBN 978-3-030-74640-7 (eBook)
https://doi.org/10.1007/978-3-030-74640-7

This Springer imprint is published by the registered company Springer Nature Switzerland AG
The registered company address is: Gewerbestrasse 11, 6330 Cham, Switzerland

Foreword

One of the appeals of optimization is its applicability across countless fields of interest; another is the opportunity it affords to interface with, and benefit from, allied disciplines.

Optimization methodologies have been transformed dramatically since I was drawn to their allure 50 years ago. Notable advances, that I have had a particular interest in, include those that have been made in global optimization, multi-objective optimization, robust optimization, and black box and data-driven optimization.

These developments have greatly enhanced optimization's capability to satisfactorily address real-world problems, thereby strengthening users' confidence. Through global optimization, we are released from the hazards of focusing on specific, local areas of the search space. Multi-objective optimization satisfies the dilemma of how to balance competing multiple objectives; it offers increased transparency by providing a family of trade-off solutions to enable domain experts to select the desired compromise solution. In real-world applications, it is almost inevitable that uncertainties will arise from either problem formulation or the implementation environment or both. For example, these might occur through deficiencies in the fidelity of a model, the precision of an objective or the ability to perfectly realize the proposed solution in practice. Robust optimization methods enable practitioners to manage these uncertainties. Evolutionary computing has been a catalyst that has promoted progress in these developments.

But what if, in a design optimization, there is no analytic representation for a particular objective or objectives? Instead, an objective (or objectives) might rely on the outcome of computer simulations, such as dynamic systems modelling, finite element analysis or the modelling of computational fluid dynamics. Or, the objectives may even depend on results obtained from experiments. Further, the collected data may be incomplete and noisy or vary in content. It will not come as a surprise, then, to learn that evolutionary computing has an important role in dealing with such data-driven problems as well.

Data-driven optimization forms the central focus of this textbook. Allying the population-based metaheuristics of evolutionary computing with the use of surrogate models and machine learning algorithms, the authors demonstrate a formidable array of approaches to address a range of data-driven optimization challenges.

As I started out on my adventures with optimization research (and these resulted in a lifetime fascination), a textbook that I regularly turned to, and still do from time to time, was Edgar and Himmelblau's *Optimization of Chemical Processes*, not because I was a chemical engineer (I was not) but because of the blend of its clarity of exposition, its practical approach and the authors' evident experience of applying their knowledge to real-world applications. *Data-Driven Evolutionary Optimization* has the same appeal to me; it is written by extremely experienced researchers who have a clear view of the importance of tailoring methodologies to address practical problems and here they provide a sound grounding in the assorted required strategies and their underpinning methodologies.

Sheffield, UK
March 2021

Peter Fleming

Preface

I started working on fitness approximation in evolutionary optimization when I moved back to Germany from the USA in 1999 to take up a research scientist position at the Honda Research Institute Europe in Offenbach to complete my second Ph.D. degree. The motivation then was to use evolutionary algorithms, in particular evolution strategies to optimize the aerodynamic performance of a turbine engine by designing the geometry of the stator or rotor blades. To accomplish this task, time-consuming computational fluid dynamics simulations must be performed, which prevents one from using tens of thousands of fitness evaluations, as typically done in evolutionary optimization. To reduce the time consumption for evolutionary aerodynamic optimization, fitness approximation using machine learning models, also called meta-models or surrogates, came into play. I was very much attracted by this research topic, since it provides a nice platform to integrate evolutionary computation with neural networks, two topics I was interested in. Out of this reason, I made much effort to promote this new area in the evolutionary computation community and continued working on this topic whenever there was an opportunity after I moved to the University of Surrey in 2010.

Over the past twenty years, fitness approximation in evolutionary optimization, also known as surrogate-assisted evolutionary optimization, has grown and developed into a fascinating research field, which is now termed data-driven evolutionary optimization. Data-driven evolutionary optimization focuses on a class of real-world optimization problems in which no analytical mathematical functions can be established for the objectives or constraints. Generally speaking, data-driven optimization may include the following situations: first, simulation-based optimization, where the quality of a solution is assessed by an iterative, computationally intensive process, ranging from numerically solving a large set of differential equations, to training a deep neural network on a large data set, and second, physical experiment-based optimization, where the objective or constraint values of a candidate solution can be evaluated only by performing physical or human experiments. This is usually because a high-quality computer simulation of the whole system is not possible, either because it is computationally prohibitive, e.g. numerical simulations of the aerodynamics of the whole aircraft, or because it is intractable because the process is not yet fully understood, e.g. the working mechanism of the human decision-making process. Finally, purely data-driven optimization, where only data collected in real-life is available for optimization and neither user-designed computer simulations nor physical experiments are allowed. For example, optimization of a complex industrial process or social system. In all the above cases, the amount of the collected data may be either small or big, and the data may be heterogeneous, noisy, erroneous, incomplete, ill-distributed or incremental.

Clearly, data-driven evolutionary optimization involves three different but complementary scientific disciplines, namely evolutionary computation, machine learning and deep learning, and data science. To effectively and efficiently solve a data-driven optimization problem, data must be properly pre-processed. Meanwhile, machine learning techniques become indispensable for handling big data, data paucity and various degrees of uncertainty in the data. Finally, solving the optimization problem becomes extremely demanding when the optimization problem is high-dimensional or large-scale, multi-objective and time-varying.

This book aims to provide researchers, including postgraduate research students, and industrial practitioners a comprehensive description of the state-of-the-art methods developed for data-driven evolutionary optimization. The book is divided into 12 chapters. For the self-containedness of the book, a brief introduction to carefully selected important topics and methods in optimization, evolutionary computation and machine learning is provided in Chaps. 1–4. Chapter 5 provides the fundamentals of data-driven optimization, including heuristics and acquisition function-based surrogate management, followed by Chaps. 6–8, presenting ideas that use multiple surrogates for single-objective optimization. Representative evolutionary algorithms for solving multi- and many-objective optimization algorithms and surrogate-assisted data-driven evolutionary multi- and many-objective optimization are described in Chaps. 7 and 8, respectively. Approaches to high-dimensional data-driven optimization are elaborated in Chap. 9. A plethora of techniques for transferring knowledge from unlabelled to labelled data, from cheap objectives to expensive ones, and from cheap problems to expensive ones are presented in Chap. 10, with the help of semi-supervised learning, transfer learning and transfer optimization. Since data-driven optimization is a strongly application-driven research area, offline data-driven evolutionary optimization is treated in Chap. 11, exemplified with real-world optimization problems such as airfoil design optimization, crude oil distillation optimization and trauma system optimization. Finally, deep neural architecture search as a data-driven expensive optimization problem is highlighted in Chap. 12. Of the 12 chapters, § 3.5– 3.6, § 4.2, § 5.2, § 6.4– 6.5, § 7.2– 7.3, § 9.6– 9.7, § 11.1, § 11.3 and Chap. 12 are written by Handing Wang, and § 3.7– 3.8, § 5.4.1, § 5.5, § 6.2– 6.3, § 9.2– 9.3 and Chap. 10 by Chaoli Sun. Handing worked as Postdoctoral Associate during 2015–18 and Chaoli at first as Academic Visitor during 2012–13 and then as Postdoctoral Associate during 2015–17 in my group at Surrey.

To make it easier for the reader to understand and use the algorithms introduced in the book, the source code for most data-driven evolutionary algorithms presented in Chaps. 5–12 is made available (http://www.soft-computing.de/DDEO/DDEO.html) and all baseline multi-objective evolutionary algorithms introduced in this book are implemented in PlatEMO, an open-source software tool for evolutionary multi-objective optimization (https://github.com/BIMK/PlatEMO).

This book would not have been possible without the support of many previous colleagues, collaborators and Ph.D. students of mine. First of all, I would like to thank Prof. Dr. Bernhard Sendhoff and Prof. Dr. Markus Olhofer, with both of whom I closely worked at the Honda Research Institute Europe during 1999–2010. After I joined Surrey in 2010, Markus and I still maintained close collaboration on a number of research projects on evolutionary optimization. I would also thank Prof. Kaisa Miettinen from the University of Jyväskylä, Finland, with whom I worked closely as *Finland Distinguished Professor* during 2015–17 on evolutionary multi-objective optimization. Thanks go to Prof. Tianyou Chai and Prof. Jinliang Ding from Northeastern University, China, with whom I also collaborate with on evolutionary optimization as *Changjiang Distinguished Professor*. The following collaborators and previous or current Ph.D. students of mine have contributed to part of the work presented in this book: Prof. Yew-Soon Ong, Prof. Jürgen Branke, Prof. Qingfu Zhang, Prof. Xingyi Zhang, Prof. Aimin Zhou, Prof. Ran Cheng, Prof. Xiaoyan Sun, Dr. Ingo Paenke, Dr. Tinkle Chugh, Mr. John Doherty, Dr. Dan Guo, Dr. Cuie Yang, Dr. Ye Tian, Dr. Cheng He, Dr. Dudy Lim, Dr. Mingh Nhgia Le, Dr. Jie Tian, Dr. Haibo Yu, Dr. Guo Yu, Dr. Michael Hüsken, Ms. Huiting Li, Ms. Xilu Wang, Ms. Shufen Qin, Mr. Hao Wang, Ms. Guoxia Fu, Mr. Peng Liao, Mr. Sebastian Schmitt, Ms. Kailai Gao, Dr. Jussi Hakanen, Dr. Tatsuya Okabe, Dr. Yanan Sun, Dr. Jan O. Jansen, Mr. Martin Heiderich, Dr. Yuanjun Huang and Dr. Tobias Rodemann. I would also like to take this opportunity to thank Prof. Xin Yao, Prof. Gary Yen, Prof. Kay Chen Tan, Prof. Mengjie Zhang, Prof. Richard Everson, Prof. Jonathon Fieldsend, Prof. Dr. Stefan Kurz, Prof. Edgar Körner and Mr. Andreas Richter for their kind support over the past two decades. Finally, financial support from EPSRC (UK), TEKES (Finland), National Natural Science Foundation of China, Honda Research Institute Europe, Honda R&D Europe and Bosch Germany is gratefully acknowledged.

Guildford, UK
February 2021

Yaochu Jin

Contents

Acronyms

ACO	Ant colony optimization
AdaBoost	Adaptive boosting
AIC	Akaike information criterion
APD	Angle penalized distance
AUE2	Accuracy updated ensemble algorithm
Bagging	Bootstrap aggregating
BEO	Bayesian evolutionary optimization
BO	Bayesian optimization
BOA	Bayesian optimization algorithm
CAL-SAPSO	Committee-based surrogate-assisted particle swarm optimization
CART	Classification and regression tree
CFD	Computational fluid dynamic
CGA	Compact genetic algorithm
CMA-ES	Covariance matrix adaptation evolutionary strategies
C-MOGA	Cellular multi-objective genetic algorithm
CNN	Convolutional neural network
COBRA	Constrained optimization by radial basis function interpolation
CPF	Coverage over the Pareto front
CSEA	Classification-based surrogate-assisted evolutionary algorithm
CSO	Competitive swarm optimizer
CSSL	Co-training semi-supervised learning
DAG	Directed acyclic graph
DE	Differential evolution
DSE	Data stream ensemble
EA	Evolutionary algorithm
EBO	Evolutionary Bayesian optimization
ECT	Electricity consumption for a ton of magnesia
EDA	Estimation of distribution algorithm
EGO	Efficient global optimization
EI	Expected improvement
ES	Evolution strategies
ESD	Estimated standard deviation

FCM	Fuzzy clustering method
FDA	Factorized distribution algorithm
FLOPs	Floating point operations
GA	Genetic algorithm
GAN	Generative adversarial network
GD	Generational distance
G-MFEA	Generalized multifactorial evolutionary algorithm
GP	Genetic programming
GP-MOEA	Gaussian process-assisted multi-objective evolutionary algorithm
GS-MOMA	Generalized surrogate-assisted multi-objective memetic algorithm
GS-SOMA	Generalized surrogate-assisted single-objective memetic algorithm
HeE-MOEA	Heterogeneous ensemble assisted multi-objective evolutionary algorithm
HSA-MSES	Hierarchical surrogate-assisted multi-scenario evolution strategy
HV	Hypervolume
IBEA	Indicator-based evolutionary algorithm
IGA	Interactive genetic algorithm
IGD	Inverted generational distance
KnEA	Knee point-driven evolutionary algorithm
KNN	K-nearest neighbours
LCB	Lower confidence bound
LHS	Latin hypercube sampling
LSTM	Long short-term memory
MaOP	Many-objective optimization problem
MBN	Multidimensional Bayesian network
MFEA	Multi-factorial evolutionary algorithm
MIC	Multi-objective infill criterion
MLP	Multi-layer perceptron
MOEA	Multi-objective evolutionary algorithm
MOEA/D	Multi-objective evolutionary algorithm based on decomposition
MOP	Multi-objective optimization problem
MSE	Mean squared error
MTO	Multitasking optimization
NAS	Neural architecture search
PBI	Penalty boundary intersection
PBIL	Population-based incremental learning
PCA	Principal component analysis
PD	Pure diversity
PoI	Probability of improvement
PR	Polynomial regression
PSO	Particle swarm optimization
QBC	Query by committee
RBF	Radial basis function
RBFN	Radial basis function network
RF	Random forest

RL	Reinforcement learning
RM-MEDA	Regularity model-based multiobjective estimation of distribution algorithm
RMSE	Root mean square error
ROOT	Robust optimization over time
SA-COSO	Surrogate-assisted cooperative swarm optimization algorithm
SBX	Simulated binary crossover
SGD	Stochastic gradient descent
SL-PSO	Social learning particle swarm optimization
SOM	Self-organizing map
SOP	Single-objective optimization problem
SP	Spacing
SQP	Sequential quadratic programming
SVDD	Support vector domain description
SVM	Support vector machine
TLSAPSO	Two-layer surrogate-assisted particle swarm optimization
TSP	Travelling salesman problem
UCB	Upper confidence bound
UMDA	Univariate marginal distribution algorithm

Symbols

$\mathcal{R}$	Real number space
$\mathcal{N}$	Integer space
Ω	Feasible region
$\mathcal{D}$	Data set
x	Decision variable or input variable (feature)
$\mathbf{x}$	Decision vector or input vector (solution)
y	Label or output variable
$\mathbf{y}$	Label or output vector
n	Number of decision variables or dimension of input
d	Number of data
m	Number of objective functions
$f(\mathbf{x})$	Objective function
$g(\mathbf{x})$	Inequality constraint function
$h(\mathbf{x})$	Equality constraint function
$\hat{f}(x)$	Approximated function
$F(\mathbf{x})$	Function vector
φ	Fidelity of approximate objective function
$\mathbf{s}$	Scenario
$\mathbf{z}^{\text{ideal}}$ or $\mathbf{z}^{\text{å}}$	Ideal point of multi-objective optimization problem
$\mathbf{z}^{\text{nadir}}$	Nadir point of multi-objective optimization problem
$\mathbf{z}^{\text{utopian}}$	Utopian point of multi-objective optimization problem
P	Parent population
O	Offspring population
N	Population size
N_{FE}	Number of fitness evaluations
α	Learning rate

Chapter 1
Introduction to Optimization

1.1 Definition of Optimization

Optimization problems can be found everywhere in engineering, economics, business, and daily life. For instance, design of a turbine engine intends to minimize the pressure loss, while design of airfoil or a race car usually aims to minimize the drag. In process industries, maximization of product quality and quantity is desired, and in automated trading or business investment, one is often interested in maximizing the profit. Even in daily life, many decision-making processes such as purchasing a personal computer or mobile phone requires to minimize the costs. In general, optimization involves making the best use of given situation or resources.

1.1.1 Mathematical Formulation

Formally, optimization, also known as mathematical programming, is a mathematical discipline that aims to find of the minimums and maximums of an objective function, subject to certain constraints (Kochenderfer & Wheeler, 2019). An optimization problem consists of three main components (Nocedal & Wright, 1999):

- *Objective function*: This defines the mathematical representation of the measure of performance in terms of the decision variables. For example, a linear objective function may look like:

$$\text{minimize} f(x_1, x_2) = 4x_1 - x_2; \tag{1.1}$$

- *Decision variables*: Each decision variable represents a quantifiable decision that needs to be made. By changing the decision variables, the objective value will also be changed. In Eq. (1.1), x_1 and x_2 are decision variables, and the value of the objective function $f(x_1, x_2)$ will be changed by changing x_1 or x_2.

Y. Jin et al., *Data-Driven Evolutionary Optimization*,
Studies in Computational Intelligence 975,
https://doi.org/10.1007/978-3-030-74640-7_1

- *Constraints*: Constraints are mathematical equations or inequalities that indicate restrictions or conditions on the values the decision variables can take. For example in the above example, the optimal solution should satisfy the following conditions:

$$x_1 + x_2 = 10.0, \tag{1.2}$$

$$x_1 - x_2 < 5.0 \tag{1.3}$$

$$0 \leq x_1, x_2 < 10.0 \tag{1.4}$$

Of the above two constraints, Eq. (1.2) is called an equality constraint, Eq. (1.3) is called an inequality constraint, and Eq. 1.4 is known as boundary constraints that defines the lower and upper bounds of the decision variables. For an optimization problem, one combination of the decision variables is known as a solution. For example, for the optimization problem in (1.1), ($x_1 = 5.0$, $x_2 = 5.0$) is solution of the problem, and since this solution satisfies both constraints, it is called a feasible solution. By contrast, ($x_1 = 1.0, x_2 = 8.0$) does not satisfy the equality condition, and therefore it is known as a non-feasible solution. The task of the above optimization problem is to find a feasible solution that minimizes the objective function (1.1).

Generally, a minimization problem can be formulated as

$$\text{minimize} \quad f(\mathbf{x}) \tag{1.5}$$

$$\text{subject to} \quad g_j(\mathbf{x}) < 0, \quad j = 1, 2, \ldots, J. \tag{1.6}$$

$$h_k(\mathbf{x}) = 0, \quad k = 1, 2, \ldots, K. \tag{1.7}$$

$$\mathbf{x}^L \leq \mathbf{x} \leq \mathbf{x}^U. \tag{1.8}$$

where $f(\mathbf{x})$ is the objective function, $\mathbf{x} \in \mathcal{R}^n$ is the decision vector, $\mathbf{x}^L$ and $\mathbf{x}^U$ are the lower and upper bounds of the decision vector, n is the number of decision variables, $g_j(\mathbf{x})$ are the inequality constraints, and $h_k(\mathbf{x})$ are the equality constraints, J,K are the number of inequality and equality constraints, respectively. If a solution $\mathbf{x}$ satisfies all constraints, it is called a feasible solution, and the solutions that achieve the minimum value are called the optimal solutions or minimums. For a differentiable objective function where all its partial derivatives equal zero, the solution is called a stationary point, which may be a local minimum, a local maximum, or a saddle point, as illustrated in Fig. 1.1.

Problem formulation in solving real-world optimization problems is of extreme importance but challenging. The main reasons include

- In optimization of complex systems, there are large numbers of decision variables, objectives and constraints and it is unpractical to optimize the overall system at once. For example, deigning a car may be divided into conceptual design, design of components and design of parts. Note that the amount of time and cost allowed for these different design stages are different.
- Even for a subsystem, it is not easy for the user to determine whether a particular requirement should be considered as an objective or as a constraint. In addition, some objectives or constraints are added only after some preliminary optimization

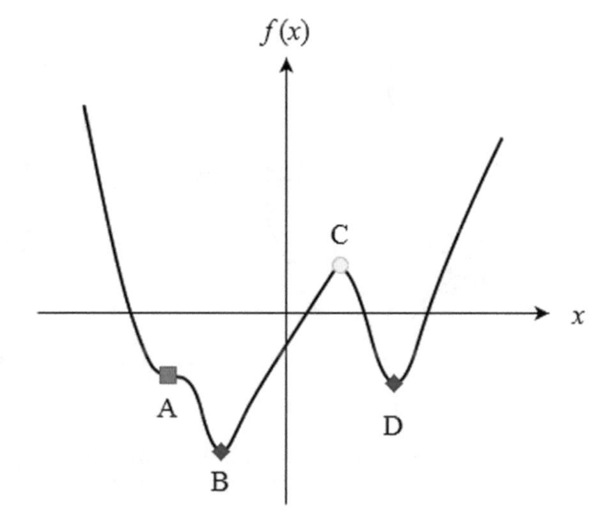

Fig. 1.1 Illustration of stationary points of a one-dimensional function, where solution 'A' is a saddle point, 'B' and 'D' are local minimums, and 'C' is a local maximum

runs are performed. For example in designing a turbine engine, minimization of the pressure loss is the primary concern. However, after a preliminary design is optimized, it is found that the variance of flow rate along the outlet is too big, which is detrimental. Thus, a second objective to minimize the variance, or an additional constraint on the variance must be included in the problem formulation.

1.1.2 Convex Optimization

Definition 1.1 The optimization problem in Eq. (1.5) is called convex, if for all $\mathbf{x}_1$, $\mathbf{x}_2 \in \mathcal{R}^n$, and for all $\alpha + \beta = 1$, $\alpha \geq 0$, $\beta \geq 0$, the following condition holds:

$$f(\alpha \mathbf{x}_1 + \beta \mathbf{x}_2) \leq \alpha f(\mathbf{x}_1) + \beta f(\mathbf{x}_2). \tag{1.9}$$

If the optimization problem in Eq. (1.5) is convex, the objective function is a convex function and the feasible solution set is a convex set. If a function $f(\mathbf{x})$ is convex, then $-f(\mathbf{x})$ is concave.

An example of a convex function is given in Fig. 1.2a. Visually, the function is convex if all points on the line connecting any two points (here x_1 and x_2) are above the function.

A convex optimization problem may have none, one or multiple optimal solutions. However, if the problem is strictly convex, then the problem has at most one optimal solution. Note that a complex optimization problem can be solved in polynomial time, whilst a non-convex optimization problem is generally NP-hard.

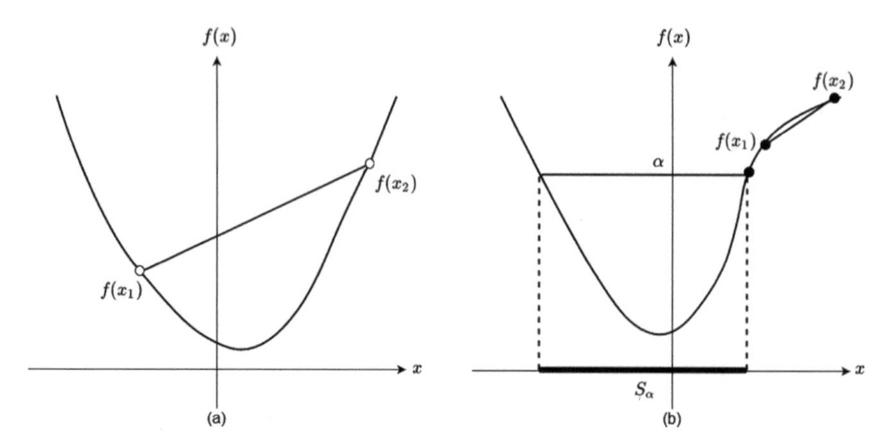

Fig. 1.2 **a** An example of a convex function. **b** An example of quasi-convex function, which is convex for $x \in \mathcal{S}_\alpha$. However, $f(x)$ is concave, when e.g., $x_1 \leq x_2$

1.1.3 Quasi-convex Function

Definition 1.2 A function $f(\mathbf{x})$ is said to be quasi-convex on $\mathcal{S}$, where $\mathcal{S}$ is a convex subset of $\mathcal{R}$, the following condition satisfies:

$$f(\lambda \mathbf{x}_1 + (1-\lambda)\mathbf{x}_2) \leq \max\{f(\mathbf{x}_1), f(\mathbf{x}_2)\}. \tag{1.10}$$

for all $\mathbf{x}_1, \mathbf{x}_2 \in \mathcal{S}$, and all $\lambda = [0, 1]$.

Alternatively, function $f(\mathbf{x})$ is quasi-convex on a sublevel set $\mathcal{S}_\alpha = \{\mathbf{x} | f(\mathbf{x}) \leq \alpha\}$, if $\mathcal{S}_\alpha$ is convex.

An example of quasi-convex function is given in Fig. 1.2b, where $f(x)$ is convex on $\mathcal{S}_\alpha$. However, it is concave on the upper right part, as indicated by the line connecting $f(x_1)$ and $f(x_2)$, where all points on the line connecting $f(x_1)$ and $f(x_2)$ are below $f(x)$.

Note that a strictly quasi-convex function is also known as *unimodal*. A unimodal function has one local optimum, which is also the global optimum. By contrast, functions having multiple local optimums are called *multi-modal* functions.

1.1.4 Global and Local Optima

Consider the following minimization problem:

$$\text{minimize} \quad f(\mathbf{x}) \tag{1.11}$$

$$\text{subject to } \mathbf{x} \in \mathcal{F} \tag{1.12}$$

where $f : \mathcal{R}^n \rightarrow \mathcal{R}$, n is the number of decision variables, $\mathcal{F} \subset \mathcal{R}^n$ is the feasible region. Then we have the following definitions.

Definition 1.3 A point $\mathbf{x} \in \mathcal{F}$ is a global optimum of the optimization problem in Eq. (1.11), if

$$f(\mathbf{x}^*) \leq f(\mathbf{x}) \tag{1.13}$$

$\forall \mathbf{x} \in \mathcal{F}$.

Definition 1.4 A point $\mathbf{x} \in \mathcal{F}$ is a local optimum of the optimization problem in Eq. (1.11), if there exists $\rho > 0$ such that

$$f(\mathbf{x}^*) \leq f(\mathbf{x}) \tag{1.14}$$

$\forall \mathbf{x} \in \mathcal{F}$ and $\| \mathbf{x} - \mathbf{x}^* \| < \rho$.

1.2 Types of Optimization Problems

Optimization problems can be categorized into various classes, typically according to the nature of the decision variables, the number of objectives to be optimized, whether constraints need to be considered, or whether uncertainty is present in the optimization problems. Finally, there also optimization problems whose objective functions are unknown or cannot be described by analytic mathematical equations.

1.2.1 Continuous Versus Discrete Optimization

A large number of real-world optimization problems continuous problems, in which the decision variables are real numbers and they can be changed continuously in the feasible regions. For example, geometry design of a turbine engine is a continuous optimization problems, where the decision variables describing the shape of the turbine are real numbers.

In contrast to continuous optimization problems, all or part of decision variables of discrete optimization problems can take on discrete sets of values only, often integers. For instance, number of people, number of cities or locations, and the sequence in which a manufacturing process is organized. Discrete problems can further be divided into mixed-integer problems, in which part of the decision variables are integers, integer problems, and combinatorial problems. Combinatorial optimization can be seen as a special class of integer optimization, where the task is to find an optimal combination of a finite set of objects. Representative combinatorial problems include the travelling salesman problem (TSP), the knapsack problem, and the minimum spanning tree problem.

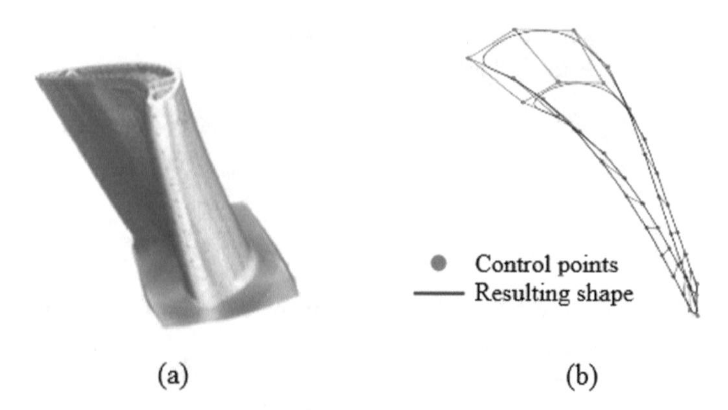

Fig. 1.3 An example of a continuous optimization problem. **a** Illustration of a jet engine turbine blade. **b** A 3D B-spline representation of a blade using two 2D sections

Consider a simplified turbine engine optimization problem, where the task to design the geometry of the turbine blades for energy efficiency. The problem can be formulated as follows:

- *Decision variables*. Parameters defining the geometry of the turbine blade. Note that many different methods can be used to represent a geometry, e.g., based on parameterization or B-splines, as shown in Fig. 1.3. Here, the decision variables are related to the blade geometry and therefore are continuous. So this is a continuous optimization problem. There is typically a trade-off between completeness and compactness in representing a design. A complete representation is able to represent all different structures, may may have a large number of decision variables. By contrast, a compact representation has a small number of decision variables, but may be limited in presenting complex structure. Other constraints may include causality and locality (Jin & Sendhoff, 2009).
- *Constraints*. Here the constraints may include the mechanical constrains, among others.
- *Objectives*. Usually, minimization of the pressure loss is the main target in design of a turbine blade, which is closely related to energy efficiency. However, objectives such as deviation from a desired inflow angle, or the variance of the flow rate at the out let, may also be included in optimization. Note that the pressure loss cannot be calculated using an analytic mathematical function and can be evaluated using numerical simulations or wind tunnel experiments only.

The TSP is a classical combinatorial optimization problem which aims to find the shortest path connecting cities visited by a travelling salesman on his sales route. Each city can be visited only once. The TSP can be formulated as follows.

- *Decision variables*. Here the connecting order of the cities to be visited is the decision variable. For example, in the 10-city example shown Fig. 1.4, one solution is $A \rightarrow B \rightarrow C \rightarrow D \rightarrow E \rightarrow F \rightarrow G \rightarrow H \rightarrow I \rightarrow J \rightarrow A$.

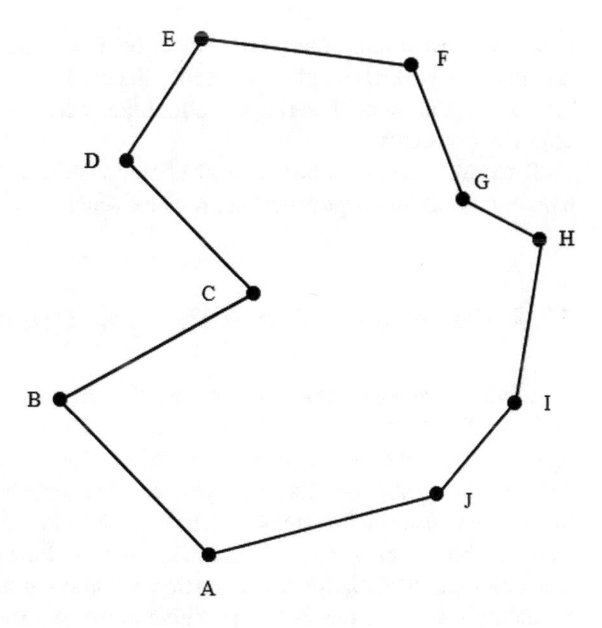

Fig. 1.4 An example of the travelling salesman problem with 10 cities showing a path starting from city A and back to A: $A \rightarrow B \rightarrow C \rightarrow D \rightarrow E \rightarrow F \rightarrow G \rightarrow H \rightarrow I \rightarrow J \rightarrow A$

- *Constraints*. The constrains are each city can be visited only once and the salesman must be back to the starting city A is he starts from there.
- *Objectives*. The objective is to minimize the length of the total travelling path.

1.2.2 Unconstrained Versus Constrained Optimization

In unconstrained optimization problems, no equality or inequality constrains need to be taken into account. By contrast, constrained optimization must satisfy equality and/or inequality constraints, making the problem-solving harder. A special case of constrained problems are *box-constrained problems*, in which only constraints on the range of the decision variables are limited. Box-constrained optimization problems reduce to unconstrained optimization when the bounds are infinite.

1.2.3 Single Versus Multi-objective Optimization

The problem in Eq. (1.5) has one objective function, therefore, such optimization problems are known as single-objective optimization problems (SOPs). In the real world, however, most problems have more than one objective to optimize, and these objectives are usually conflicting with each other. These problems are called multi-objective optimization problems (MOPs). In the evolutionary computation community, MOPs with more than three objectives are termed many-objective optimization problems (MaOPs), mainly because the hardness to solve MaOPs will become dramatically more challenging for algorithms designed for solving two- or three-objective problems.

There are also cases where the task of optimization is to find a feasible solution for a highly constrained optimization problem, which is called *constraint satisfaction*.

1.2.4 Deterministic Versus Stochastic Optimization

Usually it is assumed that a given optimization is time invariant and there is no uncertainty in the objective or constraint functions, nor in the decision variables. These optimization problems are called deterministic optimization problems. In contrast to deterministic problems, stochastic problems are subject to uncertainty in the objective and/or constraint functions, and in decision variables. The uncertainty may come from different sources, such as noise in sensors, randomness in performing numerical simulations, and changes in the operating condition. In these cases, one is interested in finding a solution that is less sensitive to the uncertainty, which is an important research topic of robust optimization and dynamic optimization to be elaborated in Sect. 1.4.

It should be noted that stochastic optimization may also refer to research methodologies in which the decision variables are treated as random variables or the search allow random changes. Here by an stochastic problem, we mean an optimization problem that is subject to various types of uncertainty.

1.2.5 Black-Box and Data-Driven Optimization

In the discussions so far, we assume that the objective of an optimization problem can be described analytically by a mathematical expression. However, this is not always possible for solving real-world problems, where the objective function is unknown or cannot be written in an analytic form. These optimization problems are sometimes called *black-box optimization*.

Typically, the objective value of a black-box optimization problem is evaluated using computer simulations. There are also cases when the objective is evaluated by performing physical or chemical experiments, although the system to be optimized is not necessarily a black-box. For example, the aerodynamic performance of a turbine engine or a vehicle is well well understood scientifically, although the aerodynamic performance of such systems can be evaluated only by solving a large number of differential equations. In addition, in the era of data science and big data, many optimization problems are solved based on collected data, which may be incomplete, noisy, and heterogeneous. Therefore, these problems are better called *data-driven optimization problems* (Jin et al., 2018), where optimization relies on data collected from computer simulations, physical or chemical experiments, and data from daily collected life.

1.3 Multi-objective Optimization

1.3.1 Mathematical Formulation

In real-life, almost all optimization problems have multiple objectives to be maximized or minimized. For example, when one wants to buy a mobile phone, one needs to consider the price that can be afforded, and the performance that is desired. For performance, there are again multiple aspects to be taken into account, e.g., the specifications of the camera, the easiness to use and many others. It is practically impossible to simultaneously optimize all objectives, because there is usually a trade-off between different objectives. For instance, as illustrated in Fig. 1.5, there is a compromise, also known as trade-off, between price minimization and the performance maximization. In other words, the better the performance, the higher the price will be. Unlike in single-objective optimization where one optimal solution can be obtained, there is always a set of trade-off solutions. Finally, one solution may be selected based on the user's preference. For example, if the user is willing to pay more for a better mobile phone, he or she will select solution C. If the user wants to have a good price/performance ratio, the user may select solution B. If the user put more importance on the minimization of the cost, he or she will select solution A.

Mathematically, a multi-objective optimization problem can be formulated as follows (Miettinen, 1999):

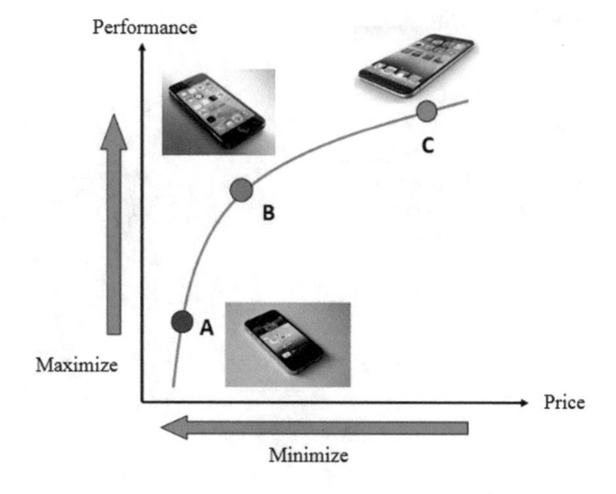

Fig. 1.5 Purchase of a mobile phone is a bi-objective optimization problem that maximizes the performance and minimizes the price. A set of optimal trade-off solutions will be obtained instead of a single ideal solution

$$\text{minimize} \quad f_i(\mathbf{x}) \quad i = 1, 2, \ldots, m. \tag{1.15}$$

$$\text{subject to} \quad g_j(\mathbf{x}) < 0, \quad j = 1, 2, \ldots, J. \tag{1.16}$$

$$h_k(\mathbf{x}) = 0, \quad k = 1, 2, \ldots, K. \tag{1.17}$$

$$x_l^L \leq x_l \leq x_l^U, \ l = 1, 2, \ldots, n. \tag{1.18}$$

where $m > 1$ is the number of objectives. If $m = 1$, it reduces to the single-objective optimization problem described in Eq. (1.5).

1.3.2 Pareto Optimality

The comparison between solutions is different from that in single-objective optimization, as more than one objective needs to be taken into account. For this reason, a concept of Pareto dominance is introduced and is the most important definition in multi-objective optimization.

Definition 1.5 For the minimization problem in Eq. (1.15), solution $\mathbf{x}_1$ Pareto dominates, denoted by $\mathbf{x}_1 \prec \mathbf{x}_2$, solution $\mathbf{x}_2$ if

$$\forall i \in \{1, 2, \ldots, m\}, \ f_i(\mathbf{x}_1) \leq f_i(\mathbf{x}_2), \ \text{and} \tag{1.19}$$

$$\exists k \in \{1, 2, \ldots, m\}, \ f_k(\mathbf{x}_1) < f_k(\mathbf{x}_2). \tag{1.20}$$

For example in Fig. 1.6, solution A dominates solutions C and D, while solution B is not dominated by solution A. On the other hand, solution B does not dominate A either. Therefore, solutions A and B are non-dominated, and similarly, solutions B, C and D, do not dominate one another. If solutions A and B are not dominated by any feasible solutions, they are Pareto optimal solutions.

Definition 1.6 A solution $\mathbf{x}^*$ is called Pareto optimal, if there does not exist any feasible solution $\mathbf{x} \in \mathcal{R}^n$ such that $\mathbf{x} \prec \mathbf{x}^*$.

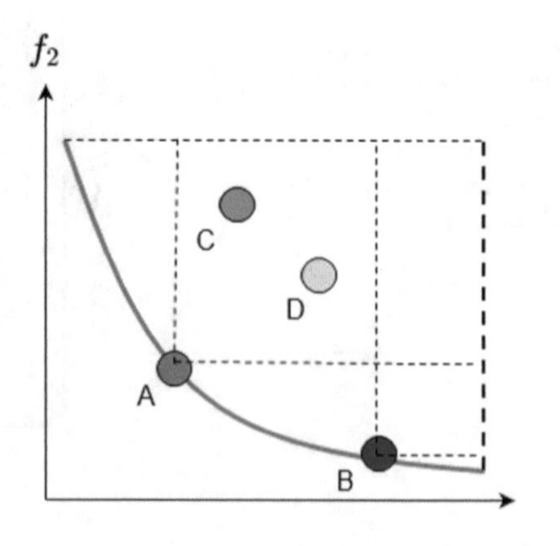

Fig. 1.6 Solution comparison in multi-objective optimization, where solution A dominates C and D, but does not dominate solution B

The image of all Pareto optimal solutions in the objective space is called the Pareto front or Pareto frontier, and the set of all solutions in the decision space is the Pareto optimal set consisting of two connected piece-wise linear curves, as shown in Fig. 1.7. In the figure, a few particular points are often used in optimization, which are the ideal point, the nadir point and the knee point.

Definition 1.7 The ideal point, denoted by $\mathbf{z}^{\text{ideal}}$, is constructed from the best objective values of the Pareto set.

Definition 1.8 The nadir point, denoted by $\mathbf{z}^{\text{nadir}}$, is constructed from the worst objective values over the Pareto optimal set.

In practice, a utopian point is often defined as follows:

$$z_i^{\text{utopian}} = z_i^{\text{ideal}} - \epsilon, \text{ for all } i = 1, 2, \ldots, m \tag{1.21}$$

where ϵ is a small positive constant.

Definition 1.9 A knee point is defined to be the one having the maximum distance from the convex hull of individual minimums to the hyperplane constructed by the extreme points.

In Fig. 1.7, solutions A and B are called the boundary points of the Pareto front, the line connecting these two solutions is the hyperplane, and the intersections of the hyperplane with the axes, here points D and E are called the extreme solutions. Solution C has the maximum distance to the hyperplane, and therefore is the knee point. Knee points of a Pareto front are of interest since they need a large compromise in at least one objective to gain a small improvement in other objectives.

Note that there are also other definitions for knee points (Deb & Gupta, 2011). In addition to the above distance based definition, there are also angle-based approaches measuring the angle between a solution and its two neighbors. Although angle-based approaches are straightforward, they are only applicable to bi-objective optimization problems. By contrast, distance-based approaches can handle problems with more than two objectives. Nevertheless, specifying knee points is notoriously difficult in high-dimensional objective spaces, because a Pareto front may have a large number of knee points.

In addition to the standard Pareto dominance relation, other dominance relationship have also been suggested by modifying the Pareto dominance. In the following, we introduce the ϵ-dominance (Laumanns et al., 2002) and α-dominance (Ikeda et al., 2001) that are often used in evolutionary multi-objective optimization.

Definition 1.10 A solution $\mathbf{x}_1$ ϵ-dominates solution $\mathbf{x}_2$ for some $\epsilon > 0$, if

$$\forall i \in \{1, 2, \ldots, m\}, (1 - \epsilon) f(\mathbf{x}_1) \leq f(\mathbf{x}_2). \tag{1.22}$$

Definition 1.11 Given two solutions $\mathbf{x}_1, \mathbf{x}_2 \in \mathcal{R}^n$ and $\mathbf{x}_1 \neq \mathbf{x}_2$, solution $\mathbf{x}_1$ α-dominates $\mathbf{x}_2$, denoted by $\mathbf{x}_1 \prec_\alpha \mathbf{x}_2$, if

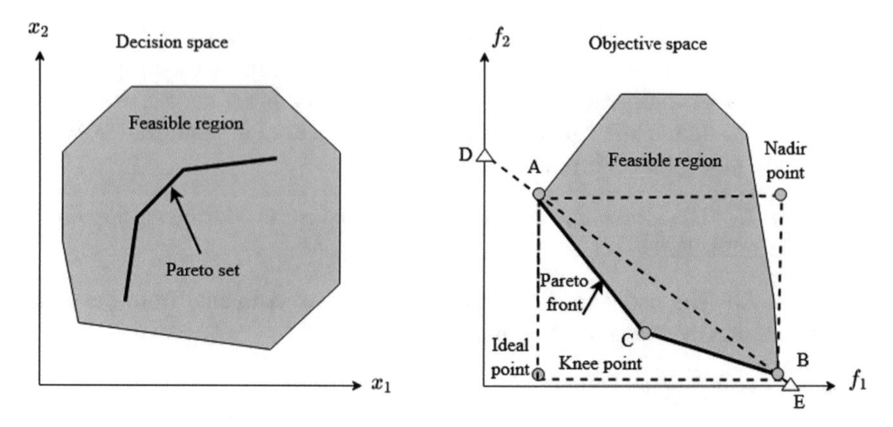

Fig. 1.7 Illustration of the Pareto set (left panel) and Pareto front (right panel), together with the ideal point, nadir point, knee point (C), boundary points (A and B) and extreme points (C and D)

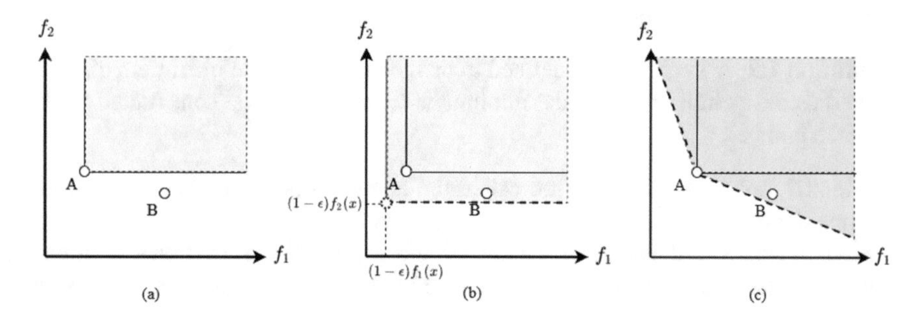

Fig. 1.8 Illustration of dominance relations. **a** Pareto dominance, **b** ϵ-dominance, and **c** α-dominance

$$\forall i \in \{1, 2, \ldots, m\},\ g_i(\mathbf{x}_1, \mathbf{x}_2) \leq 0, \text{ and} \tag{1.23}$$

$$\exists j \in \{1, 2, \ldots, m\},\ g_j(\mathbf{x}_1, \mathbf{x}_2) < 0, \tag{1.24}$$

where

$$g_i(\mathbf{x}_1, \mathbf{x}_2) = f_i(\mathbf{x}_1) - f_i(\mathbf{x}_2) + \sum_{j \neq i}^{m} \alpha_{ij}(f_j(\mathbf{x}_1) - f_j(\mathbf{x}_2)), \tag{1.25}$$

and a_{ij} is a parameter defining the trade-off rate between the i-th and j-th objectives.

From the above definitions, we can see that both ϵ-dominance and α-dominance strengthen the Pareto dominance relation, enabling a solution to dominate more solutions.

An illustration of Pareto dominance, ϵ-dominance and α-dominance relations is given in Fig. 1.8. In Fig. 1.8a, the region solution A Pareto dominates is shaded and therefore, solution B is not dominated by A. Figure 1.8b shows the region solution A ϵ-dominates, which is larger than that is dominated by the Pareto dominance.

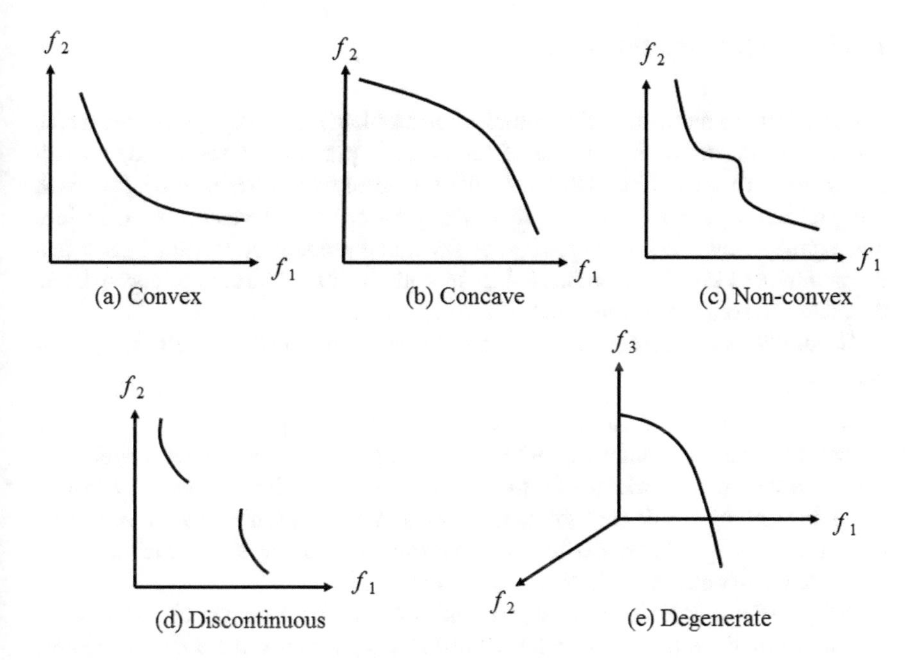

Fig. 1.9 Different types of Pareto optimal fronts

Nevertheless, solution A still cannot dominate B. The region solution A α-dominates is shown in Fig. 1.8c, where we can see that now solution B is dominated solution A. From these examples, we can see that both ϵ-dominance and α-dominance relations can dominate more the standard Pareto dominance, therefore, they can accelerate the convergence of an optimization algorithm that uses the dominance relation for comparing solutions.

The Pareto front of a multi-objective optimization problem may have very different natures, which heavily influences the hardness of problem-solving. Figure 1.9 illustrates a few typical types of Pareto fronts. Note that the definition of convexity of Pareto fronts is similar to the convexity of functions. For example, in Fig. 1.9a, the Pareto front is convex, because all points on the line connecting any two points on the Pareto front are dominated by the Pareto optimal solutions between these two points. By contrast, all solutions connecting two points on a concave Pareto front are infeasible solutions. In Fig. 1.9e the Pareto front is called degenerate, as for a three-objective optimization problem, the dimension of a Pareto front is usually a two-dimensional surface. Here, it degenerates to a one-dimensional curve.

1.3.3 Preference Modeling

The solution to a multi-objective optimization problem is a Pareto optimal set, which may consist of an infinite number of solutions. In practice, however, only a small subset of the Pareto optimal solutions will be adopted for implementation, for which user preferences are needed. Two important questions must be answered to achieve this, namely, how the user's preferences should be properly represented and when the preferences should be embedded in an optimization algorithm to approximate the preferred solutions (Wang et al., 2017b).

Generally speaking, user preferences can be represented using the following methods.

- *Goals*. The most straightforward way to articulate user preferences is to provide goal information (Gembicki, 1974), typically in terms of reference solutions or reference points. In order words, the user can describe their preferences by providing one or multiple desired solutions. For example in optimizing a diesel engine, the user may specify the desired fuel consumption and emission, which might be a certain percentages better than the baseline design.
- *Weights*. For a completely new problem, it can be hard for the user to provide desired solutions (goals) for different objectives. However, the user may be able to assess the importance of different objectives. Since there is always a trade-off between different objectives, it might be easier for the user to assign different levels of importance to different objectives by using a weight vector $\mathbf{w} = \{w_1, \ldots w_m\}$, where m is the number of objectives. Given the weights, a multiple objective optimization problem can then be scalarized into a single-objective optimization problem. The most widely used scalarizing is the linear scalarization, also known as the weighted aggregation method (Miettinen, 1999):

$$g(\mathbf{x}) = \sum_{i=1}^{m} w_i f_i(\mathbf{x}) \tag{1.26}$$

where $w_i > 0, i = 1, 2, \ldots, m$ and optimal solutions of the above single-objective optimization problem are Pareto optimal solutions of the multi-objective optimization problem in (1.15).
The linear scalarizing method is relatively easy to implement, however, has several issues. The biggest problem is that one is not able to achieve Pareto optimal solutions on the concave part of the Pareto front by solving the single-objective problem in (1.26), no matter how the weights are tuned.
Another widely used scalarizing method is called the Chebyshev scalarizing function, which is expressed as follows (Miettinen, 1999):

$$g(\mathbf{x}) = \max_{1 \le i \le m} \{w_i f_i(\mathbf{x})\}, \tag{1.27}$$

where $f_i(\mathbf{x})$ is the i-th objective and w_i is the i-th weight.

Similar to goals, however, it is hard for the user to to specify accurate weights without a good understanding the problem. It is particularly tricky when the Pareto front is non-convex, or non-uniform, discrete, or degenerate.

- *Reference vectors*. Reference vectors (Wang et al., 2017b) are similar to goals and weights in that they are meant to provide an expectation of the objectives. Different to the goals (reference points), reference vectors provide information about the directions in the objective space in which solutions are desirable. Similar to weights, reference vectors can also be used to convert a multi-objective optimization problem into single-objective optimization problems.
- *Preference relation*. Another natural way of preferences over different objectives is to describe the relative importance between pairs of objectives using human linguistic variables. For example, one may indicate that *objective 1 is more important than objective 2*, or *objective 1 and objective 2 are equally important*.
  In optimization, a preferred order of the objectives can be converted into weights or weight intervals. However, one main disadvantage is that the preference relation cannot handle non-transitivity.
- *Utility functions*. Utility functions can be used to represent user, where the preference information is implicitly involved in the objective function to rank solutions. Unlike preference relations, the utility function ranks solutions rather than the objectives. For example, given N solutions $\mathbf{x}_1, \mathbf{x}_1, \ldots, \mathbf{x}_N$, the user is required to give his or her preferences over the solutions by ranking them in a order. Then, an imprecisely specified multi-attribute value theory formulation is employed to infer the relative importance of the objectives. However, utility functions are based on a strong assumption that all attributes of the preferences are independent, thereby being unable to handle non-transitivity.
- *Outranking*. Neither preference relations based on the importance of objectives nor the utility functions based on solutions are able to handle non-transitivity. An alternative is so-called outranking that allows for non-transitivity. To determine an outranking, a preference and indifference thresholds for each objective are given by a preference ranking organization method for enrichment evaluations (Brans et al., 1986), according to which every two solutions are compared. Consequently, a preference ranking is obtained, which can be used for search of preferred solutions. However, the outranking based methods require a large number of parameter settings, which is non-trivial for the user for problems having a large number of objectives (Brans et al., 1986).
- *Implicit preferences*. Articulation of preferences is always challenging for the user when there is little knowledge about the problem to be solved. In this case, knee solutions, around which a small improvement of any objective causes a large degradation of others, are always of interest to to the user. In addition to knee points, extreme points or the nadir point can work as a special form of preferences. With the help of extreme points or the nadir point, the user can acquire knowledge about the range of the Pareto front so as to describe their preferences more accurately.

1.3.4 Preference Articulation

Once the user preferences are properly modeled, they must be included in the optimization process to obtain the preferred solutions, although this is not straightforward. In the following, we elaborate various preference articulation methods widely used in multi-objective optimization.

- *No-preference*. The simplest case is that no particular preferences are included in the optimization and all objectives are treated equally important. For example, the following scalarizing function can be used:

$$\text{minimize} \quad \| f(\mathbf{x}) - \mathbf{z}^{\text{ideal}} \|, \tag{1.28}$$

$$\text{subject to} \quad \mathbf{x} \in \mathcal{R}^n. \tag{1.29}$$

  where $\| \cdot \|$ is usually L_1, L_2 or L_∞. Usually, the objectives need to be normalized into a uniform scale so that no strong bias is included in the optimization.
- *A priori approaches*. *A priori* approaches to multi-objective optimization embed the preferences in the objective functions before the optimization starts so that a small subset of preferred Pareto optimal solutions are obtained. Most preference models, such as goals (reference points), weights, and reference vectors can be adopted to guide the search. In addition to the above-mentioned linear and Chebyshev scalarization methods that use weights for modeling preferences, we also have the augmented weighted Chebyshev scalarizing method:

$$\min \max_{i=1,2,\ldots,m} w_i[|f_i(\mathbf{x}) - z_i^{\text{utopian}}|] + \rho \sum_{i=1}^{m} |f_i(\mathbf{x}) - z_i^{\text{utopian}}|, \tag{1.30}$$

$$\text{subject to:} \quad \mathbf{x} \in \mathcal{R}^n. \tag{1.31}$$

  where $\rho > 0$ is a small constant.
  Alternatively, if the user can provide a reference point $\bar{\mathbf{z}}$, then the achievement scalarizing function can be used:

$$\min \max_{i=1,2,\ldots,m} \left[\frac{f_i(\mathbf{x}) - \bar{z}_i}{z_i^{\text{nadir}} - z_i^{\text{utopian}}} \right] + \rho \sum_{i=1}^{m} \frac{f_i(\mathbf{x})}{z_i^{\text{nadir}} - z_i^{\text{utopian}}}, \tag{1.32}$$

$$\text{subject to:} \quad \mathbf{x} \in \mathcal{R}^n. \tag{1.33}$$

  In principle, most classical optimization methods that need to convert a multi-objective optimization problem into a single-objective problem using a scalarizing function belong to *a priori* methods. *A priori* approaches suffer from two main weaknesses. On the one hand, it is non-trivial for the user to provide informative and reasonable preferences prior to the optimization. On the other hand, an optimization algorithm may not be able to find the preferred solutions even if the user's preferences are properly modelled.

- *A posteriori approaches*. In the *A posteriori* approaches, the user is allowed to articulate his or her preferences after a set of non-dominated solutions has been obtained. To this end, the algorithms shall be able to approximate a *representative* subset of the whole Pareto optimal front so that the user is able to select the preferred ones from the obtained solutions. Since mathematical programming methods can get one Pareto optimal solution in one run, they need to be run for a sufficient large number of times to get a solution set. In contrast to mathematical programming methods, evolutionary algorithms and other population based metaheuristics are able to obtain a set of Pareto optimal or non-dominated solutions in one single run. However, since the theoretical Pareto front is unknown beforehand, these algorithms aim to promote the diversity of the obtained solutions in an effort to get a representative subset of Pareto optimal solutions.
  It should be pointed out that the assumption that a population based optimization algorithm is able to obtain a representative subset of Pareto optimal solutions may practically fail to hold given limited computational resources, in particular when the Pareto front is discrete, degenerate, or when the number of objectives becomes large.
- *Interactive approaches*. Considering the difficulty in precisely modeling the user preferences in the *a priori* approach, and the challenges in acquiring a representative subset of Pareto optimal solutions, interactive approaches, also known as progressive approaches, aim to progressively embed the user's preferences in the course of the optimization so that the user is able to refine or even change his or her preferences according to the solutions obtained.
  One main challenge of the interactive approaches is that the human user needs to be directly involved in the optimization, which may become infeasible due to human fatigue. One possible solution to address this issue is to partly automatize the preference refinement or modification process using machine learning techniques.

1.4 Handling Uncertainty in Optimization

Most real-world optimization problems are subject to different degree of deterministic and probabilistic uncertainty (Jin & Branke, 2005; Liu & Jin, 2021). One most widely seen uncertainty is introduced in evaluating the objective functions introduced by various sources, such as measurement noises or numerical errors. In the real-world, optimization is often subject to more systematic uncertainty. For example, design of a turbine engine must consider its performance in various operating conditions when the velocity, weight and altitude if the aircraft changes. These time-varying factors are not decision variables, but will influence the objective functions. On the other hand, there might be small errors introduced in manufacturing the turbine blades or the blades may slowly worn out in the lifetime. Ideally, the performance of the turbine blade should be relatively insensitive to such changes in the operating condition (called environmental parameters hereafter) or in the decision variables. Without loss of generality, the objective function of a non-constrained single-objective optimization problem in the presence of uncertainty can be described as follows:

$$\min f(\mathbf{x} + \Delta\mathbf{x}, \mathbf{a} + \Delta\mathbf{a}) + z, \tag{1.34}$$

where $\Delta\mathbf{x}$, $\Delta\mathbf{a}$ are perturbations in the decision variables and environmental parameters, respectively, and $z \sim \mathcal{N}(0, \sigma^2)$ is additive noise, σ^2 is the variance of the noise.

Different methods have been proposed to handle different uncertainty in optimization. For additive noise in fitness evaluations, the typical method is to evaluate the fitness multiple times to reduce the influence of the noise (Rakshit et al., 2017). For addressing the non-additive perturbations in the decision variables and environmental parameters, one approach is to find a solution that is robust to changes in environmental parameters or in the decision variables, which is typically known as robust optimization. However, if the operating conditions change significantly, a more realistic and effective approach will be multi-scenario optimization. If the changes cannot be captured by a probability distribution, for example, there are continuous or periodic changes in the environment or decision variables, then dynamic optimization may be more effective. Finally, robust optimization over time, which aims to make a best compromise between performance and robustness, as well as the cost that may occur in switching the designs.

1.4.1 Noise in Evaluations

The most straightforward method to cancel out the addition noise in fitness evaluations is to do averaging either over time or over space, meaning sampling multiple times of the same solution, or sampling multiple solutions around the solution, respectively, and then use the average of the sampled objective values as the final objective value (Liu et al., 2014).

Since evolutionary algorithms are population based search methods, other approaches to noise reduction have also been proposed, including using a large population size together with proportional selection, introducing recombination, and introducing a threshold in selection. Since fitness evaluations can be computationally intensive, noise reducing based on re-sampling or using a large population can be impractical. Thus, efficiently re-sampling during the optimization becomes important. For example, adapting the sample size is an effective approach. Typically, one can use a smaller sample size in the early search stage and a larger sample size in the later stage; alternatively, a large sample size can be used for a more promising solution, and a small sample size is used for a poor solution. Apart from adaptive sampling, use of local regression models to estimate the true fitness value is also possible.

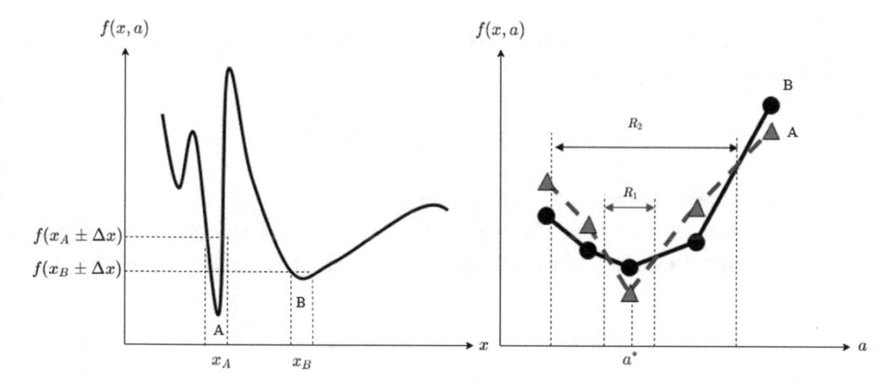

Fig. 1.10 Illustration of robust solutions against changes in the decision variable (left panel) and the environmental parameter (right panel)

1.4.2 Robust Optimization

The robustness of the solutions against perturbations in the decision variables and environmental parameters is a basic requirement in practice. Figure 1.10 illustrates two different situations of robustness. In the left panel, *A* and *B* are two local optima of the objective function. Without considering the robustness of the solutions, solution *A* has better performance than solution *B*, if we are considering a minimization problem. However, when there is a perturbation Δx in the decision variable, $f(x_A)$ will be much worse than $f(x_B)$. Therefore, we consider solution *B* is more robust to solution *A*. By contrast, in the right panel, solution *A* has the best performance at the normal operating condition when $a = a^*$. However, when a becomes larger or smaller, the performance worsens very quickly. In contrast solution *A*, solution *B* has worse performance than solution *A* at the normal operating condition, however, its performance degrades more slowly. As a result, solution *B* performs better than *A* over $a \in R_2$ except for $a \in R_1$. Thus, solution *B* is more robust than solution *A* against changes in the environmental parameter a. Before we discuss methods for obtaining robust optimal solutions, we first introduce a few widely used definitions for robustness in the context of optimization (Beyer & Sendhoff, 2007; Jin & Branke, 2005).

- *Robust counterpart approach.* For the optimization problem in the presence of uncertainty in the decision variables, the robust counterpart function $F(\mathbf{x}, \epsilon)$ is defined by

$$F(\mathbf{x}, \epsilon) = \sup_{\xi \in R(\mathbf{x}, \epsilon)} f(\xi), \tag{1.35}$$

  where $R(\mathbf{x}, \epsilon)$ is a neighborhood of x of size ϵ. Typically, the neighborhood of $\mathbf{x}$ of size ϵ is defined by $\mathbf{x} \pm \epsilon$. Therefore, this definition is the worst case scenario, which is also known as a *minimax* optimization problem.

If we consider the perturbations to the environmental parameters, then the robust counterpart function is given by

$$F_W(\mathbf{x}, \epsilon) = \sup_{\mathbf{a} \in A(\epsilon)} f(\mathbf{x}, \mathbf{a}), \tag{1.36}$$

where $A(\epsilon)$ is a neighborhood of $\mathbf{a}$ of size ϵ.

- *Expected objective function.* The expected objective function considering perturbations in both decision variables and environmental parameters, we have

$$F(\mathbf{x}) = \int f(\mathbf{x} + \delta, \mathbf{a}) p(\delta, \mathbf{a}) d\delta d\mathbf{a}, \tag{1.37}$$

If we consider the perturbations in the decision variables only, the expected objective function will be:

$$F_E(\mathbf{x}) = \int f(\mathbf{x} + \delta, \mathbf{a}) p(\delta) d\delta. \tag{1.38}$$

- *Dispersion based robustness measure.* The robustness can also be defined using the dispersion of the objective function in the presence of uncertainty: e.g.

$$F_D(\mathbf{x}) = \int (f(\mathbf{x} + \delta) - f(\mathbf{x}))^2 p\delta d\delta, \tag{1.39}$$

Methods for search of robust optimal solutions are based on the above robustness definitions and their variants. That is, if one is interested in finding the robust optimal solution, then, one of the above robustness measure can be used to replace the original objective function. Since estimation of the robustness, no matter which of the above definitions is used, requires to evaluate the objective value multiple times for one solution, several ideas for reducing the computational costs have been proposed by taking advantage of the population based search algorithm. For example, for the expected objective function, the *explicit averaging* approach can calculate the mean objective value to approximate the expected objective:

$$\bar{f}(\mathbf{x}) = \frac{1}{N} \sum_{i=1}^{N} f(\mathbf{x} + \delta), \tag{1.40}$$

where $\delta \sim \mathcal{N}(0, \sigma^2)$ is a Gaussian noise, N is the sample size. Thus, $N - 1$ additional function evaluations are needed for evaluating each solution. To reduce the computational cost, an *implicit averaging* method has been suggested, in which each solution in the currently population is evaluated only once:

$$\bar{f}(\mathbf{x}) = f(\mathbf{x} + \delta), \tag{1.41}$$

where $\delta \sim \mathcal{N}(0, \sigma^2)$ is a Gaussian noise. Actually, the implicit approach is a special case of the explicit averaging approach, where the sample size equals one. This implicit works better when the population size of a population based search algorithm is larger.

In the expected objective approach, it is assumed that the uncertainty is a Gaussian noise, which requires to specify the level of the noise (variance). This is not easy, and a wrong noise level used for calculating the robustness may either find a too conservative solution, or a solution that is not robust enough. Thus, the worst case scenario removes the assumption of Gaussian noise, but still requires the definition of the neighborhood size and a large number of extra function evaluations to find out the worst objective in the neighborhood. A variant of the worst scenario robustness, called inverse robustness that aims to avoid the specification of the neighborhood size, has been suggested. In the inverse robust optimization, the user can specify the maximum tolerable performance degradation, based on which the maximum allowed change in the decision variable will be found out. In practice, specification the maximum tolerable performance degradation is more realistic than specifying the noise level.

As can be seen in Fig. 1.10, there is usually a trade-off between the normal objective value and the degree of robustness (Jin & Sendhoff, 2003). Thus, robustness optimization has often been addressed using the multi-objective optimization approach. For example, one can minimize the expected objective function and minimize the variance:

$$F_E(\mathbf{x}) = \frac{1}{N}\sum_{i=1}^{N} f(\mathbf{x}_i) \tag{1.42}$$

$$F_D(\mathbf{x}) = \frac{1}{N}\sum_{i=1}^{N} [f(\mathbf{x}_i) - F_E(\mathbf{x})]^2 . \tag{1.43}$$

where N is the sample size.

Other combinations can also be used as the two objectives, such as the normal objective function and the expected objective function, or the normal objective function together with the variance.

Once a set of non-dominated robust solutions are obtained, the user eventually needs to select one solution a posteriori once any knowledge about the noise or the tolerable performance degradation is available.

1.4.3 Multi-scenario Optimization

In practice, the performance of a design in several operating conditions must be considered. For example for aircraft or turbine design, one must consider at least three scenarios, take-off, cruise, and landing. For designing a vehicle, the dynamics

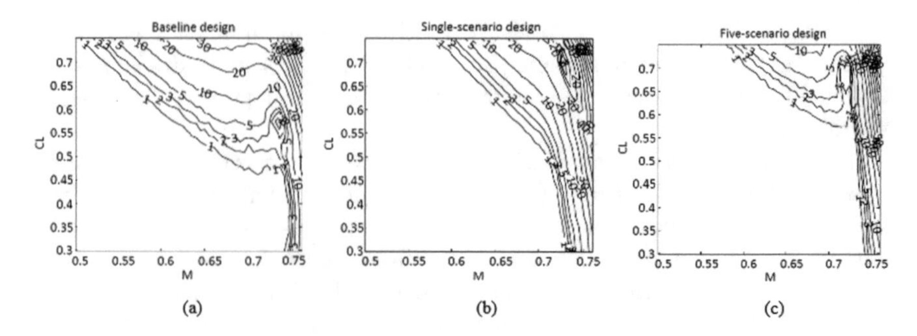

Fig. 1.11 The wave drag coefficient distribution of three RAE5225 airfoil designs obtained by single- and multi-scenario optimization methods

of the vehicle in multiple driving scenarios must be considered, such as driving in a city, on highway, on a curve, or passing a bridge. In a more extreme case, an infinite number of scenarios must be considered, for example in designing a high-lift airfoil system, where one aims to maximize the lift and minimize the drag for a given range of flying velocity. This is also known as full scenario optimization (Wang et al., 2018a). Figure 1.11 shows the drag coefficient distribution over a range of velocity (from 0.5 mach number to 0.75 mach number) of three optimized designs of airfoil RAE5255. From these figures, we can see different design philosophy may lead to very different performances.

Several approaches can be adopted to deal with multi-scenario optimization. The simplest approach is to aggregate the objectives in all considered scenarios. Alternatively, the objectives in each scenario are considered as separate objectives, and consequently, the number objectives can become very large when multi-scenarios are taken into account. In addition, the final decision making will also be difficult without any preference over the scenarios.

Robustness optimization can also be used to tackle multi-scenario optimization, where the averaging performance in multiple scenarios is optimized. However, this will not lead to good performance if the number of scenarios is large.

1.4.4 Dynamic Optimization

Handling uncertainty by finding an optimal solution that is relatively insensitive to uncertainty is effective if the level of the uncertainty is limited and can be described by a certain type of Gaussian or uniform noise. However, this approach will fail if the level of uncertainty is deterministic, continuous and large. In this case, the target becomes to dynamically track the changing optimum or the changing Pareto optimal solutions, which is known as dynamic optimization.

Generally, a dynamic optimization problem can be described as follows:

$$\min f(\mathbf{x}, \mathbf{a}(t)), \tag{1.44}$$

where $\mathbf{a}(t)$ is a set of time-varying environmental parameters. In other words, dynamic optimization is mainly concerned with uncertainty in the environmental parameters. Although $\mathbf{a}(t)$ may change continuously over time, one basic hypothesis in dynamic optimization is that the optimization problem can be seen as a sequence of stationary problems (Branke, 2002; Jin & Branke, 2005):

$$< f(\mathbf{x}, \mathbf{a}_1), f(\mathbf{x}, \mathbf{a}_2), \ldots, f(\mathbf{x}, \mathbf{a}_L) > \tag{1.45}$$

where $f(\mathbf{x}, \mathbf{a}_i)$, $i = 1, 2, \ldots, L$ is a time-invariant (stationary) function in the i-th environment, L is the total number of environments. The total time interval for the l environments is assumed to be $[0, T]$. If the frequency of environmental change is τ, then each $\lceil L = T/\tau \rceil$.

The most straightforward strategy of solving the above dynamic optimization problem is to restart the optimization once an environmental change is detected. However, the restarting strategy is uninteresting and inefficient. Thus, another basic assumption in dynamic optimization is that the optimization problem $f(\mathbf{x}, \mathbf{a}_i)$ in the i-th environment is more or less related to the one in the $i-1$-th environment or those in more than one previous environment. Thus, knowledge about the problems in the previous environment is helpful in solving the problem in the current environment more efficiently.

Many evolutionary dynamic optimization algorithms for transferring knowledge in previous environments into the current environment have been proposed, which can be divided into the following categories (Branke, 2002; Nguyen et al., 2012):

- Maintain or introduce diversity in the population. In solving stationary single-objective optimization problems, an optimizer, e.g., an evolutionary algorithm, will typically converge to one point once it finds an optimum. Loss of diversity reduces the search capability of an optimizer if there is a change in the location of the optimum. Thus, a straightforward strategy is to maintain a certain degree of diversity in the population so that once an environmental change occurs, the population can react quickly to find the new optimum. Alternatively, diversity is introduced into the population once a change is detected.
  Maintaining diversity in the genotype while achieving fast convergence is an interesting phenomenon that can also be found in biology. It has been revealed that this capability in biological systems can be partly attributed to phenotypic plasticity in the lifetime.
- Use implicit memory. One idea is to use multiple sub-populations in solving a dynamic optimization problem so that each sub-population can separately track one of the optima and a new sub-population can be generated if there is a change in the environment.
  Alternatively, ideas inspired from diversity maintenance mechanisms in biology can be used. Natural evolution has to deal with frequent environmental changes. Thus, many genetic mechanisms have been evolved in biological systems to han-

dle uncertainty in the environment. One mechanism is called multiploid, which uses multiple copies of chromosome to encode one same phenotype. Thus, different copies of the chromosome will 'memorize' the best phenotype in different environments, and once an environmental changes occurs, the corresponding chromosome will be activated and expressed, thereby quickly responding to the new environment.

- Use explicit memory. Since it is assumed that there is certain degree of correlation between the optima in the previous and current environment, it is useful to store this information, in the form of solutions (the whole genome) or pieces of genomes. Thus, these previous promising solutions or partial solutions can be added in the new population for the new environment.
  Alternatively, a model, either a probabilistic model or an associative model can be built to explicitly store the knowledge of the promising solutions in the previous environments to guide the search in the current environment.
- Predict the new optimum (Zhou et al., 2014). One most popular approach to dynamic optimization is to predict the new positions of the optimum using different models. For single-objective optimization one optimum is predicted, while for multi-objective optimization, the Pareto front in the new environment are predicted, based on individual solutions, the whole population or a model for describing the Pareto front. Figure 1.12a illustrates the idea of predicting the center of a moving Pareto front, and Fig. 1.12b is an example of predicting the moving Pareto front using a polynomial model.
- Use multi-task optimization or knowledge transfer between the different environments (Jiang et al., 2017; Yang et al., 2020). Since the problems in different environments are closely related, techniques such as multi-task optimization, which was inspired from multi-task machine learning and simultaneously optimize multiple related optimization problems, are well suited for dynamic optimization. In addition, transfer machine learning can be used for data-driven dynamic optimization problems.

The performance of a dynamic optimization algorithm can be assessed by the best or average performance over the generations, or the best performance averaged over the environments. Other related measures include performance drop and convergence speed, i.e., how fast an algorithm reacts to an environmental change. A most recent comprehensive survey of evolutionary dynamic optimization can be found in Yazdani et al. (2021a, 2021b).

1.4.5 Robust Optimization Over Time

Robust optimization and dynamic optimization are two very different philosophies for handling uncertainty in optimization. While robust optimization assumes that one single robust optimal solution can handle all uncertainty to be experienced in the lifetime of the design, dynamic optimization hypothesizes that the algorithm is

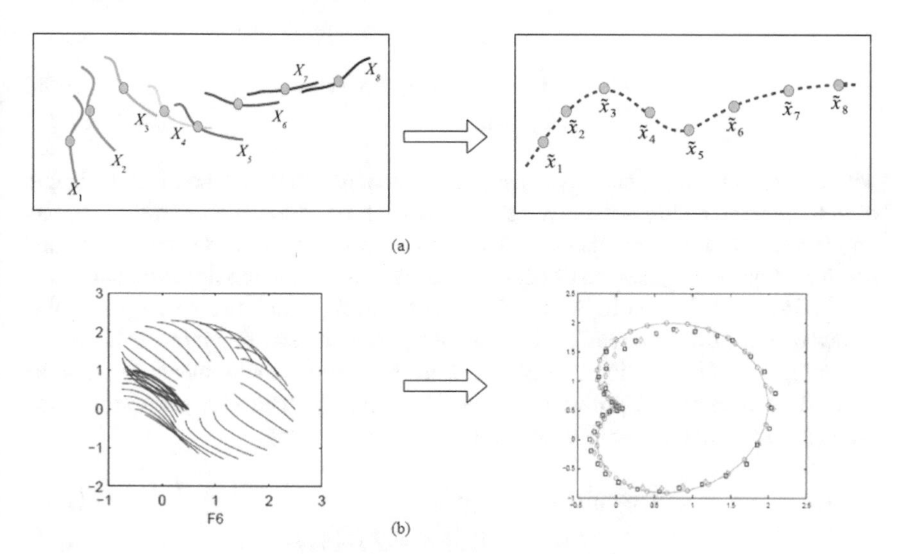

Fig. 1.12 **a** Illustration of a moving Pareto set (left), and a model predicting its center (right). **b** The moving Pareto set of a dynamic multi-objective optimization problem (left), the real trajectory of the center of the Pareto set denoted by red circles, and the predicted trajectory of the center in different environments denoted by green diamonds and blue squares (right)

always able to timely and rapidly track the moving optimum and that the change of solution is not subject to any additional cost. Apparently, robust optimization and dynamic optimization are two extreme situations and none of the basic assumptions made by them can hold in many real-world situations.

To bridge the gap between robust optimization and dynamic optimization, robust optimization over time (ROOT) was suggested (Jin et al., 2013; Yu et al., 2010). The main hypotheses of ROOT can be summarized as follows:

- An optimal solution should not be switched to a new solution even if an environmental change is detected, so long as its performance is not worse than the worst performance the user can tolerate.
- Switching solutions is subject to additional cost.
- Once the performance of a solution is worse than the worst performance the user can accept, a new optimal solutions should sought. However, the algorithm will not search for the solution having the best performance in the current environment; instead, the algorithm searches for the optimal solution that might be acceptable in a maximum number of environments to reduce the number of solution change.

From the above hypotheses, one can see that ROOT can be seen as a compromise between robust optimization and dynamic optimization. In contrast to the robustness definition in Eq. (1.37), robustness over time can be generically defined as follows:

$$F_T(\mathbf{x}) = \int_{t=t_0}^{t=t_0+T} \int_{-\infty}^{+\infty} f(\mathbf{x}, \mathbf{a}) p(\mathbf{a}(t)) d\mathbf{a}(t) dt, \tag{1.46}$$

where $p(\mathbf{a}(t))$ is the probability density function of $\mathbf{a}(t)$ at time t, and T is the length of time interval, and t_0 is the given starting time. From the above definition, we can find that ROOT not only takes into account uncertainties in the decision space and parameter space, but also the effect of these uncertainties in the time domain.

The ROOT definition in Eq. (1.46) is very generic, which is the average performance over the time interval T. However, it needs to be rewritten if one is interested in finding a ROOT solution that can be used in as many environments as possible. Given the sequence of L problems as described in (1.45), the following optimization problem can be defined to find the robust solution over time:

$$\text{maximize} \quad R = l, \tag{1.47}$$

$$\text{s.t.} \; f(\mathbf{x}, \mathbf{a}(t)) \leq \delta, \; t \in [t_c, t_c + l] \tag{1.48}$$

where δ is the worst performance the user can accept, t_c is the starting time (the time instant when the solution is to be adopted), and l is the number of environments the solution will be used. In other words, the robustness is simply defined as the number of environments the solution can be used, which is to be maximized.

Note that similar to the conventional robust optimization, there will also be a trade-off between the robustness defined in (1.47) and the average fitness over the whole time period $[0, T]$, or between the robustness and the switching cost. Thus, ROOT can also be formulated as a bi-objective optimization problem (Huang et al., 2017):

$$\text{maximize} \quad R = l, \tag{1.49}$$

$$\text{minimize} \quad C = ||\mathbf{x} - \mathbf{x}^*|| \tag{1.50}$$

$$\text{s.t.} \; f(\mathbf{x}, \mathbf{a}(t)) \leq \delta, \; t \in [t_c, t_c + l] \tag{1.51}$$

where C is the cost for switching the previous optimal solution $\mathbf{x}^*$ to the new optimal solution $\mathbf{x}$, which is defined to be the Euclidean distance between the two solutions. Of course, other definitions for the switching cost are possible.

As a result, a set of Pareto optimal solutions will be found for each environment. In practice, one of these solutions should be chosen to be implemented, e.g., the one that maximizes the ratio between robustness and switching cost (Huang et al., 2020).

It should be pointed out that finding out the ROOT solution is non-trivial as this requires to predict the performance of a solution in the future based on history data. An alternative is to find out all optima in the current environment using a multi-population strategy and then choose the ROOT solution from them according to the properties of these solutions before and after the environmental change.

1.5 Comparison of Optimization Algorithms

Comparison of optimization algorithms is of importance not only for academic research on developing new optimization algorithm, but also for choosing an optimization algorithm for solving real-world problems. It often involves the comparison of the efficiency of the optimization algorithms, and the quality of the obtained solutions. It is not an easy task to make a fair and unbiased comparison of optimization algorithms, since it involved many subtle considerations, including parameter setting, measures for comparison of the algorithm and solution quality, allowed computational resources, and benchmark problems used for comparison.

There are also differences when comparing optimization algorithms designed for single and multi-objective optimization. Table 1.1 lists the main considerations for comparing single- and multi-objective optimization algorithms. From the table, there are clear differences between single- and multi-objective optimization in evaluating the solution quality and the hardness of the problems We will discuss these points in the next subsections.

1.5.1 Algorithmic Efficiency

The efficiency of an optimization algorithm usually refers to the computational effort required to achieve a solution or a set of solutions that satisfy user's expectation. Note that in practice, it is nearly impossible to obtain the global optimum or a set of global Pareto optimal solutions for a complex problem. By computational effort, more specifically, we mean the computational complexity of an algorithm in theory, or the number of floating point operations (FLOPs), or the runtime in practice.

- *Computational complexity*. The theoretical computational complexity uses the so-called big O notation. For example, given input data size N, an algorithm with time complexity $O(N)$ is a linear time algorithm, an algorithm with time complexity

Table 1.1 Comparison of single and multi-objective algorithms

	Single-objective optimization	Multi-objective optimization
Target	The global optimum	A representative subset of Pareto optimal solutions
Performance assessment	Accuracy, efficiency	Accuracy, diversity, efficiency
Problem hardness	Fitness landscape: multi-modality, deceptiveness, ruggedness, correlation between the decision variables	Fitness landscape: multi-modality, deceptiveness, ruggedness, correlation Shape of the Pareto front: convexity, degeneration, multi-modality, regularity

is $O(\log(N))$, and one with time complexity $O(N^2)$ is proportional to the square of the size of the input data set. Generally, for a constant $k > 1$, if the complexity is $O(N^k)$, the algorithm is a polynomial time algorithm.

- *Floating point operations*. Complexity of an algorithm can also be expressed in terms the FLOPs required to find a solution, where a FLOP is a basic unit of computation such as an addition, subtraction, multiplication or division of two floating point numbers. In practice, however, the running time performance is often used to indicate the computational efficiency. Running time is usually measured by either CPU time or wall clock time and heavily depends on the implementation of the algorithm as well as the computer environment in which the algorithm is executed.
- *Memory usage*. The use of memory resources, such as registers, cache, RAM, and virtual memory, while the algorithm is being executed, is also used to measure the performance of an algorithm. Often known as the space complexity of an algorithm, the memory usage includes the amount of memory needed to hold the code for the algorithm, the input and output data, and the memory needed as working space during the optimization. Memory usage becomes increasingly important for measuring the efficiency of an algorithm, in particular when the algorithm needs to handle a huge amount of data. In this case, CPU time becomes less accurate for measuring the runtime performance, since accessing the memory will also take a lot of time.
- *Number of fundamental evaluations*. Finally, the efficiency of an optimization can be evaluated by the number of fundamental evaluations, including the number of objective function evaluations, the number of constraint function evaluations, or other information needed for optimization, such as gradient evaluations and Hessian evaluations. For many real-world optimization problems, the number of fundamental evaluations is important, particularly for expensive problems that are either computationally very expensive to evaluate, or those that can be evaluated by physical or chemical experiments only.

1.5.2 Performance Indicators

For single-objective optimization, the quality of a solution obtained by an optimization algorithm is fairly trivial to assess. For an optimization problems, the smaller the solution, the better. If a desired solution is known or given by the user, the difference between the desired and the obtained solutions can be used for evaluating the solution quality. For constrained optimization, the degree of constraint violation, or the number of violated constraints may also be used to assess the quality of the solutions, if no feasible solution is obtained. As seen in Table 1.1, it is much trickier to evaluate the quality of solutions for multi-objective optimization, where both accuracy and diversity of the solutions need to be assessed so that they can be representative of the whole Pareto front. Accuracy in the context of multi-objective optimization is also known as the convergence property of an algorithm, whilst diversity includes dis-

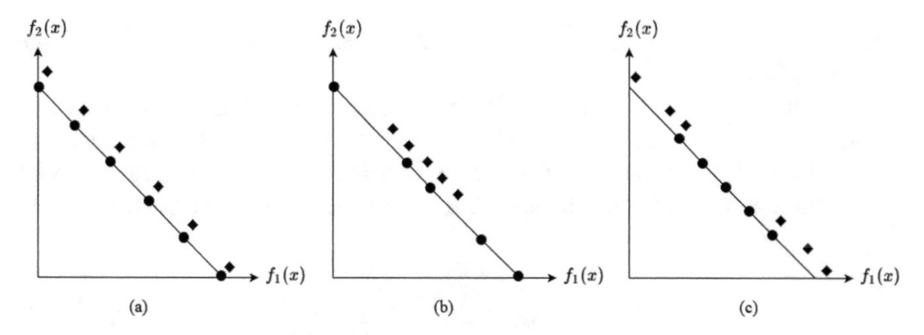

Fig. 1.13 Illustration of various qualities of the solution sets in terms of accuracy and diversity. **a** The solution set denoted by dots has good accuracy, good distribution, and good spread, while the solution set denoted by diamonds has poor accuracy, good distribution and good spread. **b** The solution set denoted by dots has good accuracy, poor distribution but good spread, while the solution set denoted by diamonds has poor accuracy, good distribution but poor spread; **c** The solution set denoted by dots has good accuracy, good distribution but poor spread, while the solution set denoted by diamonds has poor accuracy, poor distribution but good spread

tribution (also known as evenness of uniformity) and spread of the solution set. For example in Fig. 1.13, in the following, we present a number of frequently used performance indicators for measuring the quality of solution sets obtained by different algorithms (Okabe et al., 2018).

- *Distance based indicators*. Several distance based performance indicators have been suggested to account for both the accuracy and diversity of a set of solutions, assuming that a set of reference points that are uniformly sampled from the theoretical Pareto front. Given a set of solutions $A = \{\mathbf{a}_1, \mathbf{a}_2, \ldots, \mathbf{a}_{|A|}\}$ consisting of $|A|$ non-dominated solutions, and a reference set $R = \{\mathbf{r}_1, \mathbf{r}_2, \ldots, \mathbf{r}_{|R|}\}$ consisting of $|R|$ solutions, the generational distance (denoted by GD) is calculated as follows (Van Veldhuizen & Lamont, 1998):

$$GD(A) = \frac{1}{|A|}\left(\sum_{i=1}^{|A|} d_i^p\right)^{1/p}, i = 1, 2, \ldots, |A|, \tag{1.52}$$

  where $d_i = \sum_{k=1}^{m} \sqrt{(a_{i,k} - r_{j,k})^2}, k = 1, 2, \ldots, m, r_j, j \in \{1, 2, \ldots, |R|\}$ is the closest solution to a_i in R, m is the number of objectives, and p is an integer parameter. Note that for GD and its variants to be introduced below, the smaller the value, the better the quality of the solution set.
  A more widely used distance based performance indicator for measuring the quality of non-dominated solution set is called inverted generational distance, IGD for short, which is defined as follows:

$$IGD(A) = \frac{1}{|R|}\left(\sum_{j=1}^{|R|} d_j^p\right)^{1/p}, j = 1, 2, \ldots, |R|, \tag{1.53}$$

where $d_j = \sum_{k=1}^{m} \sqrt{(r_{j,k} - a_{i,k})^2}$, $k = 1, 2, \ldots, m$, $a_i, i \in \{1, 2, \ldots, |A|\}$ is the closest solution to r_j in A.
However, neither GD nor IGD is Pareto compliant. In other words, given two solutions $\mathbf{x}_1$ and $\mathbf{x}_2$, $GD(\mathbf{x}_1) < GD(\mathbf{x}_2)$ does not guarantee that $\mathbf{x}_1$ dominates $\mathbf{x}_2$. To address this issue, variants of GD and IGD, called GD^+ and IGD^+, respectively, have been proposed (Ishibuchi et al., 2015), both of which are Pareto compliant.

$$GD^+(A) = \frac{1}{|A|}\left(\sum_{i=1}^{|A|}(d_i^+)^p\right)^{1/p}, \tag{1.54}$$

$$d_i^+ = \sqrt{\left(\sum_{k=1}^{m}\max\{a_{i,k} - r_{j,k}, 0\}\right)^2}, \tag{1.55}$$

where $r_j, j \in \{1, 2, \ldots, |R|\}$ is the closest solution to a_i in R.
Similarly, IGD^+ is written by

$$IGD^+(A) = \frac{1}{|R|}\left(\sum_{j=1}^{|R|}(d_j^+)^p\right)^{1/p}, \tag{1.56}$$

$$d_j^+ = \sqrt{\left(\sum_{k=1}^{m}\max\{r_{j,k} - a_{i,k}, 0\}\right)^2}, \tag{1.57}$$

where $a_i, i \in \{1, 2, \ldots, |A|\}$ is the closest solution to r_j in A.
Note that calculation of GD, IGD and their variants requires a reference set, typically uniformly sampled from the theoretical Pareto front. In practice, however, the theoretical Pareto front is unknown. Nevertheless, this class of performance indicators can still be used for comparing the quality of the solution sets obtained by different algorithms. As illustrated in Fig. 1.14, the solution set denoted with circles is obtained by Algorithm 1 and the solution set denoted with diamonds is acquired by Algorithm 2. To compare the relative quality of these two solution sets, one can combine them and find out the non-dominated solutions, indicated by filled circles or filled diamonds, as the reference set.

- *Volume based accuracy indicators*. Hypervolume is one popular performance indicators for measuring both accuracy and diversity of a solution set, in particular when the theoretical Pareto front is unknown. Hypervolume, as the name suggests, calculates the area (in two-objective case), volume (in three-objective case) or hypervolme covered by the set of solutions related to a reference point, as shown Fig. 1.15. One can use the nadir point as the reference point, although there are also other ways of specifying the reference point. For the hypervolume, the larger the value, the better the solution set is.
  The hypervolme is a very useful performance indicator for measuring the quality of a solution set, including accuracy and diversity. However, it will becomes

computationally prohibitive when the number of objectives becomes large. Consequently, several ideas to simply the calculation of the hypervolume have been proposed, main by using e.g., Mont Carlo sampling or part of the solutions.

- *Diversity indicators*. Diversity plays an important role in multi-objective optimization, since achieving a set of representative solutions is one pivotal requirement for evolutionary multi-objective optimization. Diversity assessment involves both distribution (or evenness) and spread of the solutions, which is challenging. In the following, a few performance indicators considering distribution (evenness) or spread will be introduced, followed by a description of a diversity performance indicator inspired from biodiversity. Finally, a metric that can account for both distribution and spread will be given.

  Spacing (SP) is a distance based distribution indicator that measures the evenness in the distribution of the solutions. For a non-dominated solution set A containing $|A|$ solutions,

$$SP(A) = \sqrt{\frac{1}{|A|-1}\sum_{i=1}^{|A|}(d_i - \bar{d})^2}, \tag{1.58}$$

  where d_i is the minimum distance of the i-th solution to other solutions in A, and $\bar{d}$ denotes the mean of d_i, $i = 1, 2, \ldots, |A|$. Apparently, SP measures the standard deviation of the minimum distances of each solution to the others. This, SP mainly accounts for the evenness of a solution set without considering its spread.

  A variant of distance-based distribution is called Δ'. At first, the Euclidean distance (d_i) between two consecutive solutions in A is calculated. Then, Δ' can be calculated as follows:

$$\Delta'(A) = \sum_{i=1}^{|A|-1} \frac{|d_i - \bar{d}|}{|A|-1}, \tag{1.59}$$

  where $\bar{d}$ is the mean of d_i.

  Assessment of diversity in a high-dimensional space will become even more challenging, considering that the solutions will be very sparse in such high-dimensional space. To address this challenge, a biodiversity metric called the pure diversity (PD) proposed for measuring the diversity of biological species has been suggested as a diversity performance indicator (Wang et al., 2017a):

$$PD(A) = \max_{\mathbf{p}\in A} PD(A \setminus \{\mathbf{p}\}) + \min_{\mathbf{q}\in A\setminus\{\mathbf{p}\}} ||f(\mathbf{p}) - f(\mathbf{q})||_p, \tag{1.60}$$

  where $||f(\mathbf{p}) - f(\mathbf{q})||_p$ is the the L_p-norm based distance between two solutions $\mathbf{p}$ and $\mathbf{q}$ in objective space, and $\mathbf{p}$ and $\mathbf{q}$ are two solutions in solution set A.

  However, the above distribution and diversity measures may fail to properly characterize the degree of diversity of the solutions, in particular when the number of objectives is larger than two. Figure 1.16 shows two set of solutions, from which we can see that the one in Fig. 1.16a has a better overall diversity than that in Fig. 1.16a. However, according to the above two diversity indicators, SP and PD, the latter will have a better value than the former. To assess the overall diversity more precisely, one metric, called coverage over the Pareto front (CPF) (Tian et al., 2019), projects the solutions of an m-objective problem onto a $(m-1)$-dimensional unit simplex and a $(m-1)$-dimensional unit hypercube. Assuming the theoretic Pareto front is known, from which a reference set R is sampled. Then for a given solution set A, the main steps to calculate the $CPF(A)$ metric are as follows:

1. Replace each solution in A by its closest reference points in R to get solution set A';
2. Normalize the points in A' and R by

$$a_i = \frac{a_i - \min_{\mathbf{r} \in R} r_i}{\max_{\mathbf{r} \in R} r_i - \min_{\mathbf{r} \in R} r_i}, \tag{1.61}$$

   where a_i, r_i, $i = 1, 2, \ldots, m$, are the i-th objective of each point in A' and R, respectively.
3. Project the points in A and R to a unit simplex;
4. Project the points in A and R to a unit hypercube;
5. Calculate the side length of each monopolized hypercube R by

$$l_{\mathbf{r}} = \min_{\mathbf{s} \in R \setminus \mathbf{r}} \max_{i=1,2,\ldots m-1} |r_i - s_i|. \tag{1.62}$$

6. Shrink the size length of each monopolized hypercube in R by

$$l_{\mathbf{r}} = \min(1, r_i + l_{\mathbf{r}}/2) - \max(0, r_i - l_{\mathbf{r}}/2), \tag{1.63}$$

   where $l_{\mathbf{r}}$ is the size length of $\mathbf{r}$ in the i-th dimension.
7. Calculate the 'volume' of R by:

$$V(R) = \sum_{\mathbf{r} \in R} l_{\mathbf{r}}^{m-1}. \tag{1.64}$$

8. Calculate the side length of each monopolized hypercube in A' by

$$l_{\mathbf{a}} = \min_{\mathbf{b} \in A' \setminus \mathbf{a}} \max_{i=1,2,\ldots m-1} |a_i - b_i|. \tag{1.65}$$

9. Restrict the side length of each monopolized hypercube in A' by

$$l_{\mathbf{a}} = \min \left\{ l_{\mathbf{a}}, \left(\frac{V(R)}{|A'|} \right)^{\frac{1}{m-1}} \right\} \tag{1.66}$$

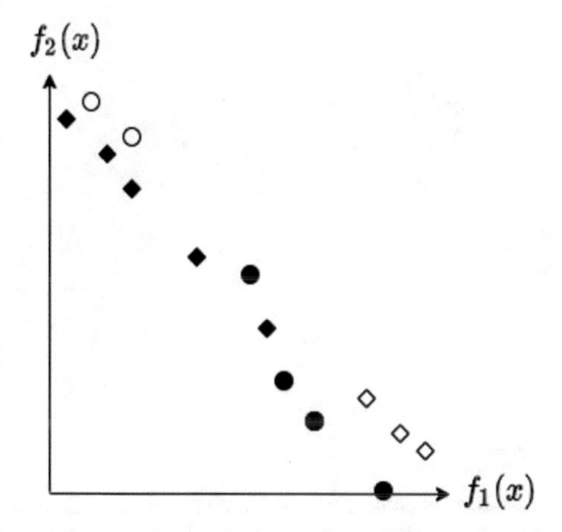

Fig. 1.14 Illustration of getting a reference set for calculating GD, IDG and their variants when the theoretical Pareto front is unknown. Filled and unfilled circles are solutions obtained by one algorithm, diamonds are solutions obtained another algorithm. Then, the non-dominated solutions of the combined solutions, denoted by filled circles and filled diamonds, can be used as the reference set to calculating GD and IGD

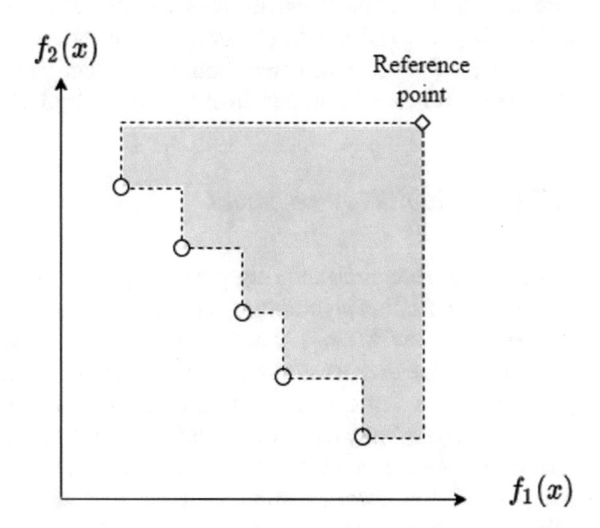

Fig. 1.15 Illustration of calculation of the hypervolume of a solution set, which is the area of the shaded region covered by the solutions with respect to the reference point

10. Calculate the 'volume' of A' by:

$$V(A') = \sum_{\mathbf{a} \; A'} l_{\mathbf{a}}^{m-1}. \tag{1.67}$$

11. Calculate $CPF = \frac{V(A')}{V(R)}$.

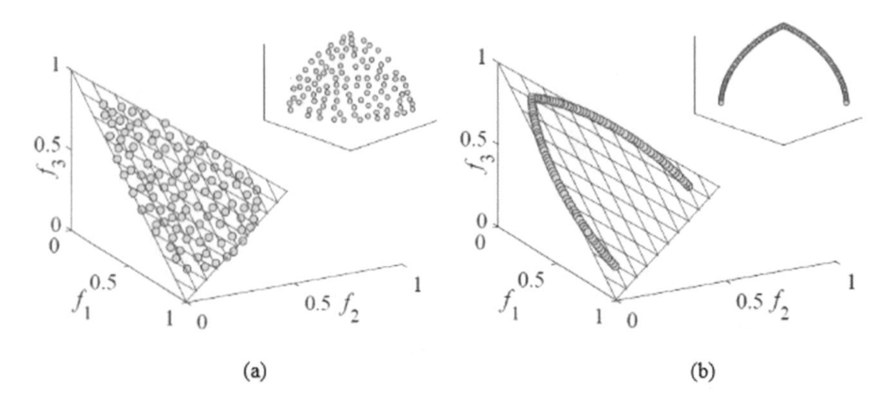

Fig. 1.16 Two different solution sets for a three-objective optimization problem, where triangle is the theoretic Pareto front, and the circles are the obtained solutions. We can visually see that the overall diversity in (**a**) is much better than (**b**)

It should be pointed out that although many performance indicators have been proposed, they are sensitive to the choice of the reference set, or the reference point, and the shape of the Pareto front. Some performance indicators may even be misleading, in particular when the Pareto front is irregular. Therefore, it is always helpful to visualize the obtained solutions in analysis of the solutions.

1.5.3 Reliability Assessment

Reliability of an optimization algorithm may refer to different aspects. For a deterministic optimization algorithm, reliability or robustness means an algorithm's performance over a wide range of optimization problems, whilst for a non-deterministic algorithm, it means robustness of the algorithm performance over multiple independent runs, and on a number of different classes of problems. Depending on the problem, the reliability may be quantitatively described by the performance, success rate, or the percentage of the global solution found. In testing algorithms, different fixed or randomly chosen starting points can be used. Note however that the all algorithms under comparison should use the same starting point for each test for the sake of fair benchmarking. Therefore, the starting points should be generated and stored before the optimization runs start.

When multiple runs on the same problem are performed, the assessment may be based on the best, average, or worst performance. When the average performance is used, the standard deviation should also be considered. A non-parametric method is the box plot, also known as box-and-whiskers plot, which visualizes the minimum and maximum values together with the quartiles, including median (50th percentile), 25th and 75th percentiles. Another benefit of the box plot is that it can also indicate the outliers in the data. Box plots display the descriptive statistic of the data without making any assumptions of the underlying statistical distribution.

1.5.4 Statistical Tests

To rigorously compare the quality of solution sets obtained by different non-deterministic optimization algorithms, one usually needs to make use of statistical analyses, which can be categorized into two classes, namely parametric and non-parametric. Parametric analyses make assumptions regarding the distribution of the data to be analyzed, including independence, normality, and homoscedasticity (or homogeneity), while non-parametric methods do not. In addition, parametric analyses usually work with categorical or nominal data whilst non-parametric methods are suited for ordinal or rank-order data.

Based on the way different data are compared, statistical tests can also be divided into pairwise and multiple comparison tests (Derrac et al., 2011). Pairwise tests are often used when one compares the quality of two sets of solutions. Otherwise, multiple comparison tests should be used. In the following, we present a number of widely used statistical tests.

- *t-test, Z-test and Wilcoxon signed-rank test.* These are three most widely used pairwise, parametric statistical tests for testing whether two data sets are statistically different. t-test can be used for normally distributed where the population standard deviation of difference is not known, while Z-test is used when the standard deviation is known. Finally, Wilcoxon signed-rank test should be used if the difference is not normally distributed. In practice, the quality difference of the solution sets obtained by non-deterministic optimization algorithms may not be described by a normal distribution.
- *ANOVA*. The analysis of variance (ANOVA) is a multiple parametric test used to determine whether there are any statistically significant differences between the means of three or more independent data sets using the F-distribution. The null hypothesis here is that the means are equal. Therefore, a significant result means that the two means are unequal. If the analysis is made on one variable (factor), it is known as one-way ANOVA, while a two-way ANOVA is able to make analysis on two independent variables.
- *Friedman's test*. The non-parametric counterpart of ANOVA is the Friedman test, which is used to test for differences between multiple data sets when the dependent variable is measured in ordinal (rank-order). Friedman's test can be used to detect significant differences between the quality of the solution sets obtained by multiple algorithms on multiple optimization problems. The null hypothesis for Friedman's test states that all algorithms perform similarly. When comparing K algorithms on N problems, we need at first to convert the original results to ranks by ranking the algorithms from the best to worst. That is, the best algorithm will have the rank 1 and the worst the K. Assume the rank of algorithm i on problem j is r_i^j, then the average rank of the i-th algorithm on all problems is

$$\bar{r_i} = \frac{1}{N} \sum_{j=1}^{N} r_i^j. \tag{1.68}$$

Then the Friedman statistic can be calculated to see if the null hypothesis is rejected:

$$Q_F = \frac{12N}{K(K+1)}\left[\sum_{i=1}^{K} \bar{r}_i - \frac{K(K+1)^2}{4}\right]. \tag{1.69}$$

Note, however, that the above statistics assumes that both K and N are sufficiently large. Empirically, K should be larger than 5, and N should be larger than 10. When the number of algorithms for comparison is small, a variant of Friedman's test, known as the Friedman Aligned Ranks test can be adopted. In this technique, the algorithms' performance are ranked from 1 to $K \times N$ in relation to each other. The ranks assigned to the aligned observations are called aligned ranks and the Friedman Aligned Ranks test statistic can be calculated as follows:

$$Q_{AF} = \frac{(K-1)\left[\sum_{i=1}^{K} R_{.i}^2 - (KN^2/4)(KN+1)^2\right]}{KN(KN+1))2KN+1)/6 - (1/K)\sum_{j=1}^{N} R_{j.}^2}, \tag{1.70}$$

where $R_{.i}$ is the rank total of the i-th algorithm, and $R_{j.}$ is equal to the rank total of the j-th problem, $i = 1, 2, \ldots, K$ and $j = 1, 2, \ldots, N$.

Non-parametric statistical tests can be applied to compare both deterministic and non-deterministic algorithms. It should be noted that pairwise methods are for comparing two algorithms only and multiple comparison methods must be used when comparing multiple algorithms. In addition, the number of problems should be larger than the number of algorithms if the Friedman's test or the Friedman Aligned Ranks test is used. Finally, the Friedman's tests assume that all problems to be compared are equally important. If the problems are of different levels of importance, then the Quade test can be used.

1.5.5 Benchmark Problems

Benchmark problems are particularly important in the field of optimization for multiple reasons, for example for demonstration of the performance of a new algorithm comparing with the state-of-the-art, for studies for determining the parameter settings in optimization algorithms, and for helping select the best algorithm for solving a real-world problem.

Constructing useful and challenging test problems is non-trivial and should follow a few basic principles. For example, the test problems should be scalable to have an arbitrary number of objective functions and an arbitrary number of decision variables. In addition, the designed properties, such as the location and number of the optimums, the shape of the Pareto front, the correlation between the decision variables should be known. Ideally, the designated hardness of the test problems should be tunable, and finally, the designed test problems should be able to reflect the hardness in real-world problems.

A large number of test suites has been proposed, which can be categorized into the following groups:

- *Singe-objective optimization test problems* (Jamil & Yang, 2013). The main purposes of these benchmark problems aim to test an algorithm's performance dealing with multi-modality, correlation between the decision variables (separability), depth and width of basins and valleys, and deceptiveness.
- *Constrained single-objective optimization test problems* (Liang et al., 2006). The hardness of constrained optimization problems to be tested may include the number of constraints, the linearity of the constraints, the connectedness of the feasible regions and the ratio of feasible regions to the overall search space.
- *Multi- and many-objective optimization test problems*. Most test suites for multi- and many-objective optimization (Deb et al., 2005; Zitzler et al., 2000) are meant for designing different shapes of Pareto fronts, such as convexity, connectedness, multi-modality, and degenerateness, and the complexity of the Pareto front. In addition, separability, the complexity in the distribution and uniqueness of the Pareto optimal solutions in the decision space, and the uniformity of the functional map from the decision space to the objective space may also be considered in designing test problems (Huband et al., 2006; Okabe et al., 2004). Test problems have also been designed for testing algorithms' capability in identifying knee points on the Pareto front, and those dedicated to problems having irregular Pareto fronts (Yu et al., 2019).
- *Dynamic optimization test problems*. Dynamic test problems aims to examine an algorithm's performance in handling various changes of the number and location of the optima in the decision and objective spaces (Farina et al., 2004; Jiang et al., 2017; Jin & Sendhoff, 2004). These test problems focus on the nature of the changes (e.g., continuous,linear, periodic, random changes among others), the severity of the changes, and the speed of the changes. In some test problems, both the objective and constraint functions may change over time.
- *Large-scale optimization test problems*. This category of test problems includes large-scale single-objective problems (Li et al., 2013; Omidvar et al., 2015) and large-scale multi- and many-objective test problems (Cheng et al., 2017a, 2017b). For single-objective large-scale test problems, the focus is mainly on the non-separability, non-uniformity of the sub-problems, degree of overlapping and importance of the sub-problems, ill-conditioning, symmetry-breaking and irregularity.
- *Multi-fidelity optimization test problems*. Some real-world problems may be evaluated using multiple levels of fidelity to approximate the original problems. For example, in design of a turbine engine, one may use two-dimensional Euler solver, or two-dimensional Navier-Stokes solving or three-dimensional solvers. There is typically a trade-off between the fidelity and computational complexity between these different solvers. So the main considerations in designing multi-fidelity test problems (Wang et al., 2018b) include resolution errors, i.e., the inconsistency between different levels of approximations, which are deterministic, stochastic errors resulting from the randomness in numerical simulations, and instability

errors caused by divergence in numerical simulations, which may happen at a probability. In addition, the ratio in the computational complexity between different levels of approximation is also a critical aspect.

- *Data-driven optimization test problems*. While nearly all test problems are white-box problems, test problems for data-driven optimization created either from white-box benchmark problems or real-world problems have also been proposed. Compared with white-box test problems, these test problems are presented in the form of data and therefore, solution of problems are based on data only (He et al., 2020).

Other test problems may include multiple aspects listed above, such as dynamic multi-objective problems, constrained multi-objective optimization problems, and test problems for robust optimization, among several others.

One major debating issue in designing test problems is whether the hardness considered in the test problems reflects the difficulties in solving real-world problems. While it is always valuable to design a range of benchmark problems to test different facets of an optimization algorithm, introducing problems from the real-world with an appropriate level of abstraction may become increasingly important.

1.6 Summary

This chapter provides a brief yet comprehensive introduction to the most important aspects related to optimization, including the definition and categorization of optimization problems, main theoretical and practical challenges optimization, assessment of the quality of approximated solutions, and evaluation of the performance of optimization algorithms. This shall offer adequate background knowledge for understanding the main topics in optimization to be detailed in this book.

References

Beyer, H.-G., & Sendhoff, B. (2007). Robust optimization—A comprehensive survey. *Computer Methods in Applied Mechanics and Engineering*, *196*(33), 3190–3218.

Branke, J. (2002). *Evolutionary optimization in dynamic environments*. Kluwer.

Brans, J.-P., Vincke, P., & Mareschal, B. (1986). How to select and how to rank projects: The PROMETHEE method. *European Journal of Operational Research*, *24*(2), 228–238.

Cheng, R., Jin, Y., Olhofer, M., & Sendhoff, B. (2017a). Test problems for large-scale multiobjective and many-objective optimization. *IEEE Transactions on Cybernetics*, *7*(12), 4108–4121.

Cheng, R., Li, M., Tian, Y., Zhang, X., Yang, S., Jin, Y., et al. (2017b). A benchmark test suite for evolutionary many-objective optimization. *Complex & Intelligent Systems*, *3*(1), 67–81.

Deb, K., & Gupta, S. (2011). Understanding knee points in bicriteria problems and their implications as preferred solution principles. *Engineering Optimization*, *43*(11), 1175–1204.

Deb, K., Thiele, L., Laumanns, M., & Zitzler, E. (2005). Scalable test problems for evolutionary multiobjective optimization. In *Evolutionary multiobjective optimization* (pp. 105–145). Springer.

Derrac, J., García, S., Molina, D., & Herrera, F. (2011). A practical tutorial on the use of non-parametric statistical tests as a methodology for comparing evolutionary and swarm intelligence algorithms. *Swarm and Evolutionary Computation*, *1*, 3–18.

Farina, M., Deb, K., & Amato, P. (2004). Dynamic multiobjective optimization problems: Test cases, approximations, and applications. *IEEE Transactions on Evolutionary Computation*, *8*(5), 425–442.

Gembicki, F. (1974). *Vector optimization for control with performance and parameter sensitivity indices* (Ph.D. thesis, Case Western Reserve University, Cleveland, Ohio).

He, C., Tian, Y., Wang, H., & Jin, Y. (2020). A repository of real-world datasets for data-driven evolutionary multiobjective optimization. *Complex & Intelligent Systems*, *6*, 189–197.

Huang, Y., Ding, Y., Hao, K., & Jin, Y. (2017). A multi-objective approach to robust optimization over time considering switching cost. *Information Sciences*, *394*, 183–197.

Huang, Y., Jin, Y., & Hao, K. (2020). Decision-making and multi-objectivization for cost sensitive robust optimization over time. *Knowlewdge-Based Systems*, *199*.

Huband, S., Hingston, P., Barone, L., & While, L. (2006). A review of multiobjective test problems and a scalable test problem toolkit. *IEEE Transactions on Evolutionary Computation*, *10*(5), 477–506.

Ikeda, K., Kita, H., & Kobayashi, S. (2001). Failure of pareto-based MOEAs: Does non-dominated really mean near to optimal? *Congress on Evolutionary Computation* (pp. 957–962).

Ishibuchi, H., Masuda, H., Tanigaki, Y., & Nojima, Y. (2015). Modified distance calculation in generational distance and inverted generational distance. In *Proceedings of the International Conference on Evolutionary Multi-criterion Optimization* (pp. 110–125).

Jamil, M., & Yang, X.-S. (2013). A literature survey of benchmark functions for global optimization problems. *International Journal of Mathematical Modelling and Numerical Optimisation*, *4*(2), 150–194.

Jiang, M., Huang, Z., Qiu, L., Huang, W., & Yen, G. G. (2017). Transfer learning-based dynamic multiobjective optimization algorithms. *IEEE Transactions on Evolutionary Computation*, *22*(4), 501–514.

Jin, Y., & Branke, J. (2005). Evolutionary optimization in uncertain environments—A survey. *IEEE Transactions on Evolutionary Computation*, *9*(3), 303–317.

Jin, Y., & Sendhoff, B. (2003). Trade-off between performance and robustness: An evolutionary multiobjective approach. In *Proceedings of Second International Conference on Evolutionary Multi-criteria Optimization* (pp. 237–251).

Jin, Y., & Sendhoff, B. (2004). Constructing dynamic test problems using the multi-objective optimization concept. In *Applications of Evolutionary Computing, LNCS 3005* (pp. 525–536). Springer.

Jin, Y., & Sendhoff, B. (2009). A systems approach to evolutionary multiobjective structural optimization and beyond. *IEEE Computational Intelligence Magazine*, *4*(3), 62–76.

Jin, Y., Tang, K., Yu, X., Sendhoff, B., & Yao, X. (2013). A framework for finding robust optimal solutions over time. *Memetic Computing*, *5*(3), 3–18.

Jin, Y., Wang, H., Chugh, T., Guo, D., & Miettinen, K. (2018). Data-driven evolutionary optimization: An overview and case studies. *IEEE Transactions on Evolutionary Computation*, *23*(3), 442–458.

Kochenderfer, M. J., & Wheeler, T. A. (2019). *Algorithms for optimization*. MIT Press.

Laumanns, M., Thiele, L., Deb, K., & Zitzler, E. (2002). Combining convergence and diversity in evolutionary multiobjective optimization. *Evolutionary Computation*, *10*(3), 263–282.

Li, X., Tang, K., Omidvar, M. N., Yang, Z., & Qin, K. (2013). *Benchmark functions for the CEC 2013 special session and competition on large-scale global optimization*. Technical report, Evolutionary Computation and Machine Learning Group, RMIT University, Australia.

Liang, J. J., Runarsson, T.P., Mezura-Montes, E., Clerc, M., Suganthan, P. N., Coello Coello, C. A., & Deb, K. (2006). *Problem definitions and evaluation criteria for the CEC 2006, special session on constrained real-parameter optimization*. Technical report, Technical Report, Nanyang Technological University.

Liu, J., & Jin, Y. (2021). Multi-objective search of robust neural architectures against multiple types of adversarial attacks. *Neurocomputing*.

Liu, J., St-Pierre, D. L., & Teytaud, O. (2014). A mathematically derived number of resamplings for noisy optimization. In *Proceedings of the Genetic and Evolutionary Computation Conference Companion* (pp. 61–62), New York, NY, USA. ACM.

Nocedal, J., & Wright, S. J. (1999). *Numerical optimization*. Springer Science & Business Media.

Miettinen, K. (1999). *Nonlinear multiobjective optimization*. Springer.

Nguyen, T. T., Yang, S., & Branke, J. (2012). Evolutionary dynamic optimization: A survey of the state of the art. *Swarm and Evolutionary Computation*, *6*(117–129), 1–24.

Okabe, T., Jin, Y., Olhofer, M., & Sendhoff, B. (2004). On test functions for evolutionary multi-objective optimization. In *Parallel problem solving from nature* (Vol. VIII, pp. 792–802). Springer.

Okabe, T., Jin, Y., & Sendhoff, B. (2018). A critical survey of performance indices for multi-objective optimization. In *Proceedings of the IEEE Congress on Evolutionary Computation (CEC)* (pp. 878–885).

Omidvar, M. N., Li, X., & Tang, K. (2015). Designing benchmark problemsfor large-scale continuous optimization. *Information Sciences*, *316*, 419–436.

Rakshit, P., Konar, A., & Das, S. (2017). Noisy evolutionary optimization algorithms—A comprehensive survey. *Swarm and Evolutionary Computation*, *33*, 18–45.

Tian, Y., Cheng, R., Zhang, X., Li, M., & Jin, Y. (2019). Diversity assessment of multi-objective evolutionary algorithms: Performance metric and benchmark problems. *IEEE Computational Intelligence Magazine*, *14*(3), 61–74.

Van Veldhuizen, D. A., & Lamont, G. B. (1998). Evolutionary computation and convergence to a pareto front. In *Late Breaking Papers of the Genetic Programmming 1998 Conference* (pp. 221–228).

Wang, H., Doherty, J., & Jin, Y. (2018a). Hierarchical surrogate-assisted evolutionary multi-scenario airfoil shape optimization. In *Proceedings of the IEEE Congress on Evolutionary Computation (CEC)*. IEEE.

Wang, H., Jin, Y., & Doherty, J. (2018b). A generic test suite for evolutionary multi-fidelity optimization. *IEEE Transactions on Evolutionary Computation*, *22*(6), 836–850.

Wang, H., Jin, Y., & Yao, X. (2017a). Diversity assessment in many-objective optimization. *IEEE Transactions on Cybernetics*, *47*(6), 1510–1522.

Wang, H., Olhofer, M., & Jin, Y. (2017b). Mini-review on preference modeling and articulation in multi-objective optimization: Current status and challenges. *Complex & Intelligent Systems*, *3*(4), 233–245.

Yang, C., Ding, J., Jin, Y., & Chai, T. (2020). A data stream ensemble assisted multifactorial evolutionary algorithm for offline data-driven dynamic optimization. *IEEE Transactions on Cybernetics* (submitted).

Yazdani, D., Cheng, R., Yazdani, D., Branke, J., Jin, Y., & Yao, X. (2021a). A survey of evolutionary continuous dynamic optimization over two decades—Part A. *IEEE Transactions on Evolutionary Computation*.

Yazdani, D., Cheng, R., Yazdani, D., Branke, J., Jin, Y., & Yao, X. (2021b). A survey of evolutionary continuous dynamic optimization over two decades—Part B. *IEEE Transactions on Evolutionary Computation*.

Yu, G., Jin, Y., & Olhofer, M. (2019). Benchmark problems and performance indicators for search of knee points in multiobjective optimization. *IEEE Transactions on Cybernetics*, *50*(8), 3531–3544.

Yu, X., Jin, Y., Tang, K., & Yao, X.(2010). Robust optimization over time—A new perspective on dynamic optimization problems. In *Congress on Evolutionary Computation* (pp. 3998–4003). IEEE.

Zitzler, E., Deb, K., & Thiele, L. (2000). Comparison of multiobjective evolutionary algorithms: Empirical results. *Evolutionary Computation*, *8*(2), 173–195.

Zhou, A., Jin, Y., & Zhang, Q. (2014). A population prediction strategy for evolutionary dynamic multiobjective optimization. *IEEE Transactions on Cybernetics*, *44*(1), 40–53.

Chapter 2
Classical Optimization Algorithms

2.1 Unconstrained Optimization

By unconstrained optimization, we usually mean the solution of continuous optimization problems where there is no restriction on the values that the decision variables can take. Consider the following unconstrained problem (Nocedal & Wright, 1999):

$$\min f(\mathbf{x}) \tag{2.1}$$

where $\mathbf{x} \in \mathcal{R}^n$ is an n-dimensional decision vector. To solve such an optimization problem, one can usually start from an initial guess $\mathbf{x}_0$ and then repeat a number of iterations of search as follows (Boyd & Vandenberghe, 2004):

$$\mathbf{x}_{k+1} = \mathbf{x}_k + \alpha_k \mathbf{d}_k, \tag{2.2}$$

where $k = 0, 1, 2, \ldots$, $\alpha_k > 0$ and $\mathbf{d}_k$ are the step size and the direction of search at the k-th iteration, respectively. The iterative search stops when either a convergence criterion is met, or the allowed maximum of iterations is reached.

If the solution sequence $\{\mathbf{x}_k\}$ remains bounded, then we say that $\mathbf{x}_k$ has an accumulation point. A sufficient condition for the existence of a unique accumulation point is

$$\lim_{k \to \infty} \|\mathbf{x}_{k+1} - \mathbf{x}_k\| = 0 \tag{2.3}$$

Different optimization algorithms differ in the determination of the search direction $\mathbf{d}_k$ and the step size α_k. In the following, we are going to present the most widely used search method for continuous optimization, the gradient based method and its variants.

Y. Jin et al., *Data-Driven Evolutionary Optimization*,
Studies in Computational Intelligence 975,
https://doi.org/10.1007/978-3-030-74640-7_2

2.1.1 The Gradient Based Method

In the iterative search methods, it is of the greatest importance to determine the right search direction. When the objective function $f(\mathbf{x})$ is differential, the gradient based method is the most widely used optimization algorithm. To illustrate the basic idea of the gradient based method, let us take a look at the one-dimensional function shown in Fig. 2.1, which has a local minimum at $x = 1$ and a local maximum at $x = -1$. From the figure, we can see that if we start from an initial point between $-1 \leq x \leq 3$, then the search direction should be towards the negative of the gradient to approach to the local minimum. On the contrary, if we want to find the local maximum, one should search towards the positive of the gradient starting from $-3 \leq x \leq 1$. Based on the above observation, the gradient method can be described as follows:

$$\mathbf{x}_{k+1} = \mathbf{x}_k - \alpha_k \nabla f(\mathbf{x}_k), \tag{2.4}$$

where $\alpha_k > 0$, and $\nabla f(\mathbf{x}_k)$ is the gradient of $f(\mathbf{x})$ at $\mathbf{x}_k$.

The main merit of the gradient based method is that it can quickly converge to a local optimum if the objective function is differentiable and the step size properly set. However, the gradient based method is easily to get stuck in a local optimum, and may also suffer from very slow convergence due to a vanishing gradient or from divergence due to an exploding gradient, as illustrated in Fig. 2.2.

To avoid stagnation on a plateau or divergence, it is extremely important to adapt the step size, which is not an easy task, though. One method for adapting the step size is called back-tracking line search, which works in the following way.

1. Given $\alpha^{(0)} = 1, 0 < \beta < 1, 0 < \rho < 1, l = 1$
2. While $f(\mathbf{x}_k) - \alpha^{(l)} \nabla f(\mathbf{x}_k) > f(\mathbf{x}_k) - \rho\alpha^{(l)} (\nabla f(\mathbf{x}_k))^T \nabla f(\mathbf{x}_k), \alpha^{(l+1)} = \beta\alpha^{(l)}$, and set $l = l + 1$.
3. Return $\alpha_k = \alpha^{(l+1)}$.

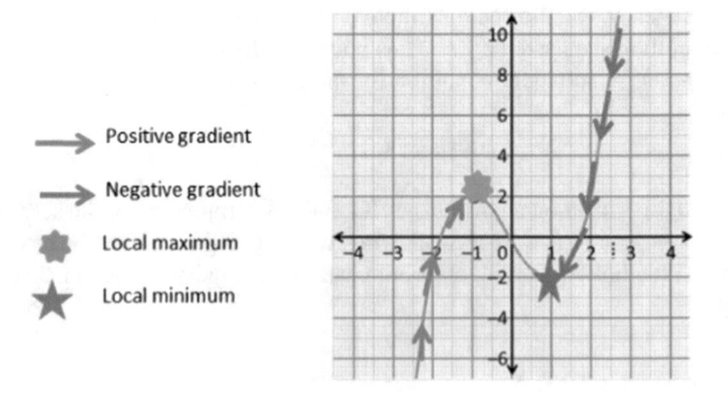

Fig. 2.1 Illustration of the gradient based method

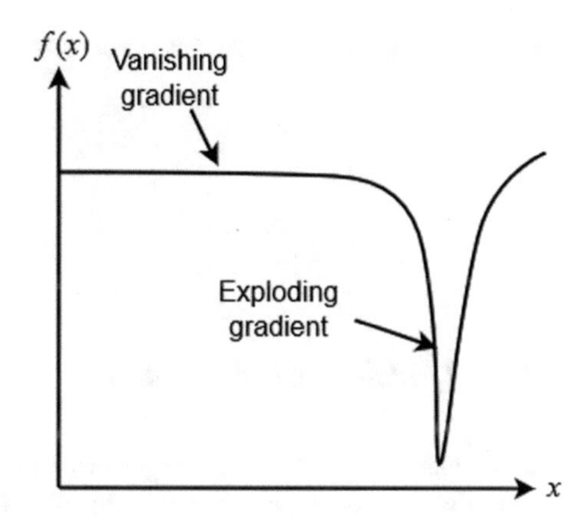

Fig. 2.2 Illustration of vanishing and exploding gradient

Another rule of thumb for adaptation of the step size is as follows:

$$\text{If } f(\mathbf{x}_{k+1}) < f(\mathbf{x}_k), \text{ Then } \alpha_{k+1} = 1.2\alpha_k \tag{2.5}$$

$$\text{If } f(\mathbf{x}_{k+1}) > f(\mathbf{x}_k), \text{ Then } \alpha_{k+1} = 0.5\alpha_k. \tag{2.6}$$

2.1.2 *Newton's Method*

The basic idea of Newton's method is to minimize a quadratic approximation of the objective function $f(\mathbf{x})$, if it is convex and twice differentiable. According to the Taylor's expansion, the quadratic approximation of $f(\mathbf{x})$ at $\mathbf{x} = \mathbf{x}_k$ is

$$f(\mathbf{x}_k + \Delta\mathbf{x}) = f(\mathbf{x}_k) + \nabla f(\mathbf{x}_k)^T \Delta\mathbf{x} + \frac{1}{2}\Delta\mathbf{x}^T \nabla^2 f(\mathbf{x}_k)\Delta\mathbf{x}, \tag{2.7}$$

Thus, the minimum of the quadratic approximation of $f(\mathbf{x})$ can be obtained by setting the gradient of $f(\mathbf{x}_k + s)$ to zero:

$$\nabla f(\mathbf{x}_k) + \nabla^2 f(\mathbf{x}_k)\Delta\mathbf{x} = 0 \tag{2.8}$$

resulting in

$$\Delta\mathbf{x} = -\frac{\nabla f(\mathbf{x}_k)}{\nabla^2 f(\mathbf{x}_k)}. \tag{2.9}$$

Thus we have the Newton's method as follows:

$$\mathbf{x}_{k+1} = \mathbf{x}_k - \frac{\nabla f(\mathbf{x}_k)}{\nabla^2 f(\mathbf{x}_k)}, \tag{2.10}$$

where $\nabla^2 f(\mathbf{x}_k)$ is also known as the Hessian matrix of $f(\mathbf{x})$ at $\mathbf{x}_k$:

$$\begin{bmatrix} \frac{\partial^2 f}{\partial x_1 \partial x_1} & \cdots & \frac{\partial^2 f}{\partial x_1 \partial x_n} \\ \vdots & \cdots & \vdots \\ \frac{\partial^2 f}{\partial x_1 \partial x_n} & \cdots & \frac{\partial^2 f}{\partial x_n \partial x_n} \end{bmatrix} \tag{2.11}$$

2.1.3 *Quasi-Newton Method*

Usually Newton's method converges faster than the gradient based method. However, Newton's method requires that the objective function be twice differentiable, which may not hold, and that the inverse of the Hessian matrix be calculated, which can be computationally intensive. To address these issues, the quasi-Newton method can be used, in particular when the number of decision variables is small to medium. The basic idea of quasi-Newton method is to approximate the Hessian matrix:

$$\mathbf{x}_{k+1} = \mathbf{x}_k - \alpha_k B_k^{-1} \nabla f(\mathbf{x}_k), \tag{2.12}$$

where B_k is an approximation of H_k, i.e., $B_k \approx H_k$. The Quasi-Newton method consists of the following steps:

1. Given starting point $\mathbf{x}_0$, and $B_0 \succ 0$, which is usually the identify matrix
2. Compute the search direction $\Delta\mathbf{x}_k = -B_k^{-1} \nabla f(\mathbf{x})$
3. Determine the step size α_k, e.g., by back-tracking line search.
4. Update $\mathbf{x}_{k+1} = \mathbf{x}_k + \alpha_k \Delta\mathbf{x}_k$
5. Update B_{k+1}.

There are different ways of updating B_k, and one most popular method is known as the Broyden-Fletcher-Goldfarb-Shanno (BFGS) algorithm, which is calculated as follows:

$$B_{k+1} = B_k + \frac{\mathbf{y}_k \mathbf{y}_k^T}{\mathbf{y}_k^T \mathbf{s}_k} + \frac{B_k \mathbf{s}_k \mathbf{s}_k^T B_k^T}{\mathbf{s}_k^T B_k \mathbf{s}_k}. \tag{2.13}$$

where $\mathbf{s}_k = \mathbf{x}_{k+1} - \mathbf{x}_k$, $\mathbf{y}_k = \nabla f(\mathbf{x}_{k+1}) - \nabla f(\mathbf{x}_k)$.

2.2 Constrained Optimization

A constrained minimization problem can be formulated as follows:

$$\min \quad f(\mathbf{x}) \tag{2.14}$$

$$\text{subject to} \quad g_i(\mathbf{x}) \le 0 \tag{2.15}$$

$$h_j(\mathbf{x}) = 0. \tag{2.16}$$

where $g_i(\mathbf{x}), i = 1, 2, \ldots, m$ is known as inequality constraints and $h_j(\mathbf{x}), j = 1, 2, \ldots, l$ are equality constraints. For the above constrained optimization problem, any solution $\mathbf{x}$ satisfying $g_i(\mathbf{x}) \leq 0$ and $h_j(\mathbf{x}) = 0$ is called a feasible solution, and the set of all feasible solutions is called the feasible solution set. In addition, a feasible local optimum of $f(\mathbf{x})$ or a feasible saddle point is called a critical point.

2.2.1 *Penalty and Barriers*

For problems with inequality constraints, one straightforward idea is to convert a constrained optimization problem into a unconstrained one, which can be realized by applying a penalty for all violated constraints (known as exterior penalty methods):

$$\min F(\mathbf{x}) = f(\mathbf{x}) + \mu \sum_{i=1}^{m} \phi_i(\mathbf{x}), \tag{2.17}$$

where $\phi_i(\mathbf{x}), i = 1, 2, \ldots, m$ are the penalty functions, and $\mu > 0$ is a hyperparameter. For example, a linear exterior penalty function may be defined as follows:

$$\phi_i(\mathbf{x}) = \max\{0, g_i(\mathbf{x})\}. \tag{2.18}$$

In the above linear penalty functions, a penalty will apply when $g_i(\mathbf{x}) > 0$. Similarly, a quadratic exterior penalty function can be adopted:

$$\phi_i(\mathbf{x}) = [\max\{0, g_i(\mathbf{x})\}]^2. \tag{2.19}$$

While a penalty increases when the violation becomes more serious, a barrier is an infinite penalty once any constraint is violated. Typically, the logarithm function can be used as the barrier function:

$$\min F(\mathbf{x}) = f(\mathbf{x}) - \mu \sum_{i=1}^{m} \log(-g_i(\mathbf{x})). \tag{2.20}$$

The penalty and barrier based methods have the following challenges in practice:

- The linear penalty function introduces non-linearity, which may cause problems for gradient based methods.
- The quadratic penalty functions usually lead to slightly infeasible solutions.
- The barrier function is not able to yield exact solutions and different hyperparameters always result in different solutions that represent trade-offs between minimizing the objective function and repelling the constraints.

2.2.2 Lagrangian Multipliers

A very popular method for solving constrained optimization problems is known as Lagrangian multipliers, which reformulates the original constrained optimization problem with inequality constraints into the following Lagrangian function:

$$\mathcal{L}(\mathbf{x}, \lambda) = f(\mathbf{x}) + \sum_{i=1}^{m} \lambda_i g_i(\mathbf{x}). \tag{2.21}$$

In this way, solution to a constrained problem is deduced by solving for both the original decision variables, called primal variables, and the optimal Lagrange multipliers, called dual variables.

The necessary conditions for a stationary point of the Lagrangian function are

$$\nabla f(\mathbf{x}) + \sum_{i=0}^{m} \lambda_i \nabla g_i(\mathbf{x}) = 0 \text{ (stationarity)} \tag{2.22}$$

$$\forall i,\, g_i(\mathbf{x}) \leq 0 \text{ (primal feasibility)} \tag{2.23}$$

$$\forall i,\, \lambda_i g_i(\mathbf{x}) = 0 \text{ (complementary)} \tag{2.24}$$

$$\forall i,\, \lambda_i \geq 0; \text{ (dual feasibility)} \tag{2.25}$$

The above conditions are also known as Karush-Kuhn-Tucker conditions (KKT) for constrained problems with inequality conditions.

Similarly, for a constrained problem with both equality and inequality constraints, the Lagrangian function is defined as:

$$\mathcal{L}(\mathbf{x}, \mu, \lambda) = f(\mathbf{x}) + \sum_{j=1}^{l} \mu_j h_j(\mathbf{x}) + \sum_{i=1}^{m} \lambda_i g_i(\mathbf{x}). \tag{2.26}$$

Then the necessary conditions for a stationary point are as follows:

$$\nabla f(\mathbf{x}) + \sum_{j=0}^{l} \mu_i \nabla h_i(\mathbf{x}) + \sum_{i=0}^{m} \lambda_i \nabla g_i(\mathbf{x}) = 0 \text{ (stationarity)} \tag{2.27}$$

$$\forall i,\, g_i(\mathbf{x}) \leq 0 \text{ (primal feasibility)} \tag{2.28}$$

$$\forall j,\, h_j(\mathbf{x}) = 0 \text{ (primal feasibility)} \tag{2.29}$$

$$\forall i,\, \lambda_i g_i(\mathbf{x}) = 0 \text{ (complementary)} \tag{2.30}$$

$$\forall i,\, \lambda_i \geq 0; \text{ (dual feasibility)} \tag{2.31}$$

2.3 Derivative-Free Search Methods

Derivative-free search methods are needed for situations when the first derivatives of the objective function are not available, or the approximation of the gradient of the objective function is impractically expensive, or the objective function is subject to noise. Classical derivative-free search methods can largely divided into direct search methods and model-based methods. Note, however, that many other methods, such as evolutionary algorithms, can also be grouped into model free derivative-free search methods, while trust region method (Powell, 1970) and Bayesian optimization (Shahriari et al., 2015) can be seen as model-based derivative-free search methods.

2.3.1 *Line Search and Pattern Search*

For an unconstrained minimization problem $\min f(\mathbf{x})$, a generic derivative-free line search method can be described as follows:

1. Choose $\mathbf{x}_0$, set $k = 0$;
2. **While** termination condition is not met, repeat
3. Choose a search direction $\mathbf{s}_k$ from $\mathbf{x}_k$
4. Calculate a step size α_k together with $\mathbf{s}_k$ such that $f(\mathbf{x}_k + \alpha_k \mathbf{s}_k) < f(\mathbf{x}_k)$
5. Set $\mathbf{x}_{k+1} = \mathbf{x}_k + \alpha_k \mathbf{s}_k$ if above step size α_k exists;
6. Otherwise set $\mathbf{x}_{k+1} = \mathbf{x}_k$ and $k = k + 1$
7. End **While**

There are different methods for performing line search, among many different approaches, the Hooke-Jeeves method is one heuristic technique that uses exploratory moves to find a good direction and then conducts a pattern move in that direction. Before the search starts, an initial start point $\mathbf{x}_0$ of the optimum and a search step size Δ are given. In the exploratory search phase, the i-th decision variable ($i = 1, 2, \ldots, n$, n is the number of decision variables) makes a change by $x_0^i = x_0^i \pm \Delta$ while keeping all other decision variables unchanged. A change is accepted if the objective value becomes better, i.e., if $f(\mathbf{x}') < f(\mathbf{x})$, where $\mathbf{x}' = \{x_0^1, x_0^2, \ldots, x_0^i \pm \Delta, \ldots, x_0^n\}$. Once the exploratory search is completed for all decision variables, the patter move phase begins by updating the start point $\mathbf{x}_0 = \mathbf{x}'$ if at least one exploratory search is successful. If none of the exploratory search is able to find a better objective value, reduce the step size by setting $\Delta = \Delta/10$.

2.3.2 *Nelder-Mead Simplex Method*

The Nelder-Mead technique is a heuristic search method that can converge to non-stationary points proposed by Nelder and Mead (1965). The Nelder-Mead method is not related to the simplex method, however, it keeps a simplex of points at each search iteration and makes use of the function values at the vertices of the simplex for guiding the search. Here, the simplex is a special polynomium type with $n + 1$ vertices in for n-dimensional problems, which could be a line segment on a line, a triangle on a plane, and a tetrahedron in three-dimensional space.

Given $\alpha \geq 1$, which is called reflection, $\beta > \alpha$, known as expansion, $\gamma \in (0, 1)$ is the contraction, and $\rho \in (0, 1)$ is the shrinkage. Given the initial simplex $X = \{\mathbf{x}^1, \mathbf{x}^2, \ldots, \mathbf{x}^{n+1}\}$ in $\mathbb{R}^n$, their function value $\{f(\mathbf{x}^1), f(\mathbf{x}^2), \ldots, f(\mathbf{x}^{n+1})\}$, and set $k = 1$. Repeat the following steps while the termination criterion is not met:

1. Order the vertices in a decreasing order: $f(\mathbf{x}^1) \leq f(\mathbf{x}^2) \leq \ldots \leq f(\mathbf{x}^{n+1})$;
2. Calculate the centroid $\bar{\mathbf{x}} = \sum_{i=1}^{n} \mathbf{x}^i$ of the n points except for $\mathbf{x}^{n+1}$.
3. Reflection: Compute $\mathbf{x}^r = \bar{\mathbf{x}} + \alpha(\bar{\mathbf{x}} - \mathbf{x}^{n+1})$. If $f(\mathbf{x}^1) \leq f(\mathbf{x}^r) < \mathbf{x}^{n+1}$, then replace the worst point $\mathbf{x}^{n+1}$ with $\mathbf{x}^r$ to form a new simplex. $k = k + 1$ and go to 1.
4. Expansion: If solution $\mathbf{x}^r$ is the best so far, i.e., $f(\mathbf{x}^r) < f(\mathbf{x}^1)$, then compute the expanded point: $\mathbf{x}^e = \bar{\mathbf{x}} + \beta(\mathbf{x}^r - \bar{\mathbf{x}})$.
   If the expanded point $\mathbf{x}^e$ is better than the reflected point $\mathbf{x}^r$, i.e., $f(\mathbf{x}^e) < f(\mathbf{x}^r)$, replace $\mathbf{x}^{n+1}$ with $\mathbf{x}^e$ to create a new simplex. Else replace $\mathbf{x}^{n+1}$ with $\mathbf{x}^r$. Set $k = k + 1$ and go to step 1.
5. Contraction: If $f(\mathbf{x}^n) \leq f(\mathbf{x}^r) < f(\mathbf{x}^{n+1})$, then compute the outside contract $\mathbf{x}^c = \bar{\mathbf{x}} + \gamma(\bar{\mathbf{x}} - \mathbf{x}^{n+1})$. If $f(\mathbf{x}^c) < f(\mathbf{x}^r)$, then obtain a new simplex by replacing the worst point $\mathbf{x}^{n+1}$ with $\mathbf{x}^c$. Set $k = k + 1$ and go to step 1. Else go to step 6.
   If $f(\mathbf{x}^r) \geq f(\mathbf{x}^n)$, compute the inside contract point $\mathbf{x}^c = \bar{\mathbf{x}} + \gamma(\mathbf{x}^{n+1} - \bar{\mathbf{x}})$. If $f(\mathbf{x}^c) < f(\mathbf{x}^{n+1})$, then obtain a new simplex by replacing the worst point $\mathbf{x}^{n+1}$ with $\mathbf{x}^c$. Set $k = k + 1$ and go to step 1. Else go to step 6.
6. Shrink: Replace all points except for the best one, i.e., for $i = 1, \ldots, n$, let $\mathbf{y}^i = \mathbf{x}^1 + \rho(\mathbf{x}^i - \mathbf{x}^1)$. For a new simplex with $X = \{\mathbf{x}^1, \mathbf{y}^1, \ldots, \mathbf{y}^n\}$.

Typically, $\alpha = 1$, $\beta = 2$, $\gamma = 0.5$, and $\rho = 0.5$. A common stop criterion is when the function values at simplex vertices are close to each other, or when the simplex has become very small.

2.3.3 Model-Based Derivative-Free Search Methods

The basic idea of model-based derivative-free search methods is to build a linear, quadratic or even probabilistic model to approximate the original non-differentiable or black-box objective function.

Given an initial solution $\mathbf{x}^0$, a number of data points $Y = \{\mathbf{y}^1, \mathbf{y}^2, \ldots, \mathbf{y}^q\}, f(\mathbf{x}^0) \leq f(\mathbf{x}^i)$, $i = 1, 2, \ldots, q$, $\xi \in (0, 1)$, and a region of size $\Delta_0 > 0$. Note that for an n-dimensional function, $q = n + 1$ points are needed for linear interpolation, and $q = (n + 1)(n + 2)/2$ points are needed for quadratic interpolation. A generic derivative-free search method based on an interpolation model is given as follows:

1. Form a linear or quadratic model $m_k(\mathbf{s})$ such that $m_k(\mathbf{s})(\mathbf{y}^i - \mathbf{x}^k) = f(\mathbf{y}^i)$, for $i = 1, 2, \ldots, q$.
2. Find the minimum of $m_k(\mathbf{s})$, $\mathbf{s}_k^*$ within the region Δ_k, that is, $\mathbf{s}_k^* \longleftarrow \min_{\mathbf{s}} m_k(\mathbf{s})$ subject to $\| \mathbf{s} \| < \Delta_k$.
3. Computer $\rho_k = \frac{f(\mathbf{x}_k) - f(\mathbf{x}_k + \mathbf{s}_k^*)}{f(\mathbf{x}_k) - m_k(\mathbf{s}_k^*)}$.
4. If $\rho_k \geq \xi$, then $\mathbf{x}_{k+1} = \mathbf{x}_k + \mathbf{s}_k^*$, $\Delta_{k+1} = \alpha \Delta_k$, $\alpha > 1$. Replace $\mathbf{y}^i \in Y$ by $\mathbf{x}^{k+1}$, update model m_k, $k = k + 1$.
5. Else if $\rho_k < \xi$, then $\mathbf{x}_{k+1} = \mathbf{x}_k$ and $\Delta_{k+1} = \beta \Delta_k, 0 < \beta < 1$. Sample a new point $\mathbf{x}'$, update model m_k, $k = k + 1$.

In model-based derivative-free search methods, what a model is to be built and how to update the model is a key issue. In trust-region, e.g., a quadratic model is usually built, which is typically obtained by interpolation. However, regression is preferred than interpolation in the presence of noise, since regression is able to smooth out the noise.

The initial data for building the model can be obtained by random sample, although design of experiments methods such as Latin hypercube sampling (LHS) (Iman et al., 1980) is usually more appealing. In case the optimum of the local model is not able to improve the solution, new solutions need to be sampled, typically using importance sampling, as done in efficient global optimization or Bayesian optimization. The trust region method and Bayesian optimization will be discussed in greater detail in Chap. 5.

2.4 Deterministic Global Optimization

2.4.1 Lipschitzian-Based Methods

Lipschitzian-based techniques are deterministic global optimization algorithms for a class of problems satisfying the Lipschitz condition over a hyperinterval with an unknown Lipschitz constant. These algorithms construct and optimize a piecewise function that underestimates the original one, thereby providing possibilities for global optimization of the original problem. Many algorithms for solving such problems have been developed, which distinguish themselves in the way of obtaining information about the Lipschitz constant and the strategy of exploration of the search space. Let $L > 0$ denote a Lipschitz constant of the objective function $f(\mathbf{x})$, then $|f(\mathbf{x}_1) - f(\mathbf{x}_2)| \leq L \| \mathbf{x}_1 - \mathbf{x}_2 \|$ for all $\mathbf{x}_1, \mathbf{x}_2 \in \mathbb{R}^n$. Among many others, the DIRECT algorithm and branch-and-bound search are two widely used Lipschitzian-based Methods.

2.4.2 DIRECT

The DIRECT algorithm was introduced by Jones et al. (1993), which stands for DIviding RECTangles. As the algorithm's name suggests, it belongs to a class of direct search algorithms. The algorithm recursively divides the search space and forms a tree of hyper-rectangles with its leaves forming a set of non-overlapping boxes, each being characterized by the function value of its base point and the size of the box. It is expected that the chance of finding an improvement inside a box is proportional to the fitness of the base point (exploitation) and to the box size (exploration, global search). DIRECT aims to select one point from current hyper-rectangles that has the best objective value for boxes of a similar size and is most likely to decrease in the objective function value. The amount of potential decrease in the current objective value is specified by a parameter that can balance the local and global search, where a larger value results in more global search. Thus, the identification of the potentially optimal boxes is basically a multi-objective problem. It is guaranteed to eventually sample a point arbitrarily close to the global optimum, if the algorithm is allowed to run for a large number of iterations and if the splitting procedure is not constrained by a maximal depth.

DIRECT consists of the following main steps:

1. Normalize the search space into a unit box, evaluate its base point.
2. While the stopping criterion is not met, repeat:
3. Identify the set of potentially optimal boxes denoted by $\mathcal{B}$.
4. For each potentially optimal box $b \in B$ with its point base being $\mathbf{c}$ do
    a. Determine the maximal side length d of box b
    b. Determine the set I of dimensions where b has the side length d.
    c. Sample the points $c \pm \frac{1}{3}d\mathbf{u}_i$ for all $i \in I$, where $\mathbf{u}_i$ are the unit vectors.
    d. Divide the box containing c into thirds along the dimensions in I: start with the dimension with the lowest $w_i = \min\{\mathbf{c} + \frac{1}{3}d\mathbf{u}_i, \mathbf{c} - \frac{1}{3}d\mathbf{u}_i\}$ and continue to the dimension with the largest w_i.
5. Return $\mathbf{x}^*$ and $f(\mathbf{x}^*)$.

DIRECT has been shown to work effectively for solving unconstrained continuous optimization problems where the derivative information is not available. In addition, DIRECT can be parallelized, making it more attractive for solving computationally expensive black-box problems.

2.5 Summary

In this chapter, we introduce a few basic mathematical programming methods for unconstrained and constrained continuous optimization. These methods perform deterministic search based on the zero order (derivative free), first order or second order derivative information of the objective function. Deterministic search methods are sometimes known as heuristics that depend on the problem to solve. They can be roughly categorized into construction heuristics that build solutions by means of iterations, and descending heuristics that seek a local optimum from a given solution. Two widely used deterministic search methods that are not introduced above are the simplex algorithm, a popular algorithm for linear programming proposed by Danzig (2009), and branch-and-bound algorithms (Morrisona et al., 2016) that adopt a tree search strategy to implicitly enumerate all possible solutions to the problem, in combination with pruning rules to eliminate unpromising regions of the search space.

In contrast to deterministic search methods, there are also many semi-deterministic or non-deterministic search methods that often known as meta-heuristics. Meta-heuristics aim to randomly scan the search area while exploring promising regions, hoping to reduce the likelihood of being trapped by a local optimum. Typical meta-heuristics include simulated annealing (Delahaye et al., 2018), tabu search (Gendreau, 2018), and evolutionary algorithms that are to be discussed in detail in Chap. 3.

References

Boyd, S., & Vandenberghe, L. (2004). *Numerical optimization*. Cambridge University Press.

Delahaye, D., Chaimatanan, S., & Mongeau, M. (2018). Simulated annealing: From basics to applications. In *Handbook of metaheuristics* (pp. 1–35). Springer.

Gendreau, M. (2018). An introduction to tabu search. In *Handbook of metaheuristics* (pp. 37–54). Springer.

Iman, R. L., Davenport, J. M., & Zeigler, D. K. (1980). Latin hypercube sampling (program user's guide). (LHC, in FORTRAN).

Jones, D. R., Perttunen, C. D., & Stuckman, B. E. (1993). Lipschitzian optimization without the Lipschitz constant. *Journal of Optimization Theory and Application*, *79*(1), 157–181.

Karloff, H. (2009). The simplex algorithm. In *Linear programming* (pp. 23–47). Birkhäuser Boston.

Morrisona, D. R., Jacobsonb, S. H., Sauppec, J. J., & Sewell, E. C. (2016). Branch-and-bound algorithms: A survey of recent advances insearching, branching, and pruning. *Discrete Optimization*, *19*, 79–102.

Nelder, J. A., & Mead, R. (1965). A simplex method for function minimization. *The Computer Journal*, *7*(4), 308–313.

Nocedal, J., & Wright, S. J. (1999). *Numerical optimization*. Springer Science & Business Media.

Powell, M. J. (1970). A new algorithm for unconstrained optimization. In *Nonlinear programming* (pp. 31–65). Elsevier.

Shahriari, B., Swersky, K., Wang, Z., Adams, R. P., & De Freitas, N. (2015). Taking the human out of the loop: A review of Bayesian optimization. *Proceedings of the IEEE*, *104*(1), 148–175.

Chapter 3
Evolutionary and Swarm Optimization

3.1 Introduction

Metaheuristics can be largely divided into two major categories, namely evolutionary algorithms that emulate natural evolution at a high abstraction level and swarm optimization algorithms that simulate swarm behaviors of social animals such as bird flocks and ant colonies. Both categories of metaheuristics are population based, stochastic search methods that are, in principle, able to solve all types of white- and black-box optimization problems. It should be noted, however, both evolutionary algorithms and swarm optimization algorithms can also be used as a generic tool for designing collective adaptive systems, or for simulating and understanding natural intelligence, e.g. in artificial life.

Evolutionary algorithms, as one area of artificial intelligence, can be traced back to 1960s, when evolutionary programming was suggested by Fogel and Fogel (1995) in the US, evolution strategies (Schwefel, 1995) were developed by Ingo Rechenberg and Hans-Paul Schwefel in Germany, and genetic algorithms were proposed by John Holland in the US and made popular by a book published by Goldberg (1989). Two classes of more recent evolutionary algorithms are genetic programming (Koza, 1992) developed by John Koza in late 1980s, and differential evolution (Storn & Price, 1996) by Storn and Price in late 1990s. Along the years, different classes of evolutionary algorithms have integrated and fused (Bäck, 1996).

The generic framework of genetic algorithms, evolution strategies, and genetic programming is very similar, as shown in Fig. 3.1. For a given representation (encoding scheme) of the problem to be solved, all evolutionary algorithms begin with an initialization of a parent population. From the optimization point of view, the initial population consists of a set of randomly initialized starting points for search. Then, crossover and mutation will be applied to the parent population to generate new candidate solutions, which are called offspring population. By mate selection, it is meant here the selection of parents for performing crossover. All offspring individuals will be evaluated using the fitness function, before environmental selection is performed. Note that environmental selection may be based on the offspring only (non-elitist),

Y. Jin et al., *Data-Driven Evolutionary Optimization*,
Studies in Computational Intelligence 975,
https://doi.org/10.1007/978-3-030-74640-7_3

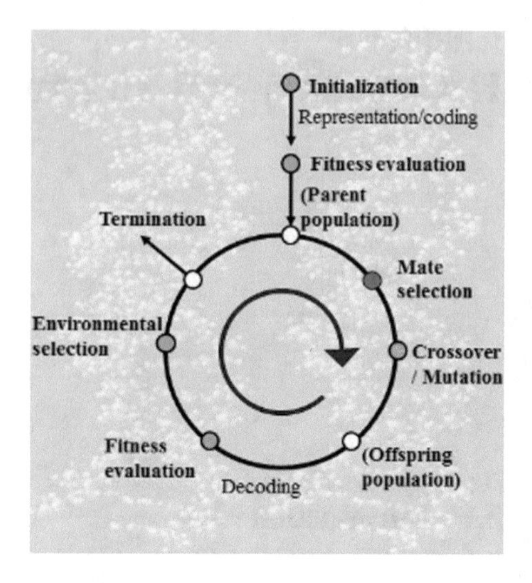

Fig. 3.1 A generic diagram of evolutionary algorithms

or based both on parent and offspring populations (elitist). The selected solutions will become the parent population of the next generation. This process repeats until a termination condition is satisfied.

Swarm optimization algorithms are another class of meta-heuristic search methods that are mainly inspired from the collective behaviors of social animals, such as bird flock and ant colony. The most popular swarm optimization algorithms are particle swarm optimization algorithms and ant colony optimization algorithms. Unlike evolutionary algorithms, swarm optimization algorithms often do not have crossover or selection operations. Note, however, that the working mechanisms of particle swarm optimization and ant colony optimization are very different, where the collective behavior of ant colonies is achieved by means of pheromone-based local communications between the ants, whereas in particle swarm optimization, particles learn from the better performing ones, resulting in a convergence of the swarm to an equilibrium.

3.2 Genetic Algorithms

3.2.1 Definitions

Genetic algorithms (GAs) aim to simulate natural evolution by modeling the genotype-phenotype mapping within each biological organism, chromosome crossover through sexual reproduction between individuals, genetic mutations at the individual level, and natural selection at the population level (Goldberg, 1989).

Most terminologies in genetic algorithms are borrowed from biology. In a genetic algorithm, a population consists of a set of individuals and the number of individuals in the population is called the population size. An individual may consist of one or multiple chromosome, each composed of genes have a particular allele. One or multiple encode a feature of the individual, e.g., the colour of a bird's feather; however, it can happen that multiple genes encode one feature (known as polygeny) or a gene may influence multiple features (pleiotropy).

Each individual is associated with a fitness value, which is, in biological systems, related to the individual's ability to survive and reproduce. In optimization, the fitness corresponds to the quality of a solution, and the mathematical description for determining the fitness value is called the fitness function.

The whole set of chromosomes is termed the genotype of the individual, while the whole set of the features is called the phenotype. Figure 3.2a illustrates a genotype-phenotype mapping consisting of eight genes and four features, while Fig. 3.2b, c shows an example of polygeny and pleiotropy, respectively.

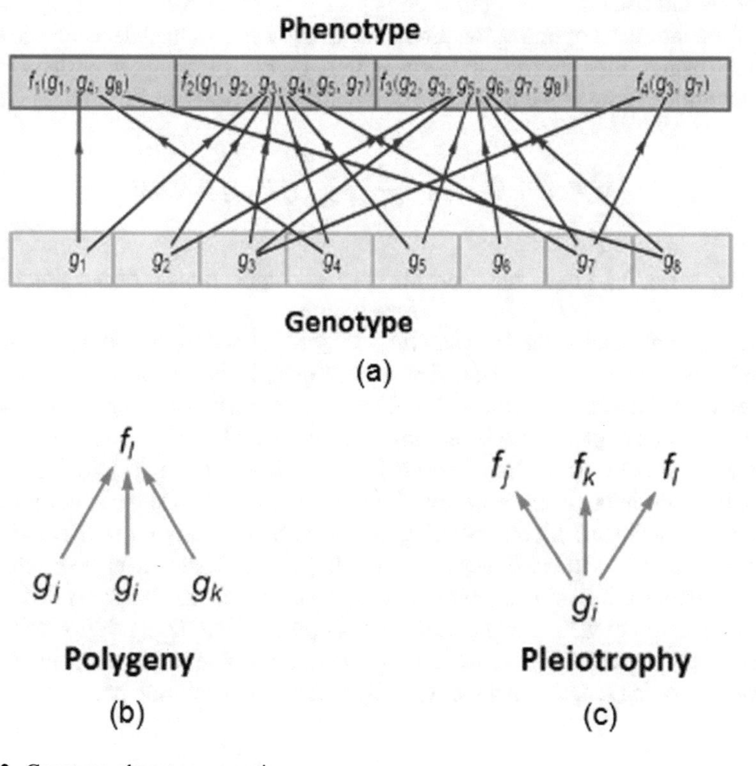

Fig. 3.2 Genotype-phenotype mappings

3.2.2 Representation

In a canonical genetic algorithm, a binary representation is adopted, simulating the DNA sequences in biological systems. For example, a 4-bit binary string is used to encode an integer decision variable, then a 12-bit string '1 0 1 1 0 1 1 0 1 1 1 0' encodes three integers, which are, from left to right, '1 0 1 1', encoding an integer of 11, '0 1 1 0', encoding 6, and '1 1 1 0', encoding 14.

More generally, the binary string as shown in Fig. 3.3 encodes n decision variables, and each variable is encoded by l-bits. Thus, if the encoded decision variables are integers, then a binary string can be decoded in the following form:

$$x_i = \sum_{j=0}^{l-1} s_j 2^j, s_j = \{0, 1\}. \tag{3.1}$$

Typically, binary coded genetic algorithms can be used for integer optimization problems and combinatorial optimization problems. In the past, they are also widely used for continuous optimization problems, where a real-valued decision variable is encoded using a binary string. In this case, the l-bit binary string can be decoded into a real number, if the range of the encoded real-valued decision variable is given:

$$x_i = a_i + (b_i - a_i)\frac{1}{2^l - 1}\sum_{j=0}^{l-1} s_j 2^j, s_j = \{0, 1\}. \tag{3.2}$$

where $x_i \in [a_i, b_i]$, i.e., a_i and b_i are the lower and upper bounds of decision variable x_i.

Note that if a real-valued number $a_i \leq x_i \leq b_i$ is encoded by l-bits, the encoded range is discretized into $2^l - 1$ equal parts and a quantization error is introduced. That is, the encoded decision variable may not be able to reach the optimum of the objective function. For example, if a decision variable between $[-1, \ 1]$ is encoded using a 3-bit string and the eight encoded numbers are $\{-1, -5/7, -3/7, -1/7, 1/7, 3/7, 5/7, 1\}$. Thus, if the optimum of the objective function is at $x = 0$, then it can never find the optimum by using a 3-bit binary string. It is clear that the larger l is, the smaller the quantization error will be. However, using a large l will lead to a huge search space, if the number of decision variables is large. Thus, in general, binary-coded genetic algorithms are not well suited to continuous optimization. One solution to address the accuracy search efficiency trade-off is to use a variable length encoding method where the coding length can vary, typically increase as evolution proceeds.

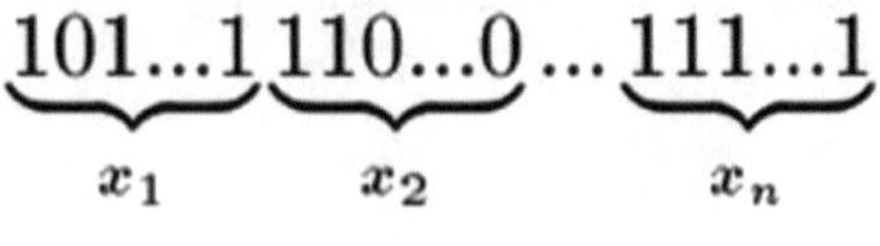

$$\underbrace{101...1}_{x_1}\underbrace{110...0}_{x_2}...\underbrace{111...1}_{x_n}$$

Fig. 3.3 A binary string encoding n decision variables, each of which is of l-bits

Another issue with binary genetic algorithms is the so-called Hamming cliff. By Hamming cliff, we mean the fact that an increment by one in the encoded decimal will require a maximum change in the binary digit. For example, a 4-bit binary string '0 1 1 1' encodes 7, while a binary string of '1 0 0 0' encodes 8. As a result, to change from 7 to 8, all four bits (a maximum change) must be changed. To resolve this issue, Gray coding code be used to replace binary coding.

3.2.3 *Crossover and Mutation*

The main driving force for changes is the crossover operator. Crossover simulates the sexual reproduction of organisms, where two parent individuals are chosen (known as mate selection) to generate new candidate solutions, called offspring individuals. During the crossover, one or multiple crossover points are randomly chosen and then the genetic materials of the two parents between the crossover points (segments of binary strings) are exchanged. Figure 3.4 illustrates three different cases of crossover, namely, one-point crossover, two-point crossover, and uniform crossover. Theoretically, for an l-bit string, one can have one-point, two-point, ..., and $(l-1)$-point crossover, and the crossover points are randomly selected. A more generic crossover method is called uniform crossover, where a binary mask of the same length as that of the chromosome is randomly generated. If a mask string bit is '1', exchange of the bits is carried out, and no exchange if the mask string bit equals '0'.

The probability of crossover is usually high, considering that it is the main genetic operator in the canonical genetic algorithm. In addition, it is also possible to perform crossover based on multiple parent individuals.

Mutation is then applied on the offspring individuals created by crossover. For binary or Gray coded genetic algorithms, mutation simply changes '1' to '0', or '0' to '1', which is also known as flip. The point mutation probability is typically low, which can be calculated by $1/l$, where l is the total length of the chromosome.

3.2.4 *Environmental Selection*

Environmental selection in genetic algorithms emulates the Darwin's theory of "survival of the fittest" in natural evolution. Through environmental selection, parents for the next generation are selected from the offspring population. A large number of environmental selection methods have been proposed, most of which in genetic algorithms are probabilistic. Widely used environmental selection strategies include fitness proportionate selection (also known as roulette wheel selection), rank proportionate selection, and tournament selection.

In fitness proportionate selection, an offspring individual is selected to be the parent of the next generation at the following probability:

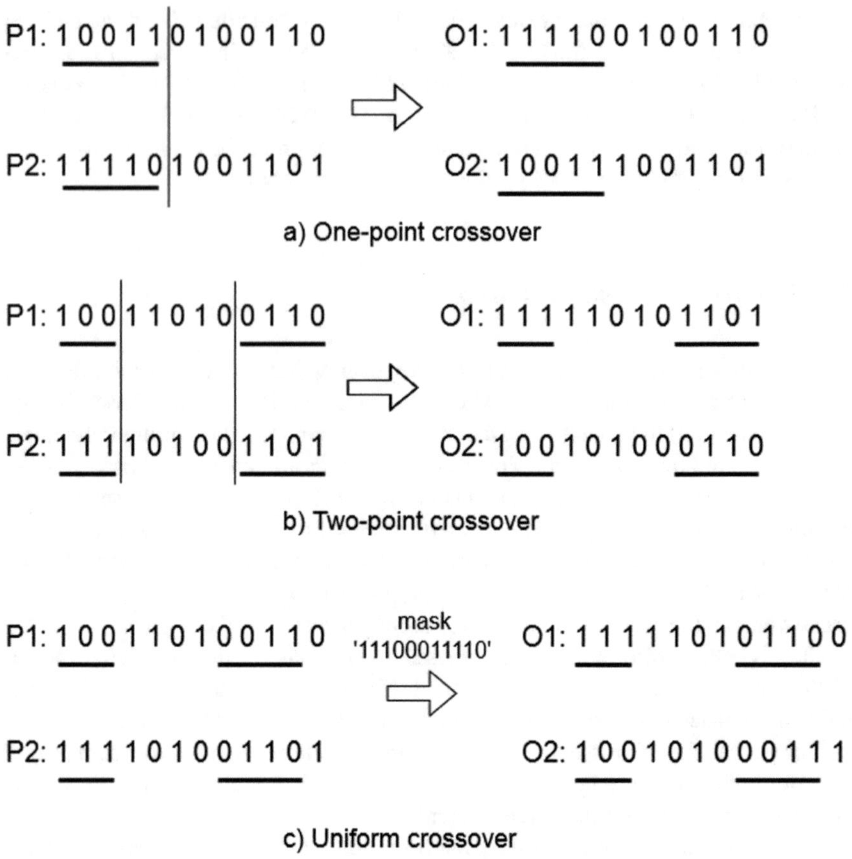

P1: 1 0 0 1 1 | 0 1 0 0 1 1 0 O1: 1 1 1 1 0 0 1 0 0 1 1 0

P2: 1 1 1 1 0 | 1 0 0 1 1 0 1 O2: 1 0 0 1 1 1 0 0 1 1 0 1

a) One-point crossover

P1: 1 0 0 | 1 1 0 1 0 | 0 1 1 0 O1: 1 1 1 1 1 0 1 0 1 1 0 1

P2: 1 1 1 | 1 0 1 0 0 | 1 1 0 1 O2: 1 0 0 1 0 1 0 0 0 1 1 0

b) Two-point crossover

P1: 1 0 0 1 1 0 1 0 0 1 1 0 mask '11100011110' O1: 1 1 1 1 1 0 1 0 1 1 0 0

P2: 1 1 1 1 0 1 0 0 1 1 0 1 O2: 1 0 0 1 0 1 0 0 0 1 1 1

c) Uniform crossover

Fig. 3.4 Illustrations of crossover. **a** One-point crossover; **b** two-point crossover; and **c** uniform crossover, where a bit of '1' in the mask string means exchange, and '0' means no exchange

$$p(i) = \frac{f_i}{\sum_{i=1}^{N} f_i}, \tag{3.3}$$

where f_i is the fitness of the i-th individual in the parent population, N is the population size. Note that for maximization problem, $f(i)$ can be the objective value of the i-th individual. For minimization, however, one needs to convert the objective value to a 'fitness' such that the smaller the objective value is, the fitter the individual will be. This can be achieved by inverting the objective value as follows:

$$f'(i) = \frac{1}{f(i) + \epsilon}, \tag{3.4}$$

where $\epsilon > 0$ is a small constant that avoids division by zero.

In fitness proportionate selection, if there is a big difference between the fitness values of the individuals, the weaker individuals will be very unlikely to be selected, leading to an overly large selection pressure. To address this problem, rank proportionate selection can be used, in which the individuals in the offspring population are ranked from the worst to the best, i.e., the worst individual has a rank of 1 while the best has a rank of N. Then the selection probability is determined by:

$$p(i) = \frac{r_i}{\sum_{i=1}^{N} r_i}, \tag{3.5}$$

where $1 \leq r(i) \leq N$ is the rank of the i-th individual. As a result, individuals whose fitness value is much smaller than others will have a larger chance to be selected. It should be pointed out that rank selection does not always reduce the selection pressure. For example, rank proportionate selection will result in a large selection pressure if the fitness values of the individuals in the population are very similar.

The tournament selection randomly picks $2 \leq k < N$ solutions from the offspring population and chooses the best of the k individual as one parent of the next generation, where k is called the tournament size. This repeats until N parents are selected. The tournament selection is called binary tournament if $k = 2$. Usually, k should be much smaller than the population size N.

None of the above three selection strategies is able to keep the best solutions in the parent population, even if they are better than all solutions in the offspring population. These selection strategies are known as non-elitism. An elitism strategy can be introduced that selects the best individual(s) together with a probabilistic selection method.

Note also that in the standard genetic algorithms, all parents are replaced by offspring individuals at each generation, which is called generational evolutionary algorithms. By contrast, one can replace the worst parent individual(s) only with the generated offspring, which is known as steady-state evolutionary algorithms.

3.3 Real-Coded Genetic Algorithms

While the canonical binary genetic algorithms can be used for integer optimization, combinatorial optimization as well as continuous optimization, their efficiency will become low if the number of decision variables is large and each one is represented with a long binary string. For example, if the optimization problem has 100 decision variables and each is represented using 20 bits, the length of the chromosome will be 2000, which forms a large search space, resulting in a low search efficiency.

3.3.1 Real-Valued Representation

In contrast to the binary coding, the real-valued coding is straightforward: For n-dimensional problem, the length of the chromosome is n, and no lower or upper bounds are needed for encoding, as shown in Fig. 3.5.

Consequently, n-point crossover or uniform crossover for binary coded genetic algorithms cannot no longer be used for real coded genetic algorithms and new variation operations must be designed. A large number of crossover operators have been designed, which can be largely divided into population centric (mean centric), meaning that the mean of the offspring generated by crossover is distributed around the center of the parent population, or parent centric, meaning that the offspring is centered around the parent individuals. Among a variety of crossover operators for real-coded genetic algorithms, the blended crossover (BLX-α) and simulated binary crossover are most widely used.

3.3.2 Blended Crossover

The blended crossover (often denoted by BLX-α) is described as follows:

1. Select two parents $\mathbf{x}_1(t)$ and $\mathbf{x}_2(t)$ from a parent population at generation t, where $\mathbf{x}_1(t) = \{x_{11}(t), x_{21}(t), \ldots, x_{i1}(t), \ldots x_{n1}(t)\}, i = 1, 2, \ldots, n, n$ is the number of decision variables.
2. Create two offspring $\mathbf{x}_1(t+1)$ and $\mathbf{x}_2(t+1)$ as follows:
3. For $i = 1$ to n do

   a. $d_i = |x_{i1}(t) - x_{i2}(t)|$
   b. Choose a uniform random real number u_1 from interval $[\min\{x_{i1}(t), x_{i2}(t)\} - \alpha d_i, \max\{x_{i1}(t), x_{i2}(t)\} + \alpha d_i]$
   c. $x_{i1}(t+1) = u_1$ (offspring 1)
   d. Choose a uniform random real number u_2 from interval $[\min\{x_{i1}(t), x_{i2}(t)\} - \alpha d_i, \max\{x_{i1}(t), x_{i2}(t)\} + \alpha d_i]$
   e. $x_{i2}(t+1) = u_2$ (offspring 2)

From the above process, we can see that BLX-α is a random interpolation or extrapolation of the two parents.

Fig. 3.5 Real-valued coding of a decision vector

x_1	x_2	...	x_n

3.3.3 *Simulated Binary Crossover and Polynomial Mutation*

Another even more popular crossover operator for real-coded genetic algorithm is called simulated binary crossover (SBX). The idea of SBX is from the factor that in binary crossover, the average of the two parents and the average of the two offspring are the same. For example, given two binary coded parents P_1: 1 0 0 1, P_2: 1 1 0 0, the decoded decimals are 9 and 12, respectively, and the average of the two parents is 10.5. If the crossover point of one-point crossover is between the second (from the left) and third bits, the two offspring will be O_1: 1 0 0 0, and O_2: 1 1 0 1, which represent 8 and 13, respectively, resulting the average of 10.5.

Thus, for a real-coded genetic algorithm, SBX aims to emulate the dynamics of binary crossover such that the mean of the offspring is the same as that of the parents. To implement this idea, a spread factor β is defined by

$$\beta = \frac{|O_1 - O_2|}{|P_1 - P_2|}, \tag{3.6}$$

where P_1, P_2, O_1, O_2 are two parents and two offspring, respectively. It can be seen that the spread factor equals 1 when the difference between the two parents and the difference between the two offspring are the same. When $\beta < 1$, the difference between the offspring is smaller than the difference between the two parents, leading to contracting effects, while $\beta > 1$ means that the difference between the two offspring are larger than the difference between the two parents, resulting in expanding effect. Offspring individuals closer to the parents are more likely to be created. Given two parent individuals, SBX generates two offspring in the following steps:

1. Generate a random number u between 0 and 1;
2. The spread factor is calculated as follows:

$$\beta(u) = \begin{cases} (2u)^{\frac{1}{\eta_c+1}}, & \text{if } u \leq 0.5 \\ \frac{1}{2(1-u)^{\frac{1}{\eta_c}}}, & \text{if } u > 0 \end{cases} \tag{3.7}$$

   where $\eta_c > 1$ is a *distribution factor*, which is a user-defined parameter.
3. Given two parents $x_{i1}(t)$ and $x_{i2}(t)$, $i = 1, \ldots, n$ is the i-th element of an n-dimensional decision vector at generation t, two offspring can be generated as follows:

$$x_{i1}(t+1) = 0.5[(1 - \beta(u))x_{i1}(t) + (1 + \beta(u))x_{i2}(t)] \tag{3.8}$$

$$x_{i2}(t+1) = 0.5[(1 + \beta(u))x_{i1}(t) + (1 - \beta(u))x_{i2}(t)]. \tag{3.9}$$

Figure 3.6 illustrates the probability density function of the offspring distribution given the parent and the distribution factor. It can be seen that density function depends on both the location of the parents and the distribution factor (η_c). The larger the distribution factor is, the more exploitative the search will be.

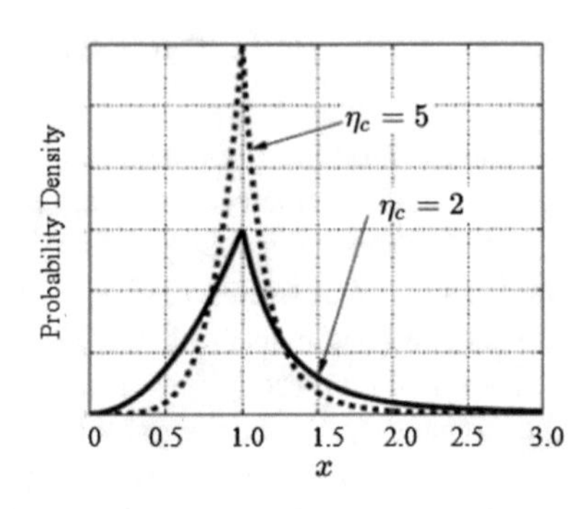

Fig. 3.6 Distribution density function of the offspring for a parent $x = 1$, and $\eta_c = 2$ and $\eta_c = 5$

Similar to the SBX, a mutation operator for real-coded genetic algorithm has also bee design, which is called polynomial mutation. Assume a decision variable of a real-coded genetic algorithm $x_i \in [x_i^L, x_i^U]$, the mutant of x_i, denoted as x_i' can be generated by:

$$x_i' = \begin{cases} x_i + \delta_L(x_i - x_i^L), & \text{if } u \leq 0.5 \\ x_i + \delta_R(x_i^U - x_i), & \text{if } u > 0.5 \end{cases} \tag{3.10}$$

where u is a randomly generated number between $[0, 1]$, δ_L and δ_R are two parameters defined as follows:

$$\delta_L = \quad (2u)^{\frac{1}{1+\eta_m}} - 1, \quad \text{if } u \leq 0.5, \tag{3.11}$$

$$\delta_R = 1 - (2(1-u))^{\frac{1}{1+\eta_m}}, \text{ if } u > 0.5, \tag{3.12}$$

where η_m is a distribution factor for mutation.

Simulated binary crossover and polynomial mutation are probably the most widely used crossover and mutation operators for real-coded genetic algorithms.

3.4 Evolution Strategies

Evolution strategies (ES) are a class of evolutionary algorithms originally designed for continuous optimization. The earliest version of evolution strategies, called $(1+1)$-ES was proposed in the 1960s, which is very similar to random search except that a rule for adaptation of the mutation strength was developed, which is known as "1/5 success rule". Notably, self-adaptation of the parameters, which is achieved by co-evolving the parameters with the decision variables, is one main difference between evolution strategies and genetic algorithms. Other main differences include that the offspring size of an evolution strategy is usually larger than the parent population size, and that Gaussian mutations are the main driving force rather than the crossover in genetic algorithms.

3.4.1 $(1+1)$-ES

As its name suggests, the $(1+1)$-ES has one parent, from which one offspring will be generated. Therefore, it is not a real population based search algorithm, however, contains the most important idea of parameter self-adaptation. For an optimization problem that minimizes a real function $f(\mathbf{x})$, where $\mathbf{x} \in \mathcal{R}^n$, the main steps of the $(1+1)$-ES can be described as below:

1. Set generation $t = 0$, initialize step size $\sigma(t)$; Randomly create one initial solution $\mathbf{x}(t) = (x_1(t), \ldots, x_n(t))$
2. For $i = 1, \ldots, n$, draw z_i from a normal distribution (Gaussian distribution)

$$x_i(t+1) = x_i(t) + z_i, z_i \sim N(0, \sigma(t+1))//\text{Mutation} \tag{3.13}$$

3. If $f(\mathbf{x}(t)) < f(\mathbf{x}(t+1))$, then $\mathbf{x}(t+1) = \mathbf{x}(t)$ // Mutation unsuccessful
4. Let $t \longleftarrow t+1$
5. Adaptation of step size $\sigma(t)$ according to the 1/5 success rule
6. If stop criterion is not met, go to step 2.

The 1/5 success rule adjusts the step size $\sigma(t)$ after every k generations by ($k \geq 5$ is defined by the user):

$$\sigma(t+1) = \begin{cases} \sigma(t)/c, & \text{if } p_s > 1/5; \text{ //increase } \sigma; \\ \sigma(t) \cdot c, & \text{if } p_s < 1/5; \text{ //decrease } \sigma; \\ \sigma(t), & \text{if } p_s = 1/5. \end{cases} \tag{3.14}$$

where p_s is the percentage of successful mutations in the past k generations, $0.8 \leq c \leq 1$. That is to say, if the percentage of success over the past k generations is larger than 1/5, the step size will be increased. If the percentage of success is less than 1/5, the step size will be decreased.

3.4.2 Evolution Strategies with One Global Step Size

Population based evolution strategies have also several variants, which differ in the number of step sizes, the way in which the step sizes are adapted, and the selection strategy. The ES with one global step size contains two chromosome, one consisting of n real-valued decision variables, and the other is a step size σ. In other words, n decision variables use the same step size for mutation. The global step size and the decision variables are mutated in the following form:

$$\sigma(t+1) = \sigma(t)\exp(\tau N(0,1)), \tag{3.15}$$

$$x_i(t+1) = x_i(t) + \sigma(t+1)N(0,1), i = 1, \ldots, n. \tag{3.16}$$

where $N(0, 1)$ denotes a random number drawn from the normal distribution, $\tau \sim 1/\sqrt{2n}$ is a constant, and n is the number of decision variables.

From the above equations, we observe that the global step σ is subject to evolution like the decision variables. Note that the order of mutation is important, i.e., the step size must be mutated before the decision variables.

3.4.3 Evolution Strategies with Individual Step Sizes

Clearly, using one global step size for all decision variables is not a good idea, especially when the the ranges of different decision variables are very different, or when the decision variables are correlated. Thus, it is more reasonable to use one step size for each decision variable. In this case, there will be n step sizes.

Similarly, the ES with individual step sizes mutates the step sizes and decision variables as follows:

$$\sigma_i(t+1) = \sigma_i(t)\exp(\tau' N(0,1) + \tau N_i(0,1)), i = 1, \ldots, n. \quad (3.17)$$

$$x_i(t+1) = x_i(t) + \sigma(t+1)N_i(0,1), i = 1, \ldots, n. \quad (3.18)$$

where $\tau' \sim 1/\sqrt{2n}$, and $\tau \sim 1/\sqrt{2\sqrt{n}}$ are two parameters, $N(0, 1)$ denotes a random number sampled from the normal distribution and used for all decision variables, while $N_i(0, 1)$ means a random number re-sampled for each decision variable. Again, the step sizes must be mutated before the decision variables.

3.4.4 Reproduction and Environmental Selection

The environmental selection in evolution strategies is very different from that in genetic algorithms in that the former adopt a deterministic selection. This is realized by producing a larger offspring population than the parent population. That is, given μ parents, then $\lambda > \mu$ offspring will be generated using the mutation strategies described. For example, for a (15, 100)-ES, $\mu = 15$ is the parent population size, $\lambda = 100$ is the offspring population size. During the reproduction, one parent will be randomly selected to generate one offspring by mutating its step size(s) at first and then the decision variables. This process is repeated for 100 times to generate 100 offspring.

There are two variants of environmental selection methods in evolution strategies, one non-elitism, denoted by (μ, λ)-ES, and the other elitism, denoted by $(\mu + \lambda)$-ES. In the non-elitism selection, the best μ individuals of the λ offspring are selected as the parents of the next generation, while in the elitist version, the parent and offspring individuals are combined, and then the best μ ones of the $\mu + \lambda$ individuals are selected as the parents of the next generation.

The overall framework of an evolution strategy can be summarized as follows:

1. Randomly initialize μ parents, where the decision variables can be randomly generated within the given search space, and the initial step size can be randomly initialized between zero and a maximum initial step size, which can empirically be set to one third of the search range;
2. Generate λ individuals using mutation in Eqs. (3.15) and (3.16) or Eqs. (3.17) and (3.18);
3. Select the μ best individuals from λ offspring (non-elitism), or from a combination of $(\mu + \lambda)$ individuals (elitism) as the parents of the next generation;
4. Go to step 2 if the stop criterion is not met.

Originally, evolution strategies do not have crossover or recombination operators. However, it has been found that if the optimum is approximately in the middle of the given search space. Since evolution strategies use real-valued coding, the following four slightly different recombination methods, local discrete or intermediate recombination, and global discrete or intermediate recombination, can be applied to both decision variables and step sizes before mutation.

The local recombination operators in evolution strategies use the same two parents for all variables. Before recombination, randomly choose two parents $\mathbf{x}_{P1}$ and $\mathbf{x}_{P2}$ from the parent population and then do the following for each decision variable:

1. Randomly choose two parents $\mathbf{x}_{P1}$ and $\mathbf{x}_{P2}$,
2. For $i = 1, 2, \ldots, n$

$$x_i' = x_{P1,i} \text{ or } x_{P2,i} \text{ (discrete recombination)} \tag{3.19}$$

$$x_i' = x_{P1,i} + \xi(x_{P2,i} - x_{P1,i}) \text{ (intermediate recombination)} \tag{3.20}$$

where $\xi \in (0, 1)$ is a parameter to be defined by the user.

By contrast, the global recombination operators may use different parents for different decision variables. That is, two parents are randomly and separately chosen for each decision variable x_i and then perform the same recombination as given in Eqs. (3.19) or (3.20).

3.4.5 Covariance Matrix Adaptation Evolution Strategy

The ES with individual step sizes provides the flexibility of separately adapting the search process along different decision variables. However, this is still insufficient when there is correlation between different decision variables. Since pairwise dependencies between the decision variables can be captured by a covariance matrix, the covariance matrix adaptation evolution strategy (CMA-ES) aims to adapt the covariance matrix to most efficiently find the optimum. The full covariance matrix $\mathbf{C}$ of the probability density function:

$$f(\mathbf{z}) = \frac{\sqrt{\det(\mathbf{C}^{-1})}}{(2\pi)^{n/2}} \exp\left(-\frac{1}{2}(\mathbf{z}^T \mathbf{C}^{-1} \mathbf{z})\right) \tag{3.21}$$

is adapted for the mutation of the decision variables, where n is the dimension of the decision vector. The derandomized CMA-ES (Hansen & Ostermeier, 1996) can be described in the following:

$$\mathbf{x}(t+1) = \mathbf{x}(t) + \delta(t)\,\mathbf{B}(t)\,\mathbf{z}, \quad z_i \sim N(0, 1) \tag{3.22}$$

where $\delta(t)$ is the overall step-size at generation t, $z_i \sim N(0, 1)$, and $\mathbf{C} = \mathbf{B}\mathbf{B}^T$. That is, $\mathbf{B}\,\mathbf{z} \sim N(\mathbf{0}, \mathbf{C})$.

The adaptation of the covariance matrix is implemented in two steps using the cumulative step-size approach, where $c_{\text{cov}} \in (0, 1)$ and $c \in (0, 1)$ determine the influence of the past during cumulative adaptation, respectively.

$$\mathbf{s}(t+1) = (1-c)\,\mathbf{s}(t) + c_u\,\mathbf{B}(t)\,\mathbf{z} \tag{3.23}$$

$$\mathbf{C}(t+1) = (1-c_{\text{cov}})\,\mathbf{C}(t) + c_{\text{cov}}\,\mathbf{s}(t+1)\mathbf{s}^T(t+1) \tag{3.24}$$

To calculate matrix $\mathbf{B}$ from $\mathbf{C}$ is non-trivial, since $\mathbf{C} = \mathbf{B}\mathbf{B}^T$ is insufficient to derive $\mathbf{B}$. Thus, we can use the eigenvectors of $\mathbf{C}$ as column vectors of $\boldsymbol{B}$, since the overall step size δ must be adapted in the last step. To achieve this, we need to know the expected length of the cumulative vector $\mathbf{s}_\delta$. Therefore, the mutation (variation) vector, which depends on $\mathbf{B}$, should obey the distribution $N(\mathbf{0}, \mathbf{1})$. This can be realized by normalizing the column vectors of $\mathbf{B}$ by the square root of the corresponding eigenvalues, yielding the matrix $\mathbf{B}_\delta$, if the column vectors of $\mathbf{B}$ are the eigenvectors of $\mathbf{C}$. Finally, $\mathbf{s}_\delta$ and δ can be adapted as follows:

$$\mathbf{s}_\delta(t+1) = (1-c)\,\mathbf{s}_\delta(t) + c_u\,\mathbf{B}_\delta(t)\,\mathbf{z} \tag{3.25}$$

$$\delta(t+1) = \delta(t)\,\exp\left(\beta\left(||\mathbf{s}_\delta(t+1)|| - \hat{\chi}_n\right)\right), \tag{3.26}$$

where $\mathbf{B}_\delta$ equals $\mathbf{B}$ with normalized columns so that $\mathbf{B}_\delta(t)\,\mathbf{z}$ follows the distribution $N(\mathbf{0}, \mathbf{1})$. $\hat{\chi}_n$ denotes the expectation of the χ_n distribution, which is the distribution of the length of an $N(\mathbf{0}, \mathbf{1})$ distributed random vector.

CMA-ES contains several hyperparameters that need to be specified. More details about determining these parameters can be found in Hansen and Ostermeier (1996).

The main differences between the ES with one global step size, ES with individual stepsizes, and CMA-ES can be illustrated in Fig. 3.7. In Fig. 3.7a, the step size of all decision variables are the same, i.e., the distribution from which the mutated samples are generated has the same variance for all decision variables (circles). By contrast, since different decision variables are allowed to have different step sizes, the distribution for different variables has different variances (ellipse), as shown in Fig. 3.7b. And since CMA-EA is able to adapt the covariance matrix, the distribution can be seen as a rotating ellipse, resulting in the most efficient search search, as

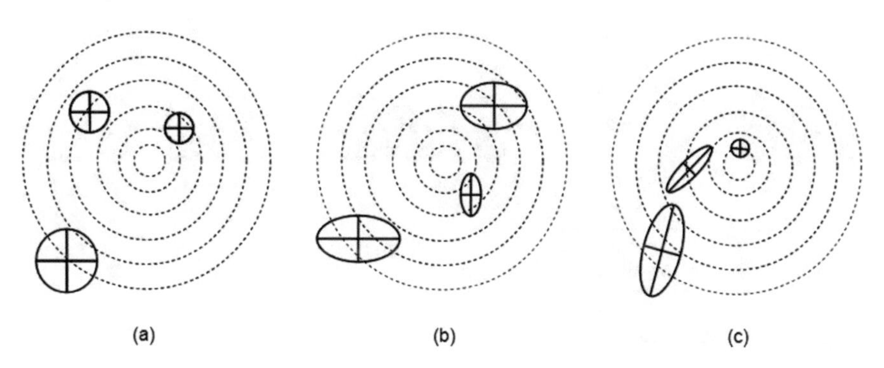

Fig. 3.7 **a** ES with one global step size; **b** ES with individual step sizes; and **c** CMA-ES

indicated in Fig. 3.7c. A large body research has been shown that CMA-ES works well for solving a large class of continuous optimization problems.

3.5 Genetic Programming

Genetic Programming (GP) (Poli et al., 2008) is a special class of EAs that automatically solve problems with a specific but high-level form, pattern, or structure. Although it can be seen as a variant of GAs using a different representation, typically a decision tree structure, GP is well suited for solving regression and feature selection problems, and therefore, can also be seen as a subset of machine learning.

GP has been used in symbolic regression (Vladislavleva et al., 2008), classification (Espejo et al., 2009), and feature selection (Muni et al., 2006). Those work can be applied to engineering design, such as antennae (Comisky et al., 2000), topology structures (Koza et al., 1996), electrical circuits (Koza et al., 1997), and software code (Forrest et al., 2009).

3.5.1 Tree-Based Genetic Programming

GP may use slightly different representations, including tree-based GP, stack-based GP, linear GP, graph GP, and strongly typed GP. Tree structure representation is the earliest and most widely used encoding method in GP (Brameier & Banzhaf, 2007). Therefore, we use tree-based GP as an example to illustrate the process of GP in this sub-section.

As shown in Fig. 3.8, GP follows the main steps of EAs: initialization, genetic variation, fitness evaluations, and environmental selection. However, GP differs from other EA techniques in the encoding (or representation) method, which leads to different initialization and variation operators.

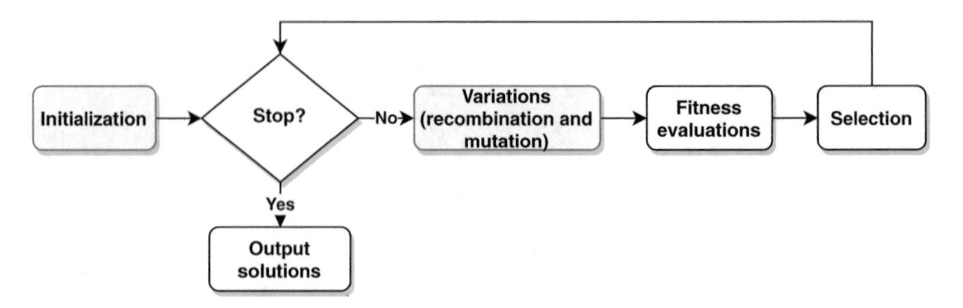

Fig. 3.8 A generic paradigm of GP

In tree-based GP, programs or mathematical expressions are represented by syntax trees rather than binary or real-valued numbers (Hoai et al., 2006). Figure 3.9 shows an example of mathematical expression $3a(3 + (y + 10))$, which is represented in the form of a decision tree. In this tree, variables or constants (3, a, y, and 10) are called leaves (or terminal nodes), and operators or functions (+) are internal nodes.

In such a tree structure, all the possible variables, constants, operators, or functions are called *primitives*, which determine the search space of GP. The *primitive set* consists of a *terminal set* for leaf nodes and a *function set* for internal nodes (Roberts et al., 2001). Depending on the domain of the problem, the terminal set can be external inputs, e.g., decision variables or attributes, and constants, the function set can be arithmetic, Boolean, mathematical, or programming functions. Table 3.1 provides some commonly primitive examples.

The choice of the primitive set is very important to the effectiveness and efficiency of GP, since it directly affects the search space. Generally, two properties need careful considerations: closure and sufficiency (Koza, 1992). The closure property consists of type consistency and evaluation safety to avoid arbitrary nodes and guarantee the feasibility of expressions during the variations. The sufficiency property means that the primitive set should contain adequate variants of expressions to represent the optimal solution of the problem to be solved.

Fig. 3.9 An example of tree structure representation of "$3a(3 + (y + 10))$"

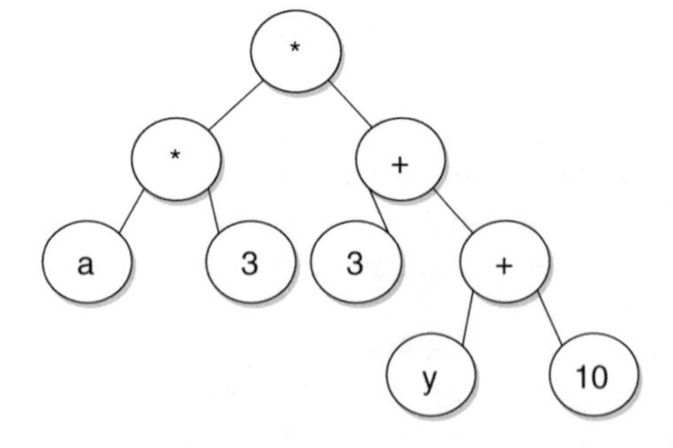

Table 3.1 Examples of primitives in GP

Category	Examples
Terminal set	
Variable	x, y, a
Constant	3, 10
.	.
.	.
Function set	
Arithmetic	$+, *, -, /,$
Boolean	AND, OR, NOT
Mathematical	sin, tan, cos, exp
Program	WHILE, FOR, IF, ELSE, END
.	.
.	.

3.5.2 *Initialization*

Because of different representations of solutions in GP, initialization of a GP population differs from that in other EAs. Two basic initialization methods (the full and grow methods) for GP are shown in Fig. 3.10, both of which randomly choose their nodes from the primitive set. The full method builds a full tree with a pre-defined depth, while the the grow method grows a tree using the depth-first search once a node reaches the the predefined depth. Their differences are elaborated in Fig. 3.10.

The above two methods are easy to be implemented, but the shapes of trees may not be sufficiently diverse. Therefore, a combination of both methods (termed ramped half-and-half) (Koza, 1992) is widely used, where half of the solutions in the initial population are generated by the full method and the other half solutions by the grow method.

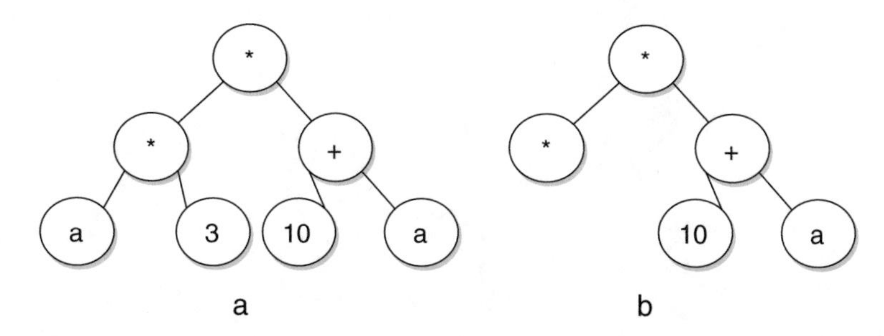

Fig. 3.10 Two basic initialization methods for GP. **a** Describes the full method, and **b** illustrates the grow method

3.5.3 *Crossover and Mutation*

Inheriting from EAs, GP applies the similar idea of crossover and mutation operators in GAs on its tree-based representations to generate new candidate solutions, termed recombination and mutation. Both operators are in the level of sub-trees as shown in Figs. 3.11 and 3.12. The recombination operator chooses one sub-tree from each parent to combine a new tree as the offspring. The mutation operator firstly chooses a node and replaces it with a randomly generated tree.

To evaluate the quality of candidate solutions, the fitness function in GP can be defined in many forms, which depends on the problem. It can be the approximation error, classification accuracy, resource or time cost for a system to reach a predefined state, or a structure incorporating user's criteria. Also, any selection mechanisms in GAs (Zames et al., 1981), such as tournament selection and roulette wheel (fitness proportionate) selection, can be applied in GP once the fitness has been properly evaluated.

3.6 Ant Colony Optimization

Ant colony optimization (ACO), as a metaheuristic inspired by the natural ant communication using pheromones (Dorigo et al., 1991, 1996), was firstly proposed to

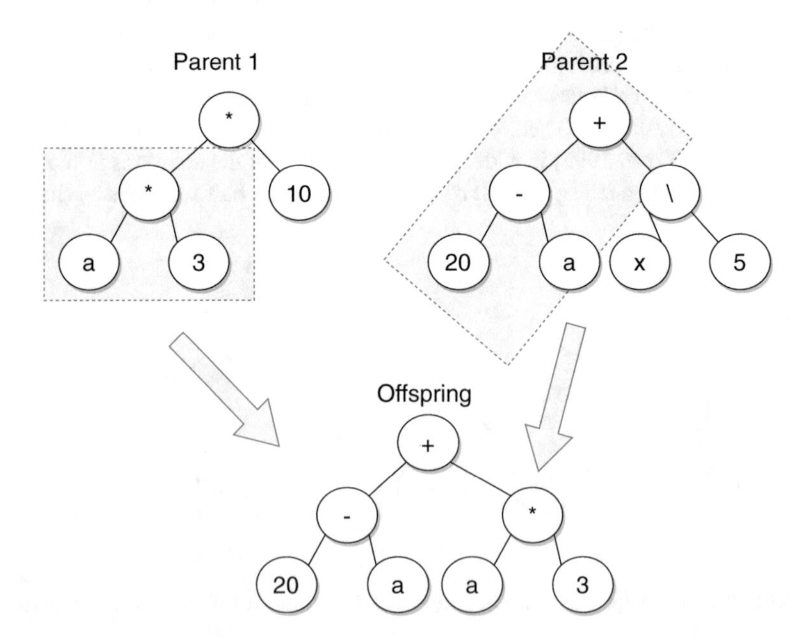

Fig. 3.11 An example of recombination in GP

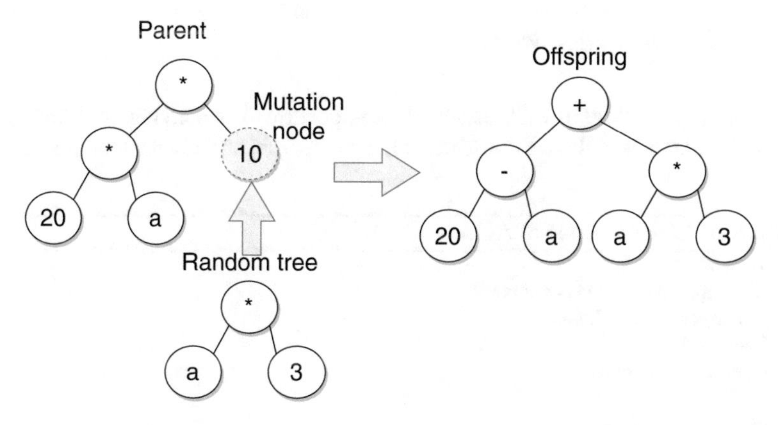

Fig. 3.12 An example of mutation in GP

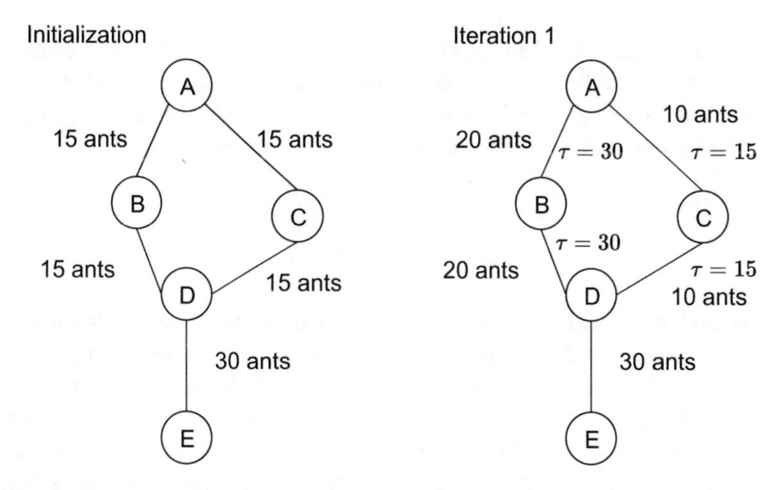

Fig. 3.13 An illustrative example of ACO

solve traveling salesman problems (TSPs) (Applegate et al., 2006). ACO employs N artificial ants (or agents) to simulate the food trailing process of an ant colony for the optimization task. As it combines the stochastic and greedy mechanisms, ACO can perform well on combinatorial optimization problems that are NP-hard (Dorigo & Stützle, 2019).

Figure 3.13 provides a simple example to illustrate the main idea of ACO, where the problem is to find the shortest path from A (starting point) to E (goal). ACO initializes each path using nearly the same number of ants, then after one iteration, ants share their pheromones τ to show the information of distance to E, and ants adjust their move strategies according to the distribution of τ. It is clear that a short distance will result in a large τ, which attracts more ants to explore the short path. At the end of ACO, ants can find the shortest path to E.

3.6.1 Overall Framework

The general procedure of ACO is shown in Algorithm 1, which has four key operations (initialization, solution construction, local search, and pheromone update).

Algorithm 1 Pseudo code of ACO.

1: Initialization.
2: **while** stopping criterion is not met **do**
3: Construct N ant solutions.
4: Local Search.
5: Update pheromones.
6: **end while**
Output: Optimal solutions.

At the very beginning of ACO, the pheromones of N ant solutions are set as τ_0. Each ant solution $\mathbf{x}$ is built based on both the heuristic and pheromone information. Assuming that the i-th dimension of $\mathbf{x}$ has a number of feasible values, the probability of the j-th feasible value is assigned as below:

$$p(x_i^j) = \frac{\tau_{ij}^{\alpha}(\eta(x_i^j))^{\beta}}{\sum \tau_{il}^{\alpha}(\eta(x_i^l))^{\beta}}, \tag{3.27}$$

where τ_{ij} and $\eta(x_i^j)$ are the pheromone and heuristic value of the j-th feasible value for x_i. Parameters α and β are pre-defined to balance the heuristic and pheromone information. After construct $\mathbf{x}$, some local search can be applied to enhance the solution, although this operation is optional. To guide the search to the optimal solution, the pheromone is updated using two mechanisms in each generation. Firstly, good solutions are collected in a set S_{upd}, if x_i^j appears in S_{upd}, τ_{ij} increases. Secondly, the pheromone evaporates over generations. Thus, the pheromone update can be expressed as follows:

$$\tau_{ij} = (1-\rho)\tau_{ij} + \sum_{\mathbf{x}\in S_{\text{upd}}|x_i^j\in\mathbf{x}} g(\mathbf{x}) \tag{3.28}$$

where ρ is a forgetting rate and $g(\mathbf{x})$ is a function to assign the pheromone increase of the component x_i^j in good solutions, which depends on the fitness function $f(\mathbf{x})$. When the stopping condition is met, ACO outputs its obtained optimal solutions.

3.6.2 Extensions

ACO has been shown to be successful in solving a wide range of combinatorial optimization problems, including routing (López-Ibáñez et al., 2009), assignment (Maniezzo & Carbonaro, 2000), job shop scheduling (Solnon, 2008), subset (Solnon & Fenet, 2006), classification (Otero et al., 2008), and haplotype inference in bioinformatics (Benedettini et al., 2008). Notably, ACO has also been adapted to continuous optimization problems too (Socha & Dorigo, 2008).

3.7 Differential Evolution

Differential evolution (DE), proposed by Price and Storn in 1996 (Storn & Price, 1996), is a class of conceptually simple and effective population-based evolutionary algorithms for solving continuous optimization problems. Given an initial population, DE searches for a global optimum of an n-dimensional problem by iteratively performing three operations, i.e., mutation, crossover, and environmental selection. The optimal solution will be the output if the method meets the stopping criterion such as the maximum number of generations or the maximum number of fitness evaluations. Different to other evolutionary algorithms, the DE algorithm is implemented by perturbing each solution using the difference vector of two randomly chosen solutions in the current population.

Figure 3.14 gives a flowchart of DE. We will give a detailed description of the four main components of DE in the following.

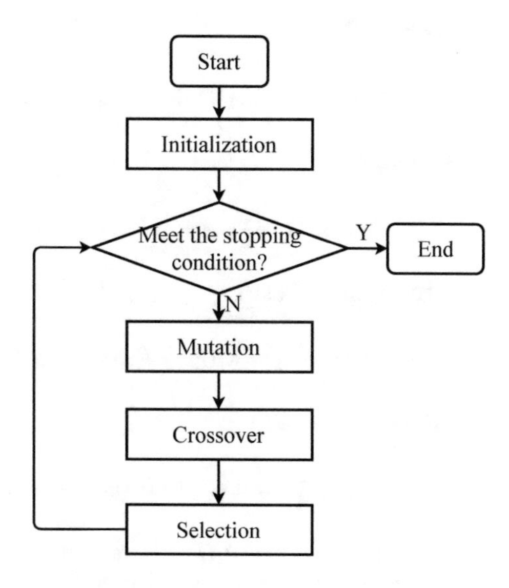

Fig. 3.14 The flowchart of DE

3.7.1 Initialization

A population P with N individuals are randomly generated in the decision space $\mathcal{R}^n$ of the optimization problem, where n is the number of decision variables. The position of any individual $\mathbf{x}_i$ on j-th dimension will be generated as follows:

$$x_{ij}(t) = x_{j,\min} + r * (x_{j,\max} - x_{j,\min}), \quad (3.29)$$

where $t = 0$ for the population initialization, r is a number randomly generated in the range $[0, 1]$. $\mathbf{x}_{\min} = (x_{1,\min}, x_{2,\min}, \ldots, x_{n,\min})$ and $\mathbf{x}_{\max} = (x_{1,\max}, x_{2,\max}, \ldots, x_{n,\max})$ are the lower and upper bounds, respectively.

3.7.2 Differential Mutation

In the context of the EAs, the differential mutation can be seen as a perturbation on decision variables. It generates a mutant, called the donor vector $\mathbf{u}$, for each individual i in the population P by adding one or more vector differences to the base vector. Some common utilized mutations are given in the following:

1. DE/rand/1

$$\mathbf{u}_i(t+1) = \mathbf{x}_{r1}(t) + F(\mathbf{x}_{r2}(t) - \mathbf{x}_{r3}(t)) \quad (3.30)$$

2. DE/best/1

$$\mathbf{u}_i(t+1) = \mathbf{x}_{\text{best}} + F(\mathbf{x}_{r1}(t) - \mathbf{x}_{r2}(t)) \quad (3.31)$$

3. DE/rand/2

$$\mathbf{u}_i(t+1) = \mathbf{x}_{r1}(t) + F(\mathbf{x}_{r2}(t) - \mathbf{x}_{r3}(t)) + F(\mathbf{x}_{r4}(t) - \mathbf{x}_{r5}(t)) \quad (3.32)$$

4. DE/best/2

$$\mathbf{u}_i(t+1) = \mathbf{x}_{\text{best}} + F(\mathbf{x}_{r1}(t) - \mathbf{x}_{r2}(t)) + F(\mathbf{x}_{r3}(t) - \mathbf{x}_{r4}(t)) \quad (3.33)$$

5. DE/current-to-best/1

$$\mathbf{u}_i(t+1) = \mathbf{x}_i(t) + F(\mathbf{x}_{\text{best}} - \mathbf{x}_i(t)) + F(\mathbf{x}_{r1}(t) - \mathbf{x}_{r2}(t)) \quad (3.34)$$

6. DE/rand-to-best/1

$$\mathbf{u}_i(t+1) = \mathbf{x}_{r1}(t) + F(\mathbf{x}_{\text{best}} - \mathbf{x}_{r1}(t)) + F(\mathbf{x}_{r2}(t) - \mathbf{x}_{r3}(t)) \quad (3.35)$$

In Eqs. (3.30)–(3.35), $\mathbf{u}_i(t+1)$ represents the vector generated by the mutation operator for individual i at generation $t+1$. The first item of the right-hand side is called the base vector. $\mathbf{x}_i(t)$ and $\mathbf{x}_{\text{best}}$ are the decision vector of individual i at generation t and the best solution found so far, respectively. $r1$, $r2$, $r3$, $r4$, and $r5$, $r1 \neq r2 \neq r3 \neq r4 \neq r5 \neq i$, are randomly selected indices and F is called the scaling factor.

3.7.3 Differential Crossover

The DE crossover is used to enhance the diversity of the population after the mutation operator. There are two commonly used crossover methods, i.e., binomial and exponential crossover, to generate a trial vector $\mathbf{v}$. In the binomial crossover, the value of j-th decision variable of a trial vector $\mathbf{v}_i$ is taken either from its parent $\mathbf{x}_i$ or the donor vectors $\mathbf{u}_i$ according to a constant in the range of [0, 1], called the crossover rate (Cr). Equation (3.36) shows the scheme of the binomial:

$$v_{ij}(t+1) = \begin{cases} u_{ij}(t+1) & \text{if } (\text{rand}() \leq Cr \text{ or } j = j_{\text{rand}}) \\ x_{ij}(t) & \text{otherwise} \end{cases} \tag{3.36}$$

where rand() represents a random number generator that generates a random number in the range of [0, 1], $j_{\text{rand}} \in \{1, 2, \ldots, n\}$ is a randomly chosen index, which ensures that at least one decision value is selected from the donor vector.

In the exponential crossover, an integer number k will be randomly selected from the range $[1, n]$, which acts as a starting point in the parent vector and L, $L < n$ consecutive elements are taken from the donor vector $\mathbf{u}$ in a circular array. The decision value will continue to be taken from donor vector until rand() $> Cr$ or the total number of elements taken from the donor vector reaches L. All other parameters of the trial vector are taken from the parent vector. Algorithm 2 gives the pseudo code of the exponential crossover.

Algorithm 2 Exponential crossover

Input: $\mathbf{x}_i$, $\mathbf{u}_i$
Output: $\mathbf{v}_i$

$\mathbf{v}_i = \mathbf{x}_i$;
Randomly choose an integer number k in the set $\{1, 2, \ldots, n\}$;
$l = 1$;
$j = k + l$;
1: **repeat**
2: $v_{ij} = u_{ij}$;
3: $l = l + 1$;
4: $j = k + l$;
5: **until** $(rand() > Cr)$ or $(l > L)$

3.7.4 Environmental Selection

Finally, the solutions which are to be passed to the next generation will be determined by fitness values of the parent and the trial vectors. The vector with no worse fitness value will replace the current parent vector, which is given in Eq. (3.37).

$$\mathbf{x}_i(t+1) = \begin{cases} \mathbf{v}_i(t+1) & \text{if } f(\mathbf{v}_i(t+1)) \leq f(\mathbf{x}_i(t) \\ \mathbf{x}_i(t) & \text{otherwise} \end{cases} \tag{3.37}$$

3.8 Particle Swarm Optimization

3.8.1 Canonical Particle Swarm Optimization

Particle swarm optimization (PSO) (Kennedy & Eberhart, 1995), simulating the bird flocking or fish schooling, was proposed by Kennedy and Eberhart in 1995. It is a population-based search method, in which a number of individuals (particles) collaborate with each other to find the optimal solution. Different to GAs or DE, in PSO, each individual i has not only a position $\mathbf{x}_i$ but also a velocity $\mathbf{v}_i$ at each generation t. Each individual will update its velocity by learning from its own personal experience and from the experience of the swarm as well, which can be given in explicit mathematical formulas as shown in Eq. (3.38):

$$v_{ij}(t+1) = \omega v_{ij}(t) + c_1 r_1 (p_{ij}(t) - x_{ij}(t)) + c_2 r_2 (g_j - x_{ij}(t)) \tag{3.38}$$

Accordingly, the position of the i-th particle can be updated as in (3.39)

$$x_{ij}(t+1) = x_{ij}(t) + v_{ij}(t+1) \tag{3.39}$$

where $\mathbf{v}_i(t) = (v_{i1}(t), v_{i2}(t), \ldots, v_{in}(t))$ and $\mathbf{x}_i(t) = (x_{i1}(t), x_{i2}(t), \ldots, x_{in}(t))$ represent the velocity and position of individual i at t-th generation. $\mathbf{p}_i(t) = (p_{i1}, p_{i2}, \ldots, p_{in})$ and $\mathbf{g}(t) = (g_1, g_2, \ldots, g_n)$ are the best positions found by the individual i itself and found by the population till now, respectively, where

$$\mathbf{p}_i(t) = \underset{k=0,1,2,\ldots,t}{\arg\min} \; f(\mathbf{x}_i(k)) \tag{3.40}$$

$$\mathbf{g}(t) = \underset{k=0,1,2,\ldots,t}{\arg\min} \; \{f(\mathbf{x}_1(k)), f(\mathbf{x}_2(k)), \ldots, \mathbf{x}_N(k)\} \tag{3.41}$$

ω is called the inertia weight, both c_1 and c_2 are positive constants which are called acceleration coefficients, where c_1 is called the cognitive parameter and c_2 is the social parameter. r_1 and r_2 are random numbers generated uniformly in the range of [0, 1].

From Eq. (3.38), we can see that velocity update of an individual involves three terms. The first part is known as the inertia, which reflects the momentum that the individual keeps to search in the previous direction. The second part is called the cognitive learning, in which the individual is encouraged to learn from the best position found so far by itself. The third part represents the social learning, in which the individual learns from the best position found so far by the population. Therefore, the best position of the population represents the result of a collaborate behavior to find the global optimal solution.

Figure 3.15 gives a simple example to show how to get the next position of an individual in a two-dimensional decision space. In Fig. 3.15, three red dot lines with arrows represent three parts of Eq. (3.38), respectively. As is shown in Fig. 3.15, the parameter ω is used to control the momentum, φ_1 and φ_2 control the degree that learn from the personal best position and the best position found so far, respectively.

Algorithm 3 gives the pseudo code of PSO. An initial population with N individuals will be firstly initialized with random position and velocity in its upper and lower bounds, respectively. The personal best position of each individual i, $\mathbf{p}_i$, will be the current position of individual i, and the best position found so far $\mathbf{g}(t)$ by all individuals will be kept. After the initialization, each individual will update its velocity and position, respectively, according to Eqs. (3.38) and (3.39), followed by the evaluation of the fitness. The personal best position of each individual i will be updated by the current position if the fitness value of the current position is better than that of the personal best position. All personal best positions are then used to update the global best position found so far. When the stopping criterion is satisfied, the best position found so far and its fitness value will be output.

The particle swarm optimization algorithm has been very popular because of its simplicity in algorithmic implementation and quick convergence. However, PSO

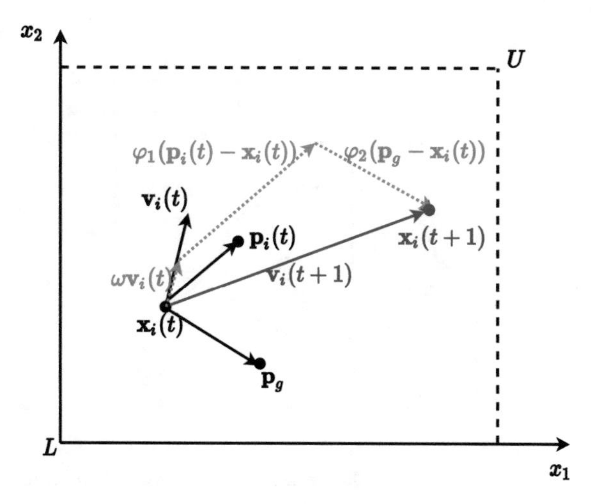

Fig. 3.15 A simple example to show the way to update the position of individual i, in which $\varphi_1 = c_1 r_1$, $\varphi_2 = c_2 r_2$

easily falls into a local optimum, especially in high-dimensional space. Therefore, a number of new learning strategies have been proposed (Cheng & Jin, 2014, 2015; Liang et al., 2006), aiming to improve the diversity of the population so as not to fall into a local optimum. In the following, we will introduce two PSO variants proposed for high-dimensional or large-scale optimization problems.

3.8.2 Competitive Swarm Optimizer

The convergence speed and global search ability are two critical properties of population-based optimization algorithms. In order to alleviate premature convergence, a competition mechanism between particles within one population is introduced in a particle swarm optimization variant, called the competitive swarm optimizer (CSO) (Cheng & Jin, 2014). Figure 3.16 gives a diagram to show the general idea of the competitive swarm optimizer. In CSO, the particles in the current population $P(t)$ will be randomly grouped into $N/2$ couples, assuming that the population size N is an even number. For example, in Fig. 3.16, there are six particles in the current population $P(t)$, then three couples will be randomly paired. Then a competition will be made between two particles in each pair of particles. Suppose $\mathbf{x}_i(t)$ and $\mathbf{x}_j(t)$ are paired. They will compete with each other according to their fitness value, and the solution with a better fitness is defined as the winner, and the other is called the loser. The winner will be directly passed to the next generation, while the loser will update its velocity and position by learning from the winner. Suppose $\mathbf{x}_i(t)$ has a better fitness value than $\mathbf{x}_j(t)$ in the given example, we can see from Fig. 3.16 that $\mathbf{x}_i(t)$ is passed the next generation, i.e., $\mathbf{x}_i(t+1) = \mathbf{x}_i(t)$. While for particle $\mathbf{x}_j(t)$, its position (and velocity) will be updated to be $\mathbf{x}_j(t+1)$ by learning from the winner $\mathbf{x}_i(t)$. So from Fig. 3.16, we can see that only $N/2$ particles in the population are updated at each generation.

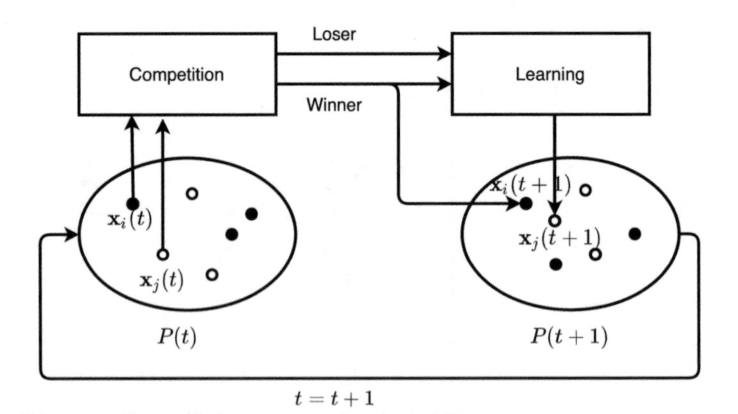

Fig. 3.16 The general idea of the competitive swarm optimizer

Algorithm 3 Particle swarm optimization

Input: N: the population size;
Output: $gbest$: the best solution found so far and its fitness;
1: Initialize a population P;
2: Evaluate the fitness of each individual i in P;
3: $t = 0$;
4: Set the current position of each individual i, $\mathbf{x}_i(t), i = 1, 2, \ldots, N$, to be its personal best position $\mathbf{p}_i(t) = \mathbf{x}_i(t)$;
5: Find the best position among the personal best solutions of all individuals and set it to the best position found so far $\mathbf{g}(t)$;
6: **while** the stopping criterion is not met **do**
7: **for** each individual i **do**
8: Update the velocity and position of individual i using Eqs. (3.38) and (3.39), respectively;
9: Evaluate the fitness value of individual i;
10: **if** $f(\mathbf{x}_i(t+1)) \leq f(\mathbf{p}_i(t))$ **then**
11: $\mathbf{p}_i(t+1) = \mathbf{x}_i(t+1)$;
12: **else**
13: $\mathbf{p}_i(t+1) = \mathbf{p}_i(t)$;
14: **end if**
15: **if** $f(\mathbf{p}_i(t+1)) \leq f(\mathbf{g}(t))$ **then**
16: $\mathbf{g}(t+1) = \mathbf{p}_i(t+1)$;
17: **else**
18: $\mathbf{g}(t+1) = \mathbf{g}(t)$;
19: **end if**
20: **end for**
21: **end while**
22: Output $\mathbf{g}(t+1)$ and $f(\mathbf{g}(t+1))$;

Let $\mathbf{x}_{w,k}(t)$, $\mathbf{x}_{l,k}(t)$, $\mathbf{v}_{w,k}(t)$ and $\mathbf{v}_{l,k}(t)$ represent the position and velocity of the winner and loser in k-th round of competition in the t-th generation, respectively, where $1 \leq k \leq N/2$. Accordingly, the velocity of the loser will be updated using the following learning strategy:

$$\begin{aligned} \mathbf{v}_{l,k}(t+1) = \; & \mathbf{R}_1(k,t)\mathbf{v}_{l,k}(t) + \mathbf{R}_2(k,t)(\mathbf{x}_{w,k}(t) - \mathbf{x}_{l,k}(t)) \\ & + \varphi \mathbf{R}_3(k,t)(\bar{\mathbf{x}}_k(t) - \mathbf{x}_{l,k}(t)) \end{aligned} \tag{3.42}$$

As a result, the position of the loser will be updated as:

$$\mathbf{x}_{l,k}(t+1) = \mathbf{x}_{l,k}(t) + \mathbf{v}_{l,k}(t+1) \tag{3.43}$$

In Eqs. (3.42) and (3.43), $\mathbf{R}_1(k,t)$, $\mathbf{R}_2(k,t)$, $\mathbf{R}_3(k,t) \in [0, 1]^n$ are three randomly generated vectors for the k-th competition at the t-th generation. $\bar{\mathbf{x}}_k(t)$ is the mean position of the relevant particles, φ is a parameter controlling the influence of the mean position $\bar{\mathbf{x}}_k(t)$. Two versions of the mean position can be utilized. One is the global version of the mean position, denoted as $\bar{\mathbf{x}}_k^g(t)$, which represents the mean position of all particles in the current population $P(t)$. The other is the local version of the mean position, denoted as $\bar{\mathbf{x}}_{l,k}^l(t)$, which is the mean position of the particles in a predefined neighborhood of the loser particle of the k-th competition at the t-th generation. The first part of Eq. (3.42) is used to ensure the stability of the search process, which is the same with the canonical PSO that keeps the momentum of the previous direction. The second part of Eq. (3.42), $\mathbf{R}_2(k,t)(\mathbf{x}_{w,k}(t) - \mathbf{x}_{l,k}(t))$, is also called cognitive component, however, different to the canonical PSO, the losing particle learns from the winner in the same couple instead of from its personal best position found so far. This is biologically more plausible in simulating the animal swarm behaviors, since it is hard to require that all particles remember their best position they have experienced in the past. The third part of Eq. (3.42), $\varphi\mathbf{R}_3(k,t)(\bar{\mathbf{x}}_k(t) - \mathbf{x}_{l,k}(t))$, is also termed social component as in the canonical PSO. However, the mean position is used instead of the best position found so far by the population, which does not require to record the best position found so far, either.

Algorithm 4 gives the pseudo code of the competitive swarm optimizer. An initial population P will be randomly generated in the decision space and evaluated using the fitness function. The best solution found so far, denoted as $gbest$, is then set to the solution having the minimum fitness value among all solutions in the current population. The following procedure will be repeated until the stopping criterion is satisfied. The current population will be randomly classified into two sub-population with an equal size $N/2$, and saved to sets S_1 and S_2, respectively. For each corresponding pair in S_1 and S_2, two solutions will be compared on their fitness values, the solution with better fitness value will be assigned to $\mathbf{x}_{w,k}(t)$ and the other will be assigned to $\mathbf{x}_{l,k}(t)$, as is shown in lines 5–16. The winner solution will be kept in set U directly, while the loser one can be able to be saved in U only after it is updated using Eqs. (3.42) and (3.43). All solutions in set U will become the next parent particles. The best solution found so far $gbest$ will be updated after all solutions in U are evaluated using the fitness function.

3.8.3 Social Learning Particle Swarm Optimizer

Learning and imitating the behaviors of better individuals in the population, known as social learning, can be widely observed in the social animals. Different from individual learning, the social learning has the advantage of allowing individuals to learn behaviors from others without incurring the costs of individual trials and errors. Thus, it is quite natural to apply the social learning mechanisms to population-based stochastic optimization. The social learning particle swarm optimizer (SL-PSO) is a PSO variant (Cheng & Jin, 2015), in which each individual learns from any better particles (also called demonstrators) in the current population instead of learning from the historical best positions. Figure 3.17 gives a general idea of the social learning particle swarm optimizer. For the current population $P(t)$, all individuals will be sorted according to their fitness in a descending order for minimization problems (in an ascending order for the maximization problems). Then the solutions in the current population with better fitness values than solution $\mathbf{x}_i(t)$ will be the demonstrators of $\mathbf{x}_i(t)$, as shown in light blue grids. After that, solution $\mathbf{x}_i(t)$ will

Algorithm 4 The pseudo code of the competitive swarm optimizer

Input: N: the population size;
Output: $gbest$: the best solution found so far
1: Initialization of a population P;
2: Evaluate the fitness of each individual i in the population;
   $t = 0$;
3: Find the best position among all solutions of P and set it to the best position found so far $gbest$;
4: **while** the stopping criterion is not met **do**
5:   $S_1 = \emptyset$;
6:   $S_2 = \emptyset$;
7:   Randomly allocate $N/2$ solutions in P to S_1 and the remainder solutions in P to S_2;
8:   $U = \emptyset$;
9:   **for** $k = 1$ to $N/2$ **do**
10:     **if** $f(S_1(i))) \leq f(S_2(i))$ **then**
11:       $\mathbf{x}_{w,k}(t) = S_1(i)$;
12:       $\mathbf{x}_{l,k}(t) = S_2(i)$
13:     **else**
14:       $\mathbf{x}_{w,k}(t) = S_2(i)$;
15:       $\mathbf{x}_{l,k}(t) = S_1(i)$
16:     **end if**
17:     $U = U \cup \mathbf{x}_{w,k}(t)$;
18:     Update $\mathbf{x}_{l,k}(t)$ using Eqs. (3.42) and (3.43);
19:     Save the updated solution into U;
20:   **end for**
21:   $P = U$;
22:   $t = t + 1$;
23:   Evaluate the fitness values of all solution in P, and update the best solution found so far $gbest$;
24: **end while**
25: Output the best solution found so far $gbest$;

update its velocity and position by learning from its its demonstrators. Note that in SL-PSO, each solution $\mathbf{x}_i(t)$, except the worst one, can serve as a demonstrator for different imitators (solutions having a worse fitness value than solution $\mathbf{x}_i(t)$). Also, each $\mathbf{x}_i(t)$, except for the best one, can learn from different demonstrators. The best particle in the current population will not be updated.

Inspired by social learning mechanism, an imitator will learn the behaviors of different demonstrators in the following manner:

$$x_{ij}(t+1) = \begin{cases} x_{ij}(t) + \Delta x_{ij}(t+1), & \text{if } p_i(t) \leq P_i^L \\ x_{ij}(t), & \text{otherwise} \end{cases} \tag{3.44}$$

where $x_{ij}(t), i = 1, 2, \ldots, N, j = 1, 2, \ldots, n$ is the j-th dimension of the behavior vector in generation t, $\Delta x_{ij}(t+1)$ is the behavior correction of particle i on the j-th dimension. Note that in social learning, the motivation to learn from better individuals may vary from individual to individual, so a learning probability P_i^L will be defined for each individual. As a result, an individual i will learn from its demonstrators

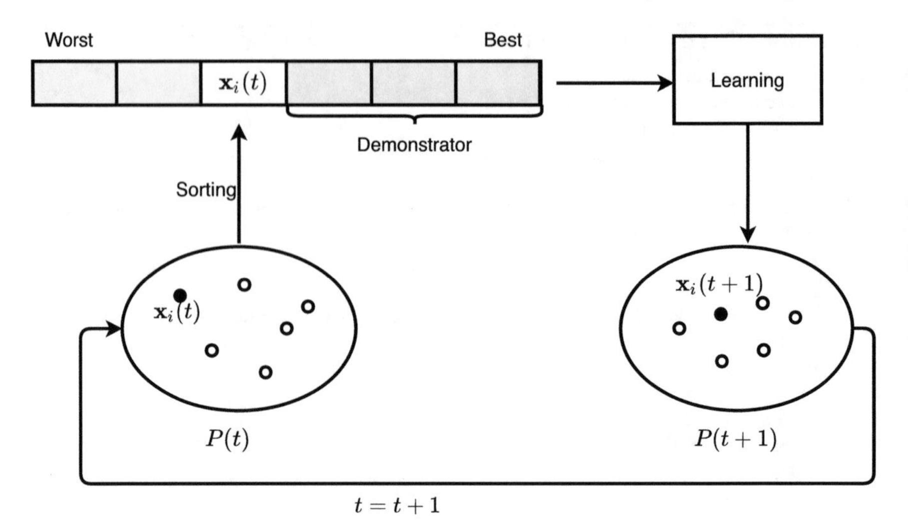

Fig. 3.17 The general idea of the social learning particle swarm optimizer

using Eq. (3.45) only if a randomly generated probability p_i satisfies the inequality condition: $0 \leq p_i(t) \leq P_i^L \leq 1$.

$$\Delta x_{ij}(t+1) = r_1(t)\, \Delta x_{ij}(t) + r_2(t)\, (x_{kj}(t) - x_{ij}(t)) + r_3(t)\, \epsilon\, (\bar{x}_j(t) - x_{ij}(t)). \tag{3.45}$$

In Eq. (3.45), $r_1(t)$, $r_2(t)$, and $r_3(t)$ are three random numbers in the range of [0, 1] in generation t, respectively. $x_{kj}(t)$ is the behavior vector of particle k on d-th dimension in generation t. $\bar{x}_j(t) = \frac{\sum_{i=1}^{N} x_{ij}(t)}{N}$ is the mean behavior of all particles in the current population of dimension j in generation t. The parameter ϵ is called as the social influence factor, which is used to control the degree of the social influence to this particle on this dimension. Similar to the velocity updating in the canonical particle swarm optimization, the velocity update in SL-PSO is also composed of three parts. The first component aims to keep the momentum of the previous direction and to ensure the stability of the search process. In the second part of Eq. (3.45), individual i will learn from its demonstrators, which is different to that in the canonical PSO, which learns from the personal best position of the individual. Note that $i < k \leq N$, and the demonstrators for individual i to learn from for different dimensions may be different. The third part of Eq. (3.45) is also different to that in the canonical PSO: instead of learning from the best position found so far by the population in PSO, in SL-PSO, individual i learns from the collective behavior of the whole population, i.e., the mean behavior of all particles in the current population.

Algorithm 5 presents the pseudo code of the social learning particle swarm optimizer. An initial population P will be generated and evaluated using the fitness function. The best position among all individuals of P will be identified and set to be the best position found so far, denoted as $gbest$. The following procedure will repeat until the stopping criterion is met. All solutions in the current population P will be sorted in a descending order according to their fitness values. For each individual in the sorted population, except for the best individual, will update its velocity by learning from its demonstrators. Thereafter, the position of the particles in the current population P will be updated. Once the demonstrators are identified for each dimension of individual i, each component of the position vector will be updated by learning from different demonstrators and the mean behavior of all particles in the current population (lines 7–14). Once their positions are also updated, all updated solutions will be evaluated based on the update position and the best solution $gbest$ will be updated accordingly.

Algorithm 5 The pseudocode of the social learning particle swarm optimizer

Input: N: the population size;
Output: $gbest$: the best solution found so far
1: Initialize a population P;
2: Evaluate the fitness of each individual i in the population;
3: $t = 0$;
4: Find the best position among all solutions of P and set it to the best position found so far $gbest$;
5: **while** the stopping criterion is not met **do**
6: Sort the current population P in a descending order according to the fitness value and update the index of individuals in the population accordingly;
7: **for** $i = 1$ to $N - 1$ **do**
8: **for** $j = 1$ to n **do**
9: Generate a random integer number k in the range of $[i + 1, N]$;
10: Randomly generate three numbers $r_1(t)$, $r_2(t)$, and $r_3(t)$ in the range of [0, 1];
11: Calculate the mean position of all particles in the current population on j-th dimension;
12: Update the behavior of particle i on j-th dimension using Eq. (3.44);
13: **end for**
14: Evaluate the fitness value of the updated particle i;
15: **end for**
16: Update the best position found so far $gbest$ using all particles in the current population;
17: $t = t + 1$;
18: **end while**
19: Output the best solution found so far $gbest$;

3.9 Memetic Algorithms

3.9.1 Basic Concepts

Memetic algorithms are a class of hybrid search methods that embed local search, also known as life-time learning, in evolutionary algorithms. This way, memetic algorithms can benefit from both genetic search, a type of more global, stochastic search,

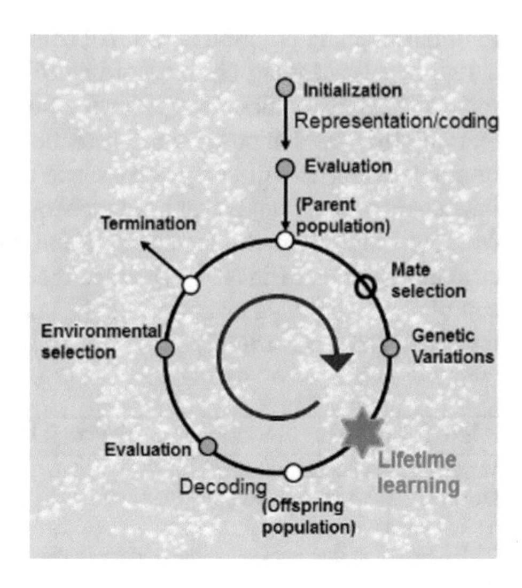

Fig. 3.18 A generic framework of memetic algorithms

with more local, and typically greedy search. A generic description of a memetic algorithm is given in Fig. 3.18, from which we can see that the main difference between a memetic algorithm and an evolutionary algorithm is that a local search will be performed before the offspring are evaluated for environmental selection.

A few questions need to be answered before a local search can be carried out. For example, which local search algorithm can be used? How many iterations of local search should be conducted? And shall local search apply on all individuals? Different settings may result in different search performance, which may also be problem specific. Although local search can be hybridized with both binary-coded and real-valued coded evolutionary algorithms in principle, it is more widely seen that a local search is combined with a binary or gray-coded genetic algorithm, since the synergistic benefits of combining global search with local search may be better exploited.

3.9.2 Lamarckian Versus Baldwinian Approaches

There are two slight different variants of memetic algorithms, one is known as Lamarckian evolution and the other Baldwinian. Note that local search is conducted in the phenotypic space, whilst genetic changes are made in the genotypic space. Thus, the main difference between the two approaches lie in whether the changes to the decision variables (phenotypic changes) in local search will directly influence the genotype, thereby making the changes heritable. Figure 3.19 illustrates the difference between the Lamarckian and Baldwinian approaches of memetic algorithms. On the left side of the figure, a local search is applied on the phenotype of solution x_1 and it changes to x_1' after the local search. In the meantime, the changed phenotype

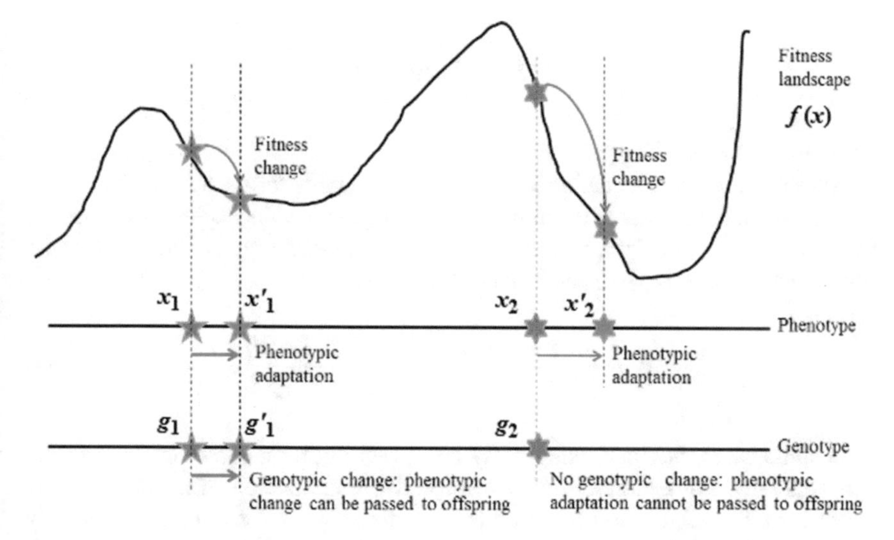

Fig. 3.19 Lamarckian (left) versus Baldwinian (right) evolution

is directly encoded back into the genotype of the individual and consequently, the genotype of this individual is changed from g_1 to g_1' after local search. In this case, a Lamarckian evolution has taken place. By contrast, on the right side of the figure, a local search is also applied to solution x_2 and it changes to x_2' after local search. However, the genotype of solution remains unchanged, which is known as Baldwinian evolution. Note that in environmental selection, the fitness of x_1 will become $f(x_1')$ and the fitness of x_2 will become $f(x_2')$. Therefore, the results of local search, regardless in Lamarckian or Baldwinian evolution, will influence the environmental selection. The only difference is that for x_1, which adopts a Lamarckian evolution, its genotype has also changed as a result of local search, while for x_2, only its fitness value has changed due to local search.

3.9.3 Multi-objective Memetic Algorithms

In single-objective optimization, the objective function used in local search is usually the same as in the genetic search. However, since most local search algorithms, such as gradient-based methods, are not suited for multi-objective optimization, a multi-objective optimization problem needs first to be converted into a single-objective one before local search can be carried out. To this end, a randomly generated weight vector may be assigned to each individual in the population so that local search can be carried out diverse directions. For example, for an individual in the population for m-objective optimization, one can use the following single-objective problem for local search:

$$f(\mathbf{x}) = \sum_{i=1}^{m} w_i f_i(\mathbf{x}), \tag{3.46}$$

where $w_i > 0$, and $\sum_{i=1}^{m} w_i = 1$.

Alternatively, one can define a set of pseudo-weights for each individual in the current population. Assume the minimum and maximum values of the i-th objective function are $f_i^{\min}$ and $f_i^{\max}$, respectively, the pseudo-weight for an individual whose i-th objective value is $f_i(\mathbf{x})$ can be calculated as follows:

$$w_i = \frac{f_i^{\max} - f_i(\mathbf{x})}{f_i^{\max} - f_i^{\min}} \Big/ \sum_{j=1}^{m} \frac{f_i^{\max} - f_i(\mathbf{x})}{f_i^{\max} - f_i^{\min}} \tag{3.47}$$

In some cases, the local search may be carried out for a subset of the decision variables only. For example, for multi-objective optimization of neural networks where both the weights and architecture are optimized, the local search is typically applied on the weights only using a gradient-based method (Jin & Sendhoff, 2008).

3.9.4 Baldwin Effect Versus Hiding Effect

Combining local search with evolutionary search aims to accelerate the search process, which is known as Baldwin effect. However, this is not always true, since local search may also slow down the evolution, which is called hiding effect. Typically, the Baldwin effect may occur if the local search reduces the differences between the innate fitness values of the individuals within the current population, while the hiding effect takes place if the local search enlarges the differences in their fitness values of the individuals in the population, as illustrated in Fig. 3.20. On the left panel, the innate fitness values of the three individuals becomes larger, and consequently the weaker individual will have a smaller probability to survive after lifetime learning, leading to a higher selection pressure, and therefore a Baldwin effect. By contrast, on the right panel, the fitness values between the three individuals becomes smaller after lifetime learning, resulting in a smaller selection pressure and therefore a hiding effect. Note that the above discussions are based on the assumption that a fitness proportionate selection strategy is adopted.

In the following, we provide a simple mathematical proof showing the condition under which Baldwin or hiding effect will be observed (Paenke et al., 2009b). Assume the innate fitness landscape (without learning) of a problem to be maximized is monotonously increasing and continuous, i.e., $f(\mathbf{x}) > 0$, its first-order derivative function $f'(\mathbf{x}) > 0$, and its first to fourth-order derivative functions $f^{(i)}(\mathbf{x}), i = 1, 2, 3, 4$ is continuous. The fitness landscape after lifetime learning is denoted as $f_l(\mathbf{x})$, which is also positive and monotonic. For a population of size N, the mean genotype is calculated as

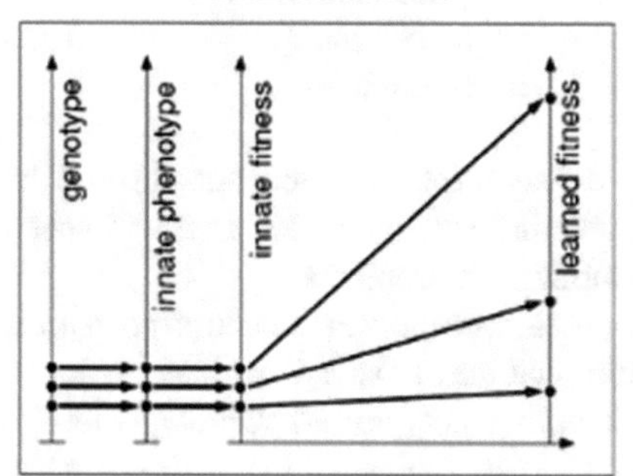

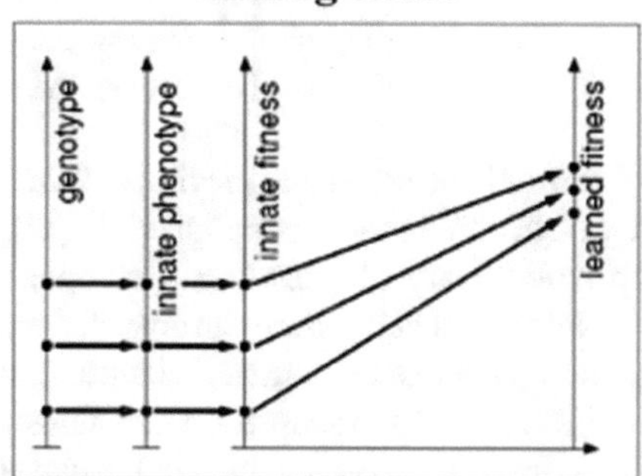

Fig. 3.20 Baldwin effect (left) versus hiding effect (right)

$$\bar{\mathbf{x}} = \frac{1}{N}\sum_{i=1}^{N} \mathbf{x}_i. \tag{3.48}$$

If a fitness proportionate selection is applied to the population, the mean genotype after selection is changed to

$$\bar{\mathbf{x}}^* = \frac{\sum_{i=1}^{N} \mathbf{x}_i f(\mathbf{x}_i)}{\sum_{i=1}^{N} f(\mathbf{x}_i)}. \tag{3.49}$$

Thus, the expected change in the genotype after selection will be:

$$\Delta\bar{\mathbf{x}} = \frac{\sum_{i=1}^{N} \mathbf{x}_i f(\mathbf{x}_i)}{\sum_{i=1}^{N} f(\mathbf{x}_i)} - \frac{1}{N}\sum_{i=1}^{N} \mathbf{x}_i. \tag{3.50}$$

Thus, evolution is accelerated (decelerated) by learning if the following expression is positive (negative):

$$\text{sign}(\Delta\bar{\mathbf{x}}_l - \Delta\bar{\mathbf{x}}) = \text{sign}\left(\frac{\sum_{i=1}^{N} \mathbf{x}_i f_l(\mathbf{x}_i)}{\sum_{i=1}^{N} f_l(\mathbf{x}_i)} - \frac{\sum_{i=1}^{N} \mathbf{x}_i f(\mathbf{x}_i)}{\sum_{i=1}^{N} f(\mathbf{x}_i)}\right). \tag{3.51}$$

For a given innate fitness function $f(\mathbf{x})$ and a learned fitness function $f_l(\mathbf{x})$, we can define the following gain function to judge whether learning is able to accelerate evolution:

$$g(\mathbf{x}) = \frac{f_l(\mathbf{x})}{f(\mathbf{x})}. \tag{3.52}$$

It can be proved that

$$g'(\mathbf{x}) = \begin{cases} > 0 \Longleftrightarrow \Delta\bar{\mathbf{x}}_l - \Delta\bar{\mathbf{x}} > 0 \text{ (accelerated)} \\ < 0 \Longleftrightarrow \Delta\bar{\mathbf{x}}_l - \Delta\bar{\mathbf{x}} < 0 \text{ (decelerated)} \\ = 0 \Longleftrightarrow \Delta\bar{\mathbf{x}}_l - \Delta\bar{\mathbf{x}} = 0 \text{ (no influence)} \end{cases} \tag{3.53}$$

It should be noted that the fitness function discussed above is very specific and the it will become much complicated to judge whether adding local search will speed up evolutionary optimization of complex optimization problems.

Note also that in nature, survival and reproduction are the main selection pressures, rather than a fast evolution. A natural question that may arise is what benefits it can bring about to have adaptation mechanisms at both genotypic and phenotypic levels. Some insight has been gained in Paenke et al. (2009a), indicating that by making use of the hiding effect, a population is able to maintain a sufficient degree of genotypic diversity, which is of particular importance for the population to quickly adapt in dynamic environments. This also implies that both Baldwin and hiding effects may be helpful for the population evolving in a frequently changing environment.

3.10 Estimation of Distribution Algorithms

Evolutionary algorithms rely on genetic variations such as crossover and mutation to search for new promising solutions. However, these operators perform random search and are not able to explicitly learn the problem structure and make use of such knowledge in the search.

Estimation of distribution algorithms (EDAs) are a class of population based search algorithms that guide the search for the optimum by building and sampling explicit probabilistic models of promising candidate solutions. A diagram of a generic estimation of distribution algorithm is given in Fig. 3.21, from which we can see that the main components of an EDA is very similar to an evolutionary algorithm, except that the genetic operators are replaced by building a probabilistic model and sampling offspring from the model.

Consequently, efficiently building probabilistic models becomes the central concern in designing EDAs. In the following, we are going to introduce a few probabilistic models widely used in EDAs.

3.10.1 A Simple EDA

The basic idea of EDAs can be illustrated using the so called onemax problem, where each decision variable is a binary bit, and the objective function can be described by:

$$\text{onemax}(x_1, x_2, \ldots, x_n) = \sum_{i=1}^{n} x_i, \tag{3.54}$$

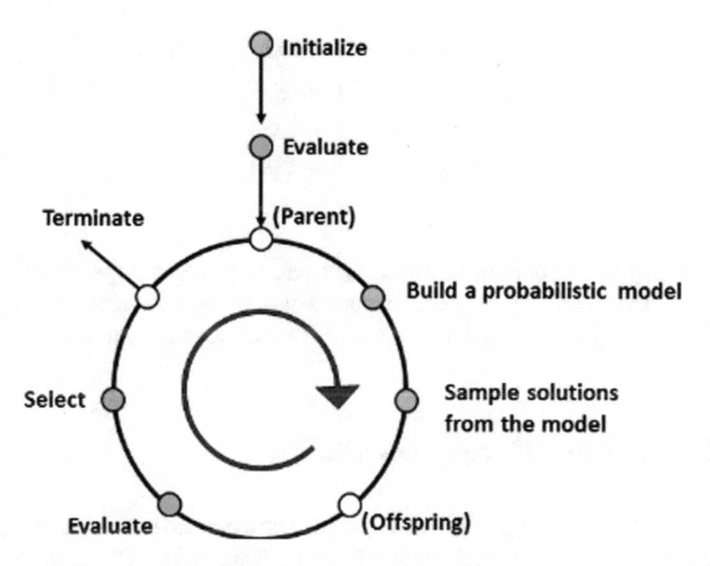

Fig. 3.21 A diagram of an estimation of distribution algorithm

where n is the number of decision variables. This problem has one global optimum where all decision variables equal 1.

Consider the *onemex* consisting of four variables. Assume that at the current generation t, the parent population contains three individuals $\mathbf{x}^1(t) = \{1, 0, 1, 1\}$, $\mathbf{x}^2(t) = \{0, 1, 1, 0\}$, and $\mathbf{x}^3(t) = \{0, 1, 1, 0\}$. Thus, we can build a probabilistic model by calculating the probability of $x_i = 1, i = 1, 2, 3, 4$, based on the three individuals:

$$\begin{aligned} p_1(x_1 = 1) &= 1/3, \\ p_2(x_2 = 1) &= 2/3, \\ p_3(x_3 = 1) &= 1, \\ p_4(x_4 = 1) &= 1/3. \end{aligned} \tag{3.55}$$

We can then sample three offspring solutions using the above probability for each decision variable $p_i, i = 1, 2, 3, 4$, say, $\mathbf{x}'^1(t) = \{0, 1, 1, 1\}$, $\mathbf{x}'^2(t) = \{1, 0, 1, 0\}$, and $\mathbf{x}'^3(t) = \{1, 1, 1, 0\}$. A selection method can then be carried out to select parents for the $(t + 1)$-th generation, say $\mathbf{x}^1(t + 1) = \{0, 1, 1, 1\}$, $\mathbf{x}^2(t + 1) = \{0, 1, 1, 1\}$, and $\mathbf{x}^3(t + 1) = \{1, 1, 1, 0\}$. Consequently, the probabilistic models will be updated as follows:

$$
\begin{aligned}
p_1(x_1 = 1) &= 1/3, \\
p_2(x_2 = 1) &= 1, \\
p_3(x_3 = 1) &= 1, \\
p_4(x_4 = 1) &= 2/3.
\end{aligned}
\tag{3.56}
$$

The above process continues until the population converges to the optimal solution $(1, 1, 1, 1)$. The algorithm is known as univariate marginal distribution algorithm (UMDA), as it does not consider any interactions between the variables.

3.10.2 EDAs for Discrete Optimization

The UMDA described in the previous section can be extended to solving more general discrete optimization problems (Hauschild & Pelikan, 2011). One idea is to incrementally update the probability models, which is called population-based incremental learning (PBIL) (Baluja, 1994):

$$p_i = (1-\alpha)p_i + \alpha\frac{1}{N}\sum_{k=1}^{N} x_i^k, i = 1, 2, \ldots, n \tag{3.57}$$

where α a learning rate to be defined by the user, $\mathbf{x}^k = \{x_1^k, x_2^k, \ldots, x_n^k\}$, N is the selected parent individuals from $M \geq N$ offspring.

A variant of PBIL is called compact genetic algorithm (CGA) (Harik et al., 1999b). The probabilistic model is initialized with $p_i = 0.5, i = 1, 2, \ldots, n$. Then two individuals are sampled according to the probabilistic model, and the probabilistic model is updated towards the best one. This process of adaptation continues until the probabilistic model converges.

While UMDA, PBIL and CGA do not take into account the interactions between the decision variables, more sophisticated EDAs use bivariate or multivariate probabilistic models that can capture interactions between two or multiple variables. For example, the extended compact genetic algorithm groups the decision variables into a number of clusters and then each cluster is modeled by one probability (Harik et al., 1999a). The factorized distribution algorithm (FDA) (Muhlenbein & Mahnig, 1999) employs a fixed factorization that the factors the distribution into conditional and marginal distributions for additively decomposed discrete functions. The Bayesian optimization algorithm (BOA) (Pelikan et al., 2000) uses a Bayesian network to model interactions between multiple decision variables using the edges between different nodes. Note that BOA is completely different from Bayesian optimization in Sect. sec5:4.

3.10.3 EDAs for Continuous Optimization

Many estimation of distribution algorithms have also been proposed for solving continuous optimization problems. Similarly, these models can be divided into univariate and multivariate probabilistic models. For the current parent population of size N, a univariate factorized Gaussian model can be described by:

$$\mu_i = \frac{1}{N}\sum_{i=1}^{N} x_i, \tag{3.58}$$

$$\delta_i = \sqrt{\frac{1}{N}\sum_{i=1}^{N}(x_i - \mu_i)^2}, \tag{3.59}$$

where N is the number of selected individuals (parents). Then, a Gaussian model can be constructed for each decision variable:

$$p_i(x_i) = \frac{1}{\delta_i\sqrt{2\pi}}\exp\left\{-\frac{(x_i - \mu_i)^2}{2(\delta_i)^2}\right\}, \tag{3.60}$$

where μ_i and δ_i are the mean and standard deviation of the Gaussian model for variable $i = 1, \ldots, n$.

Once the Gaussian model $p_i(x_i)$ is built, offspring solutions (usually more than N) can then be generated from the Gaussian distribution. After selection, the probabilistic models will be updated and new solutions will be sampled.

However, such univariate models are not able to capture the correlations between the decision variables. To address problem, a joint Gaussian distribution model can be adopted. However, building a full joint distribution model in a high-dimensional space is subject to the curse of dimensionality, and infeasible in practice. To alleviate this problem, the population can be divided into a number clusters and then a joint distribution model is built for each cluster. For k-th cluster, where $1 \le k \le K$, the following joint distribution model is constructed:

$$p^k(\mathbf{x}) = \frac{1}{(2\pi)^{n/2}|\Sigma^k|^{n/2}}\exp\left\{-\frac{1}{2}(\mathbf{x} - \Lambda^k)^T(\Sigma^k)^{-1}(\mathbf{x} - \Lambda^k)\right\}, \tag{3.61}$$

where $\mathbf{x}$ is a n-dimensional design vector, Λ^k is a n-dimensional vector of the mean value and Σ^k is an $n \times n$ covariance matrix, estimated by the individuals in the k-th cluster.

Different from the univariate factorized model, the probabilistic model of the k-th cluster can written in the following

$$p(k) = \frac{N^k}{\sum_{k=1}^{K} N^k}, \tag{3.62}$$

where N^k is the number of individuals in the k-th cluster.

3.10.4 Multi-objective EDAs

In principle, all EDAs developed for single-objective optimization can be extended to multi-objective optimization by replacing the selection scheme, however, several multi-objective EDAs have been developed that aim to model the properties of the Pareto front.

One multi-objective EDA, termed regularity modeling multi-objective evolutionary algorithm (RM-MOEA) builds a probabilistic model in the $m-1$-dimensional manifold (Zhang et al., 2008). Under mild conditions, the Pareto front of an m-objective optimization problem is an $(m-1)$-dimensional manifold. Therefore, the probabilistic model of RM-MOEA consists of a local principal curve and a number of Gaussian distribution models.

Assume a population is divided into K clusters. The solutions in the k-th cluster (denoted by C^k) can be described by a uniform distribution on a $(m-1)$-dimensional manifold M^k:

$$P^k(S) = \begin{cases} \frac{1}{V^k}, & \text{if } S \in M^k, \\ 0, & \text{else} \end{cases} \tag{3.63}$$

where S is an $(m-1)$-dimensional random vector in the latent space, V^k is the volume of M^k bounded by:

$$a_i^k \le s_i \le b_i^k, i = 1, \ldots, (m-1), \tag{3.64}$$

and

$$a_i^k = \min_{\mathbf{x} \in C^k} (\mathbf{x} - \bar{\mathbf{x}}^k)^T U_i^k, \tag{3.65}$$

$$b_i^k = \max_{\mathbf{x} \in C^k} (\mathbf{x} - \bar{\mathbf{x}}^k)^T U_i^k, \tag{3.66}$$

where $\bar{\mathbf{x}}^k$ is the mean of the points in C^k, U_i^k is the i-th principal component of the data in cluster C^k.

In addition to the uniform distribution model defined by a principal curve, an n-dimensional zero-mean Gaussian distribution is also constructed in the design space:

$$N^k(\mathbf{x}) = \frac{1}{(2\pi)^{n/2}|\Sigma^k|^{n/2}} \exp\left\{-\frac{1}{2}\mathbf{x}^T \Sigma^k \mathbf{x}\right\}, \tag{3.67}$$

where $\Sigma^k = \delta^k I$, I is an $n \times n$ dimensional identity matrix, and δ^k is calculated by:

$$\delta^k = \frac{1}{n-m+1} \sum_{i=m}^{n} \lambda_i^k, \tag{3.68}$$

where λ_i^k are the i-th largest eigenvalue of the covariance matrix of the solutions in cluster k.

The probability at which the model of cluster k is chosen to sample offspring individuals is defined by

$$p(k) = \frac{V^k}{\sum_{k=1}^{K} V^k}, \tag{3.69}$$

where V^k is the volume of the $(m-1)$-dimensional manifold. In case of a curve, it is the length of the curve.

Sampling new solutions consists of three steps. In the first step, a solution is generated on $(m-1)$-dimensional manifold M^k according to Eq. (3.63), and is then mapped onto the n-dimensional decision space. Assume S is an $(m-1)$-dimensional random vector generated in M^k for the k-th cluster, it is mapped onto the n-dimensional decision space in the following way, if the manifold M^k is a first-order principal curve:

$$\mathbf{x}_1 = \Theta_0^k + \Theta_1^k S \tag{3.70}$$

where $\mathbf{x}_1$ is a n-dimensional random vector, Θ_0^k is the mean of data in cluster $C(\mathbf{x})^k$, and Θ_1^k is $n \times (m-1)$-dimensional matrix, which is composed of the eigenvectors corresponding to the $(m-1)$ largest eigenvalues.

Next, an n-dimensional decision vector $\mathbf{x}_2$ is generated from the Gaussian model in Eq. (3.67). Finally, a new candidate solution is the sum:

$$\mathbf{x} = \mathbf{x}_1 + \mathbf{x}_2. \tag{3.71}$$

We can then repeat the above process to generate an offspring population.

While most multivariate EDAs aim to capture the interactions between the decision variables, RM-MOEA is able to model the relationship between the objectives, which has been used to deal with noisy evaluations (Wang et al., 2016). One step further is to consider interactions between the decision variables, between objectives, and between decision variables and objectives. For example in Karshenas et al. (2013) employs a multi-dimensional Bayesian network (MBN) to capture the three different interactions mentioned above. In an MBN, the nodes represent the decision variables, the arcs describe the conditional dependencies between triplets of variables, and the parameters express the conditional probability of each of its values for each variable, given different value combinations of its parent variables according to the structure of the Bayesian network. An example of MBN is shown in Fig. 3.22, where the top layer consists of three nodes representing three class variables, and the bottom layer has four nodes denoting four decision variables. As a result, the MBN is able to capture the interactions between both decision variables and objectives using the arcs (directed connections), thereby being able to build a more effective EDA for solving multi-objective optimization problems.

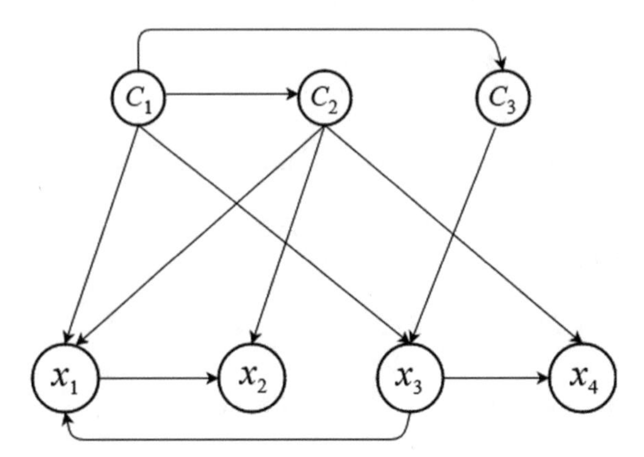

Fig. 3.22 An example of multi-dimensional Bayesian network consisting of four decision variables and three classes

3.11 Parameter Adaptation and Algorithm Selection

We have introduced so far six categories of most popular population based metaheuristic search algorithms, including genetic algorithms, evolution strategies, genetic programming, ant colony optimization, differential evolution, and particle swarm optimization. These algorithms may perform differently on different problems, and there are no clear theoretic guidelines indicating which algorithm one should use for a given problem, making it hard for practitioners to select the right algorithm without much trials and errors. In addition, most metaheuristics contain several parameters, which need to be specified by the user. In the following, we will present some typical ideas for selection and recommendation of metaheuristic search algorithms.

3.11.1 Automated Parameter Tuning

Specification of parameters for metaheuristics is non-trivial. Most parameters in metaheuristics, such as population size, crossover and mutation probabilities, number of generations, and encoding methods, have strong influence on their performance, yet no concrete guidelines are available for choosing these hyperparameters. Evolution strategies are probably an exception, since they also encode some important parameters, for example the step sizes in the chromosome and co-evolve them. In addition, there are some guidelines for determining the population size, in particular the covariance matrix adaptation evolution strategies.

Parameter adaptation in metaheuristics is challenging for several reasons. First, the parameters are not independent, and they influence the search performance together. Second, it may take a large number of generations before the algorithm is able to acquire useful knowledge to effectively tune the parameters. Finally, the optimal setting of parameters is usually problem specific.

Usually, the parameters can either be tuned intuitively, or adapted according to the feedback of the search dynamics. For example, since exploration should be encouraged in the early search stage and exploitation is more desirable in the late stage, one can control the parameters to reflect such requirements. Thus, some parameters can be controlled according to a pre-specified adaptation mechanism, typically by changing the parameters over the generations.

The other idea is to change the parameters according to the search performance or states of the algorithm. These may include the convergence properties, population diversity. More discussions about parameter adaptation can be found in Eiben et al. (2009).

Apart from parameter adaptation, representation of evolutionary algorithms can also be adapted. One basic idea for adapting the representation of binary genetic algorithm is to use a variable chromosome length instead of a fixed one (Goldberg et al., 1991). Another idea is to use a hybrid representation (Fig. 3.23), where both binary and real-valued representations are used for each decision variable and which representation is to be activated depends on the performance of the respective representation (Okabe et al., 2003). It should be noted that the two representations need to be coordinated so that they represent the same value after crossover and mutation.

3.11.2 Hyper-heuristics

Due to the none free lunch theorem, no single meta-heuristic or mathematical optimization algorithm can outperform others on all classes of problems. Naturally, it is of interest to automatically select a best performing algorithm during the search. A hyper-heuristic aims to automate the selection or combination of simpler heuristics to efficiently solve a given problem. The automated selection and generation of metaheuristics are usually realized with the help of machine learning techniques and statistical analysis. An illustration of the generic framework of hyper-heuristics is provided in Fig. 3.24.

Hyper-heuristics may be divided into online methods and offline methods. In offline methods, previous knowledge on solving various problems is extracted in terms of rules and algorithm selection can be based on case based reasoning. In online approaches, a machine learning algorithm such as reinforcement learning is applied during the search process to find out the most effective search heuristics for a given problem.

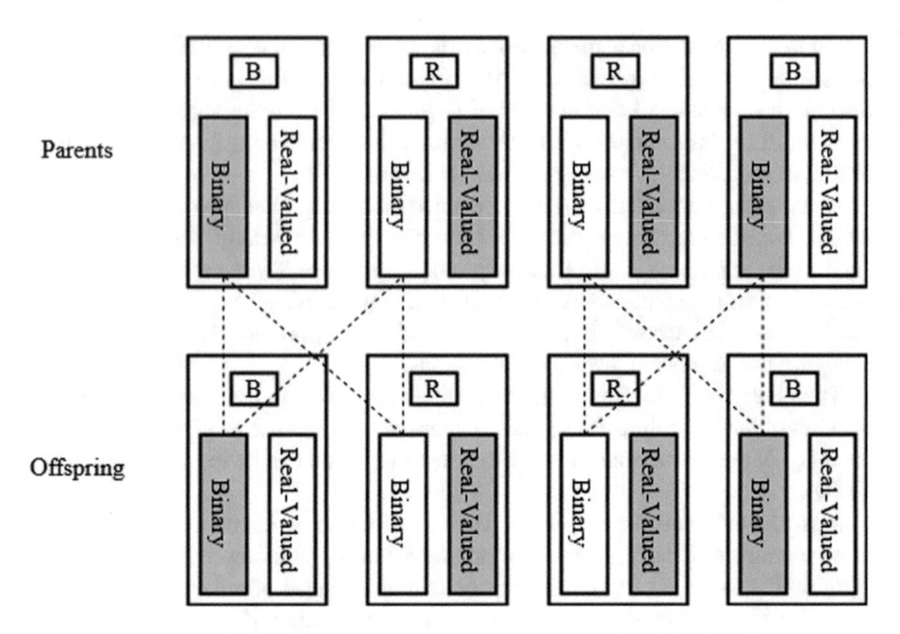

Fig. 3.23 A hybrid representation that encode each decision variable using both binary and real-valued representations. In the presentation, switch 'R' means the real-valued representation is activated while 'B' means the binary coding is activated

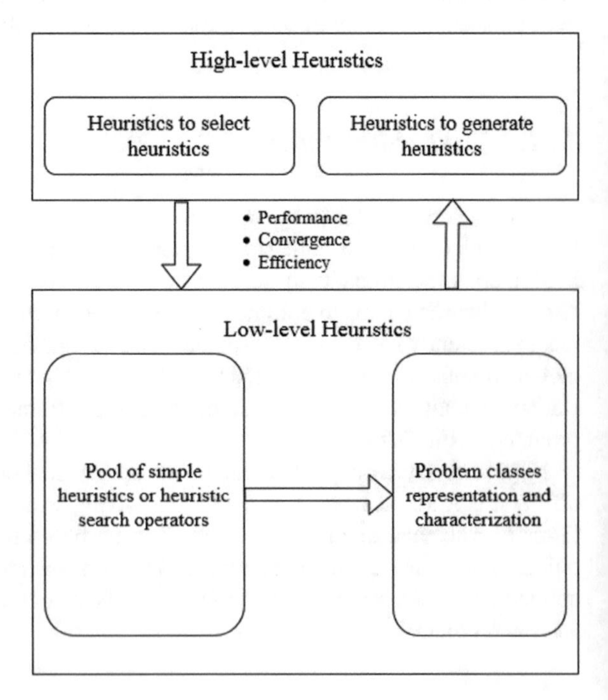

Fig. 3.24 A generic hyper-heuristic algorithm

3.11.3 Fitness Landscape Analysis

Fitness landscape analysis is another approach understand the search performance of metaheuristics on classes of optimization problems. To this end, one attempts to examine the main features that characterize the hardness in solving an optimization problem. Consequently, finding out the right features of a given problem and relating them to search algorithms or search operators will provide insight into selecting the right algorithm and right search operators. The main fitness landscape features that have widely been analyzed include (Pitzer & Affenzeller, 2012):

- *Modality*. Modality, i.e., whether the fitness landscape contains one or multiple optimums, is one of the apparent features of a fitness landscape that characterize the hardness of a problem, and determine which search operators are more effective.
- *Basins of Attraction*. This basically defines the width and depth of the minimums. Basins are grouped into strong (unconditional) and conditional (weak) attractions.
- *Ruggedness*. Ruggedness is one additional feature describing the optimums. In contrast to basin, it reflects the frequency of changes or the density of the local minimums.
- *Barriers*. This defines the minimum fitness value needed go from one minimum to another via an arbitrary path.
- *Evolvability, neutrality and epistasis*. These features are metaphors borrowed from biological networks describing the relationship between neighboring points on the fitness landscape. These are closely related to landscape walk analysis, treating the fitness landscape as a complex network.

There are several other methods for fitness landscape analysis, including fitness distance correlation (Jones & Forrest, 1991) and spectral landscape analysis (Weinberger, 1991). Information theoretic approaches have also been proposed (Vassilev et al., 2000).

3.11.4 Automated Recommendation Systems

One empirical yet promising idea is to develop a recommendation system that is able to recommend a particular metaheuristic algorithm for a given problem. In contrast to hyper-heuristics, recommendation systems aim to identify the relationship between algorithm performance and fitness landscape features offline. A recommendation system is similar to a classification system that use features representing the difficulty or structure of an optimization problem as the input and the corresponding best-performing metaheuristic algorithm as the output (label). Different search algorithms are used to solve a large number of benchmark problems, and the winning algorithm on each problem is recorded as the label. Given these data are then used as data to train the classifier. To achieve a high-performance algorithm recommendation system, it is essential to collect a sufficient number of benchmark problems, extract the most important features from the benchmark problems, and choose the representative metaheuristics to solve these problems.

While most work are based on fitness landscape analysis (Kerschke et al., 2019), one recently proposed idea is to represent an arbitrary white-box or black-box objective function with a decision tree structure consisting of basic arithmetic operators such as addition, multiplication, and base functions including square, square root, exponent, and sine, among others (Tian et al., 2020). For black-box optimization problems, symbolic regression based on genetic programming can be adopted to estimate a tree representation. The benefit of doing this is that one can achieve a uniform representation for an arbitrary optimization problem, and an infinite number of benchmark problems can be created by generating random trees based on these basic arithmetic operators and basic functions for collecting a huge amount of training data. Afterwards, a deep learning model is trained using these data so

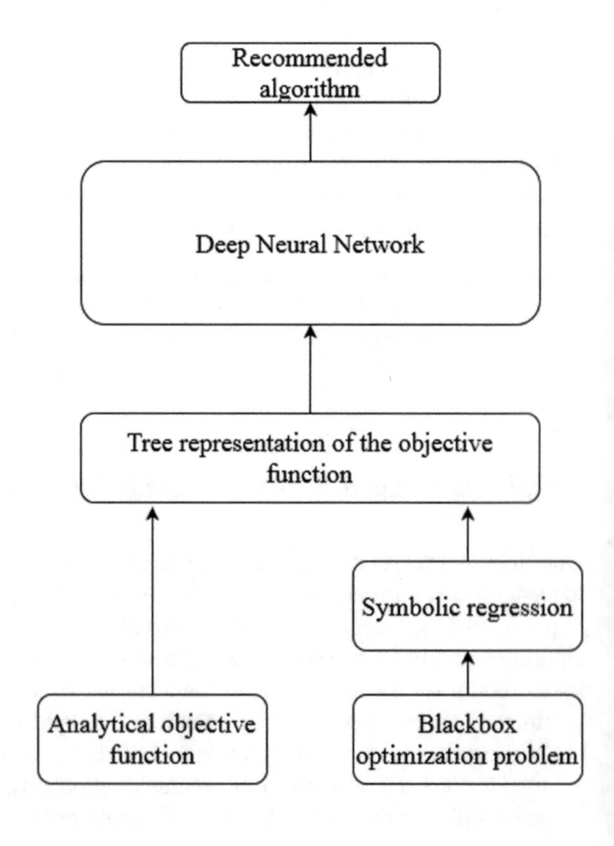

Fig. 3.25 A recommendation system for metaheuristic algorithms based on deep learning

that it can predict performance of any given function, thereby recommending the best-performing heuristic algorithm. A diagram of the recommendation algorithm is provided in Fig. 3.25.

3.12 Summary

Population based metaheuristics have been shown to be very powerful in solving a wide range of test problems that cannot be effectively solved by mathematical optimization methods. They have also enjoyed great success in solving many black-box real-world optimization problems, such as aerodynamic design optimization, optimization of industrial processes, and hybrid electric vehicle controller design. However, the power of evolutionary algorithms largely remains to be demonstrated with killer applications. To achieve this, efficient and effective data-driven evolutionary optimization algorithms will be indispensable.

References

Applegate, D. L., Bixby, R. E., Chvatal, V., & Cook, W. J. (2006). *The traveling salesman problem: A computational study*. Princeton University Press.

Bäck, T. (1996). *Evolutionary algorithms in theory and practice: Evolution strategies, evolutionary programming, genetic algorithms*. Oxford University Press.

Baluja, S. (1994). *Population-based incremental learning. A method for integrating genetic search based function optimization and competitive learning*. Technical report, Department of Computer Science, Carnegie-Mellon University, Pittsburgh, PA.

Benedettini, S., Roli, A., & Di Gaspero, L. (2008). Two-level aco for haplotype inference under pure parsimony. In *International Conference on Ant Colony Optimization and Swarm Intelligence* (pp. 179–190). Springer.

Brameier, M. F., & Banzhaf, W. (2007). A comparison with tree-based genetic programming. In *Linear Genetic Programming* (pp. 173–192).

Cheng, R., & Jin, Y. (2014). A competitive swarm optimizer for large scale optimization. *IEEE Transactions on Cybernetics*, *45*(2), 191–204.

Cheng, R., & Jin, Y. (2015). A social learning particle swarm optimization algorithm for scalable optimization. *Information Sciences*, *291*, 43–60.

Comisky, W., Yu, J., & Koza, J. (2000). Automatic synthesis of a wire antenna using genetic programming. In *Late Breaking Papers at the 2000 Genetic and Evolutionary Computation Conference*, Las Vegas, Nevada (pp. 179–186). Citeseer.

Dorigo, M., Maniezzo, V., & Colorni, A. (1991). Positive feedback as a search strategy.

Dorigo, M., Maniezzo, V., & Colorni, A. (1996). Ant system: Optimization by a colony of cooperating agents. *IEEE Transactions on Systems, Man, and Cybernetics, Part B (Cybernetics)*, *26*(1):29–41.

Dorigo, M., & Stützle, T. (2019). Ant colony optimization: Overview and recent advances. In *Handbook of metaheuristics* (pp. 311–351). Springer.

Eiben, A. E., Hinterding, R., & Michalewicz, Z. (2009). Parameter control in evolutionary algorithms. *IEEE Transactions on Evolutionary Computation*, *3*(2), 124–142.

Espejo, P. G., Ventura, S., & Herrera, F. (2009). A survey on the application of genetic programming to classification. *IEEE Transactions on Systems, Man, and Cybernetics, Part C (Applications and Reviews)*, *40*(2):121–144.
Fogel, D. B., & Fogel, L. J. (1995). An introduction to evolutionary programming. In *European Conference on Artificial Evolution: Artificial Evolution* (pp. 21–33).
Forrest, S., Nguyen, T., Weimer, W., & Le Goues, C. (2009). A genetic programming approach to automated software repair. In *Proceedings of the 11th Annual Conference on Genetic and Evolutionary Computation* (pp. 947–954).
Goldberg, D. E. (1989). *Genetic algorithms in search, optimization, and machine learning*. Addison Wesley.
Goldberg, D. E., Deb, K., & Korb, B. (1991). Do not worry, be messy. In *Proceedings of 4th International Conference on Genetic Algorithms* (pp. 24–30).
Hansen, N., & Ostermeier, A. (1996). Adapting arbitrary normal mutation distributions in evolution strategies: The covariance matrix adaptation. In *Proceedings of IEEE International Conference on Evolutionary Computation* (pp. 312–317). IEEE.
Harik, G., et al. (1999a). *Linkage learning via probabilistic modeling in the ECGA*. IlliGAL report, 99010.
Harik, G. R., Lobo, F. G., & Goldberg, D. E. (1999b). The compact genetic algorithm. *IEEE Transactions on Evolutionary Computation*, *3*(4), 287–297.
Hauschild, M., & Pelikan, M. (2011). An introduction and survey of estimation of distribution algorithms. *Swarm and Evolutionary Computation*, *1*(3), 111–128.
Hoai, N. X., McKay, R. I., & Essam, D. (2006). Representation and structural difficulty in genetic programming. *IEEE Transactions on Evolutionary Computation*, *10*(2), 157–166.
Jin, Y., & Sendhoff, B. (2008). Pareto-based multiobjective machine learning: An overview and case studies. *IEEE Transactions on Systems, Man, and Cybernetics, Part C (Applications and Reviews)*, *38*(3), 397–415.
Jones, T., & Forrest, S. (1991). Fitness distance correlation as a measure of problem difficulty for genetic algorithms. *ICGA*, 184–192.
Karshenas, H., Santana, R., Bielza, C., & Larranaga, P. (2013). Multiobjective estimation of distribution algorithm based on joint modeling of objectives and variables. *IEEE Transactions on Evolutionary Computation*, *18*(4), 519–542.
Kennedy, J., & Eberhart, R. (1995). Particle swarm optimization. In *Proceedings of ICNN'95-International Conference on Neural Networks* (Vol. 4, pp. 1942–1948). IEEE.
Kerschke, P., Hoos, H. H., Neumann, F., & Trautmann, H. (2019). Automated algorithm selection: Survey and perspectives. *Evolutionary Computation*, *27*(1), 3–45.
Koza, J. R. (1992). *Genetic programming: On the programming of computers by means of natural selection* (Vol. 1). MIT press.
Koza, J. R., Bennett, F. H., Andre, D., & Keane, M. A. (1996). Automated design of both the topology and sizing of analog electrical circuits using genetic programming. In *Artificial Intelligence in Design'96* (pp. 151–170). Springer.
Koza, J. R., Bennett, F. H., Andre, D., Keane, M. A., & Dunlap, F. (1997). Automated synthesis of analog electrical circuits by means of genetic programming. *IEEE Transactions on Evolutionary Computation*, *1*(2), 109–128.
Liang, J. J., Qin, A. K., Suganthan, P. N., & Baskar, S. (2006). Comprehensive learning particle swarm optimizer for global optimization of multimodal functions. *IEEE Transactions on Evolutionary Computation*, *10*(3), 281–295.
López-Ibáñez, M., Blum, C., Thiruvady, D., Ernst, A. T., & Meyer, B. (2009). Beam-ACO based on stochastic sampling for makespan optimization concerning the TSP with time windows. In *European Conference on Evolutionary Computation in Combinatorial Optimization* (pp. 97–108). Springer.
Maniezzo, V., & Carbonaro, A. (2000). An ants heuristic for the frequency assignment problem. *Future Generation Computer Systems*, *16*(8), 927–935.

Muhlenbein, H., & Mahnig, T. (1999). Convergence theory and applications of the factorized distribution algorithm. *Journal of Computing and Information Technology*, *7*(1), 19–32.
Muni, D. P., Pal, N. R., & Das, J. (2006). Genetic programming for simultaneous feature selection and classifier design. *IEEE Transactions on Systems, Man, and Cybernetics, Part B (Cybernetics)*, *36*(1):106–117.
Okabe, T., Jin, Y., & Sendhoff, B. (2003). Evolutionary multi-objective optimization with a hybrid representation. In *Proceedings of the IEEE Congress on Evolutionary Computation* (pp. 2262–2269).
Otero, F. E., Freitas, A. A., & Johnson, C. G. (2008). cAnt-Miner: An ant colony classification algorithm to cope with continuous attributes. In *International Conference on Ant Colony Optimization and Swarm Intelligence* (pp. 48–59). Springer.
Paenke, I., Jin, Y., & Branke, J. (2009a). Balancing population-and individual-level adaptation in changing environments. *Adaptive Behavior*, *17*(2), 153–174.
Paenke, I., Kawecki, T. J., & Sendhoff, B. (2009b). Balancing population-and individual-level adaptation in changing environments. *Artificial Life*, *15*(2), 227–245.
Pelikan, M., Goldberg, D. E., & Cantu-Paz, E. (2000). Linkage problem, distribution estimation, and Bayesian networks. *Evolutionary Computation*, *8*(3), 311–340.
Poli, R., Langdon, W. B., McPhee, N. F., & Koza, J. R. (2008). *A field guide to genetic programming*. Lulu.com.
Pitzer, E., & Affenzeller, M. (2012). A comprehensive survey on fitness landscape analysis. In *Recent Advances in Intelligent Engineering Systems* (pp. 161–191). Springer.
Roberts, S. C., Howard, D., & Koza, J. R. (2001). Evolving modules in genetic programming by subtree encapsulation. In *European Conference on Genetic Programming* (pp. 160–175). Springer.
Schwefel, H.-P. (1995). *Evolution and optimum seeking*. Wiley.
Socha, K., & Dorigo, M. (2008). Ant colony optimization for continuous domains. *European Journal of Operational Research*, *185*(3), 1155–1173.
Solnon, C. (2008). Combining two pheromone structures for solving the car sequencing problem with ant colony optimization. *European Journal of Operational Research*, *191*(3), 1043–1055.
Solnon, C., & Fenet, S. (2006). A study of aco capabilities for solving the maximum clique problem. *Journal of Heuristics*, *12*(3), 155–180.
Storn, R., & Price, K. (1996). Minimizing the real functions of the ICEC'96 contest by differential evolution. In *Proceedings of IEEE International Conference on Evolutionary Computation* (pp. 842–844). IEEE.
Tian, Y., Peng, S., Zhang, X., Rodemann, T., Tan, K. C., & Jin, Y. (2020). A recommender system for metaheuristic algorithms for continuous optimization based on deep recurrent neural networks. *IEEE Transactions on Artificial Intelligence*.
Vassilev, V., Fogarty, T., & Miller, J. (2000). Information characteristics and the structure of landscapes. *Evolutionary Computation*, *8*(1), 31–60.
Vladislavleva, E. J., Smits, G. F., & Den Hertog, D. (2008). Order of nonlinearity as a complexity measure for models generated by symbolic regression via Pareto genetic programming. *IEEE Transactions on Evolutionary Computation*, *13*(2), 333–349.
Wang, H., Zhang, Q., Jiao, L., & Yao, X. (2016). Regularity model for noisy multiobjective optimization. *IEEE Transactions on Cybernetics*, *46*(9), 1997–2009.
Weinberger, E. D. (1991). Local properties of Kauffman's n-k model, a tuneably rugged energy-landscape. *Physical Review A*, *44*(10), 6399–6413.
Zames, G., Ajlouni, N., Ajlouni, N., Ajlouni, N., Holland, J., Hills, W., et al. (1981). Genetic algorithms in search, optimization and machine learning. *Information Technology Journal*, *3*(1), 301–302.
Zhang, Q., Zhou, A., & Jin, Y. (2008). RM-MEDA: A regularity model-based multiobjective estimation of distribution algorithm. *IEEE Transactions on Evolutionary Computation*, *12*(1), 41–63.

Chapter 4
Introduction to Machine Learning

4.1 Machine Learning Problems

Machine learning aims to solve a particular type of problems through an iterative process with the help of a machine learning model and a learning algorithm. Typically, problems that can be solved by machine learning can be divided into unsupervised learning problems, such as dimension reduction and data clustering, supervised learning problems including regression and classification, and reinforcement learning problems. In addition to the above three major classes of the learning algorithms, many new learning techniques have been developed that combine the multiple learning techniques, such as semi-supervised learning, active learning, multi-task learning and transfer learning.

Figure 4.1 provides an overview of the three main categories of machine learning algorithms, together with some typical examples of machine learning problems. Behind all machine learning algorithms are typically one or more machine learning models, although there are so-called model-free reinforcement learning methods.

4.1.1 Clustering

Clustering, or clustering analysis is a process that aims to separate the given data into a number of groups so that the data in the same cluster are more similar to each other than to those in other clusters. Data in the same clusters can be represented based on connectivity, distance, distribution or density. In addition, artificial neural networks, graph networks can also be used to represent clusters.

Clustering may become very challenges since the similarity in the data must be measured using a proper proximity measure. As shown in Fig. 4.2, various clusters can be separated only using a correct model. For example, a centroid model will not be able to separate the two clusters in Fig. 4.2b. In addition, it is hard to specify the

Y. Jin et al., *Data-Driven Evolutionary Optimization*,
Studies in Computational Intelligence 975,
https://doi.org/10.1007/978-3-030-74640-7_4

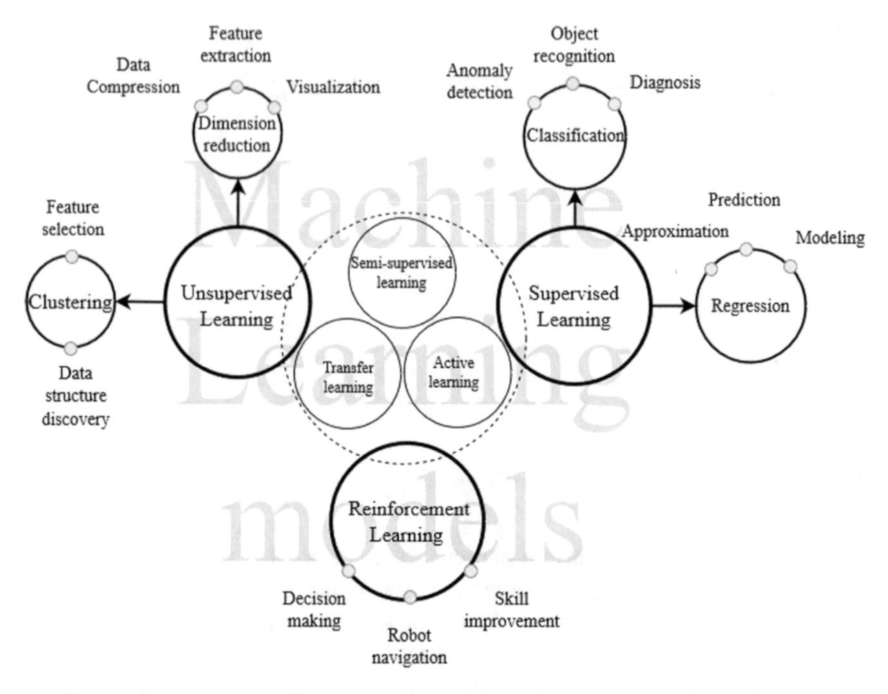

Fig. 4.1 A taxonomy of machine learning algorithms and applications

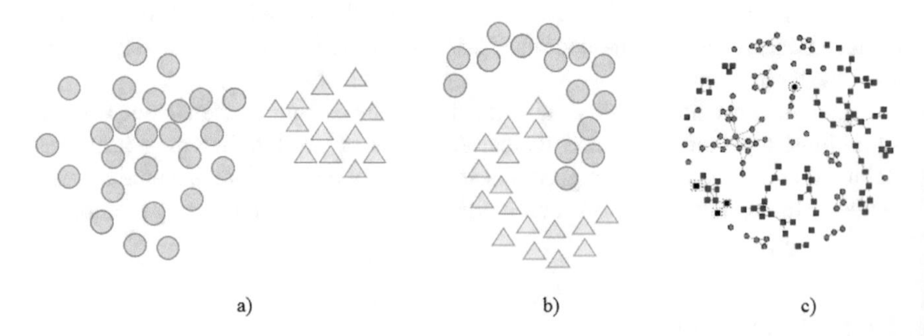

Fig. 4.2 Three examples of data clusters. **a** Two clusters can be represented by a centroid model; **b** Two clusters can be separated using a connectivity or distribution model; and **c** A large number of clusters that can be described by a graph network

number of clusters in advance, and many clustering methods require to so do. For image and text data, clustering analysis becomes even more challenging. For high-dimensional data, it is often hard to find the data structure in the overall space and clustering can only be done in some subspaces. To solve such problems, subspace clustering or biclustering (Kriegel et al., 1999) can be applied, which has achieved success in bioinformatics.

4.1.1.1 K-means Clustering

Given a set of data $\{\mathbf{x}^{(1)}, \mathbf{x}^{(2)}, \cdots, \mathbf{x}^{(d)}\}$, where each data $\mathbf{x}$ is an n-dimensional vector. The k-means clustering algorithms aims to partition the data into k groups, where $k < d$ that needs to be predefined by the user, by minimizing the within-cluster sum of squares:

$$\arg\min_{\mathbf{C}} \sum_{i=1}^{k} \sum_{\mathbf{x} \in \mathbf{C}_i} \| \mathbf{x} - \mu_i \|^2, \tag{4.1}$$

where $\mathbf{C} = \{C_1, \cdots, C_k\}$ are the k clusters of data.

In implementation, one starts with initializing k cluster centers (means) and then assign each data to the cluster with the nearest mean in terms of Euclidean distance. Update the mean of each cluster once all data are assigned. Repeat the above two steps until it converges.

Note that k-means work under the assumption that the number of clusters is known and the data structure can be captured with normal distribution in terms of the Euclidean distance, which is not always true. In addition, there is no theoretical guarantee that the global optimum can be found. Finally, the clustering result may depend on the initial mean value of each cluster.

4.1.1.2 Hierarchical Clustering

Unlike the k-means clustering algorithm, hierarchical clustering does not need to pre-define the number of clusters. Instead, it constructs a hierarchy of clusters by merging and splitting sub-clusters according to their similarity. There are two general approaches to hierarchical clustering, namely agglomerative and divisive. The agglomerative approach is a bottom-up approach, meaning that it starts by considering each data point as a cluster, and then merges similar clusters pairwise. By contrast, the divisive approach is a top-down approach that begins with putting all data in one cluster and then recursively splits the clusters hierarchy by hierarchy.

Figure 4.3 provides an illustrative example of hierarchical clustering of eight data points. As indicated by the solid line arrow, the agglomerative approach starts from the bottom by merging clusters pairwise, while the divisive approach starts from the top by splitting the clusters, as indicated by the dashed line arrow. The final number of clusters is determined by a threshold as shown in the figure, where the dashed horizontal line crosses four vertical lines, resulting in four clusters, and the dotted horizontal line passes two vertical lines, resulting two clusters.

In contrast to the k-means clustering algorithm uses Euclidean distance, hierarchical clustering can use any form of similarity or distance, making it applicable to any attribute types. One main disadvantage of hierarchical clustering is that it has $\mathcal{O}^3$ time complexity and requires $\mathcal{O}^2$ memory, which makes it too slow for medium to large data sets.

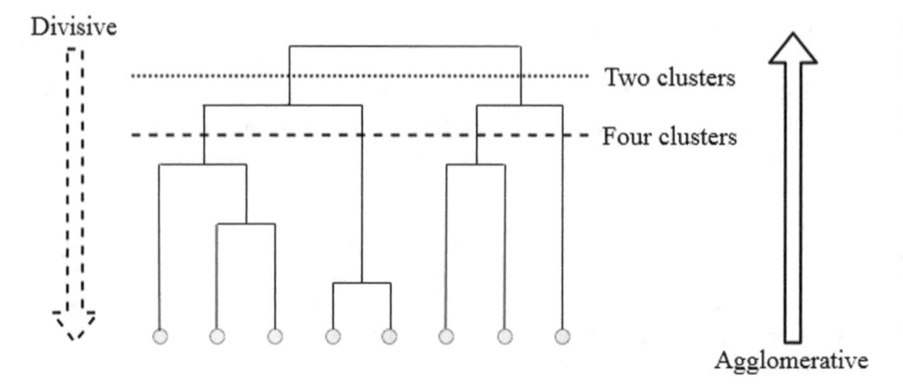

Fig. 4.3 An example of hierarchical clustering

4.1.2 Dimension Reduction

High-dimensional data often contains redundant information and is hard for visualization, analysis and learning due the curse of dimensionality. Dimension reduction is widely used in noise reduction, feature selection or feature extraction, modeling, data visualization, and data analysis.

Both feature selection and feature extraction aim to reduce the number of features or attributes, which is a typical pre-processing before a machine learning model is built. They distinguish in the fact that feature selection deletes those features that do not contribute or contribute less to the output of the model, while feature extraction maps the original features to a new, lower dimensional space where the new features are more informative and non-redundant. Both feature selection and feature extraction make it easier to build a high-quality model, in particular when the number of original features is large. Figure 4.4 shows two examples of data whose dimension can be reduced via feature extraction. On the left panel of the figure, the two-dimensional data in the x_1, x_2 coordinate system can be transferred into a new system in z_1, z_2, and since the values of z_2 of all data are close to zero, the data can be represented in one dimension (z_1) with little information loss. On the right panel of Fig. 4.4, it is clear that the Euclidean distance in the three-dimensional coordinate system is not the right proximity metrics for the data in the so-called "Swiss roll" data set. For example, data points 'A' and 'B' has a small Euclidean distance, however, their distance is much longer along the two-dimensional manifold. We can see from Fig. 4.4 that the dimension of the Swiss roll data set can be reduced to two if they can be mapped to a new coordinate system defined along the manifold.

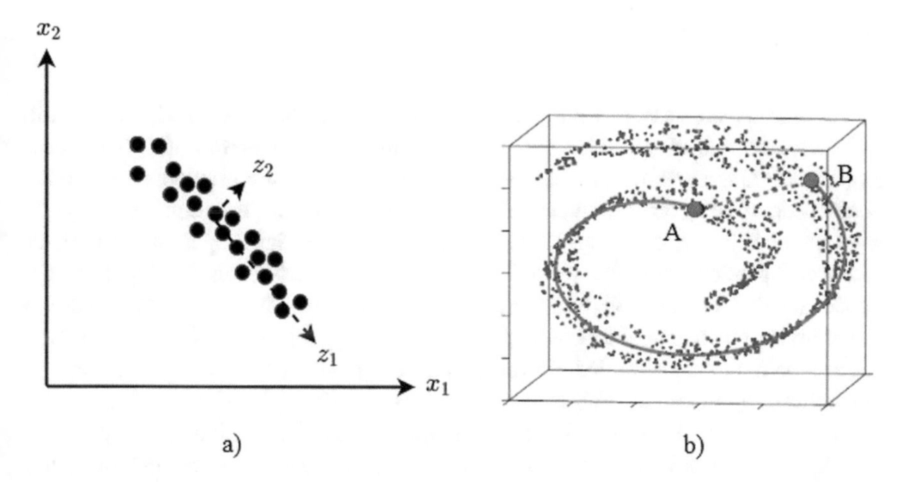

Fig. 4.4 Illustrative examples of dimension reduction of data. **a** The original two-dimensional data can be reduced to one-dimensional without loss of much information. **b** So-called Swiss roll containing three-dimensional data, which can be reduced to two-dimensional using a nonlinear dimension reduction technique

4.1.2.1 Principal Component Analysis

Principal component analysis (PCA) is a widely used technique for dimension reduction, which linear transform the n-dimensional data to a new, lower-dimensional coordinate system by maximizing the variance of the data in the new representation. Consider d n-dimensional data points $\{\mathbf{x}^{(1)}, \cdots, \mathbf{x}^{(d)}\}$, which have been centered (i.e., have zero mean), then the first principal component (vector) can be computed as follows:

$$\mathbf{w}_1 = \arg \max_{\|\mathbf{w}\|=1} \frac{1}{m} \sum_{i=1}^{m} (\mathbf{w}^T \mathbf{x}^{(i)})^2. \tag{4.2}$$

Then, the k-th ($k \leq n$) can be obtained by subtracting the first $k-1$ principal components from $\mathbf{x}^{(i)}$:

$$\mathbf{w}_k = \arg \max_{\|\mathbf{w}\|=1} \frac{1}{m} \sum_{i=1}^{m} \left\{ \left[\mathbf{w}^T \left(\mathbf{x}^{(i)} - \sum_{j=1}^{k-1} \mathbf{w}_j \mathbf{w}_j^T \mathbf{x}^{(i)} \right) \right]^2 \right\} \tag{4.3}$$

The principal components can also be obtained by means of singular value decomposition.

The main limitation of PCA is that it can mainly capture linear correlation between variables. Thus, many nonlinear PCA variants have been developed, including principal curve and manifolds and kernel PCA. In addition, neural networks can also be applied to nonlinear dimension reduction, which is going to be discussed when we introduce unsupervised learning algorithms.

4.1.3 Regression

Regression is concerned with the approximation of the functional relationship between two sets of variables using a mathematical or black-box machine learning model. One main difference between regression and the previously discussed machine learning problems such as clustering and dimension reduction is that regression involves two spaces, one is called the dependent variable space and the other is the independent variable space. The independent variables are the inputs of the model and the dependent variables are the outputs of the model. The model inputs are sometimes known as features or attributes. Depending on the problem to be solved, regression may also be called interpolation, extrapolation, function approximation, prediction, or modeling. Figure 4.5 illustrates one example of regression, where the functional relationship between the input (independent variable) x and output (dependent variable) y is to be estimated. In the figure, one linear model (solid line) and one nonlinear model (dashed line) are used, either of which is a useful estimate of the true function between the two variables. However, it is hard to judge, without the ground truth, whether the linear model or the nonlinear model is a better approximation of the true function behind the given data, as it was put by Box that "*All models are wrong, but some are useful.*" (Box, 1979). This is particularly true due to the bias-variance trade-off in training a machine learning model. For example, in Fig. 4.5, the nonlinear model has a smaller bias over the training samples, but it also introduces a larger variance in estimation.

Many statistical methods have been proposed for model selection, such as cross-validation, Akaike's information criterion, Bayesian information criterion, and minimum description length, among others. The minimum description length can be seen as a mathematical formulation of Occam's razor, the law of parsimony, meaning that the simplest model that can describe the data is the best. Note that the parsimony principle also holds for model building in classification, which is to be discussed next.

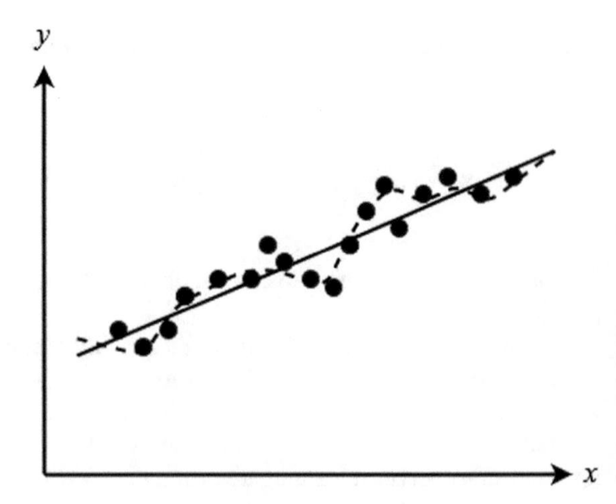

Fig. 4.5 An example of regression based on the given data describing the relationship between the independent variable x and dependent variable y

4.1.4 Classification

Classification aims to distinguish which category (or class, label) a new observation belongs to, given a set of training data containing observations (or instances) whose category (label) is known. Classification can be seen as a supervised form of clustering, since the number of classes is known and each group of the data has a given label. Classification problems having two classes are known as binary classification. By contrast, a multi-class classification problem contains more than two classes.

A multi-classification problem can be seen as multiple binary classification problems, and consequently it can be solved using binary classification methods, including the One-vs-Rest approach, which splits a multi-class classification problem into one binary classification problem per class; and the One-vs-One approach, converting a multi-class classification problem into one binary classification problem per each pair of classes. However, many classifiers are able to solve multi-class classification problems directly.

Figure 4.6 plots examples of binary classification problems. The two classes of data in Fig. 4.6a can be separated using a linear hyperplane, known as linearly separable; while those in Fig. 4.6b can be separated only using a nonlinear hyperplane. Figure 4.6c shows a multi-class classification problem containing three classes.

In addition to the problem of model selection similar regression (here the complexity of the hyperplane separating different classes), training a classifier becomes more challenging when the data is class-imbalanced, where the number of examples in the training dataset for each class label is heavily imbalanced. For example in Fig. 4.6c, the number of training data of one class is much smaller than that of the other two classes, which is very common, e.g. in fraud detection and anomaly detection. Many strategies have been developed to address the imbalanced classification problem (Wang and Yao, 2012), for example by over-sampling the the minority class to create additional synthetic data, and/or by under-sampling the majority the majority class to reduce the number of the samples of the majority classes.

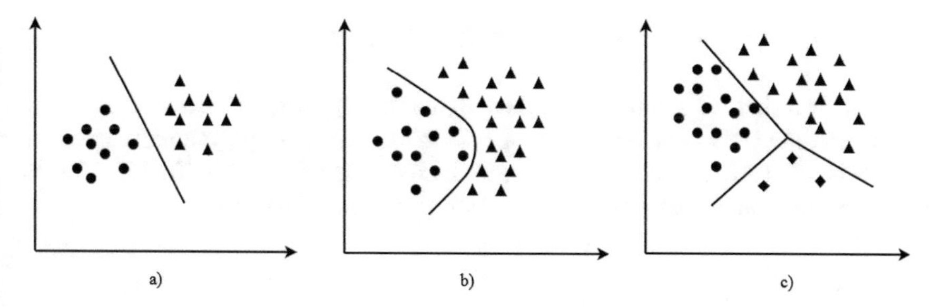

Fig. 4.6 Illustrative examples of **a** linearly separable binary classification; **b** linearly non-separable binary classification; and **c** multi-class classification

4.2 Machine Learning Models

To deal with the problems in the previous section, many machine learning models driven by data are frequently employed. In this section, we will introduce six popular models.

4.2.1 *Polynomials*

Polynomial regression (PR) models (Draper and Smith, 1998) are statistical models to analyze the the relationship between the n-dimensional input variable $\mathbf{x}$ and the output variable y, i.e., $y = f(\mathbf{x})$. As a widely used approximation method, polynomial regression models may have different orders. The first-order and second-order polynomial regression models can be written as:

$$\mathbf{y} = a_0 + \sum_{i=1}^{n} a_i x_i, \tag{4.4}$$

$$\mathbf{y} = a_0 + \sum_{i=1}^{n} a_i x_i + \sum_{i=1}^{n} \sum_{j=1}^{n} a_{ij} x_i x_j, \tag{4.5}$$

where a_i are parameters to be estimated. We can rewrite Eqs. (4.4) and (4.5) as:

$$\mathbf{y} = \mathbf{Za}, \tag{4.6}$$

where $\mathbf{a}$ are parameter vectors and n_c-dimensional $\mathbf{Z}$ are variables in different orders $\mathbf{Z} = \{1, x_1, ...\}$. To calculate the unknown parameter $\mathbf{a}$, at least n_c training data points are needed. Then, $\mathbf{a}$ can be calculated as:

$$\mathbf{a} = (\mathbf{Z}^{\mathrm{T}}\mathbf{Z})^{-1}\mathbf{Z}^{\mathrm{T}}\mathbf{y}. \tag{4.7}$$

It is clear that the complexity of the polynomial regression model increases with the increasing order. The dimension of $bf Z$ in the first-order polynomial regression model is $n + 1$, while that of the second-order model is $C_n^2 + n + 1$. Thus, the higher order polynomial regression model needs more training data point.

4.2.2 *Multi-layer Perceptrons*

The multi-layer perceptron (MLP) is one class of feedforward neural networks (Gardner and Dorling, 1998), which has been widely used for solving classification and

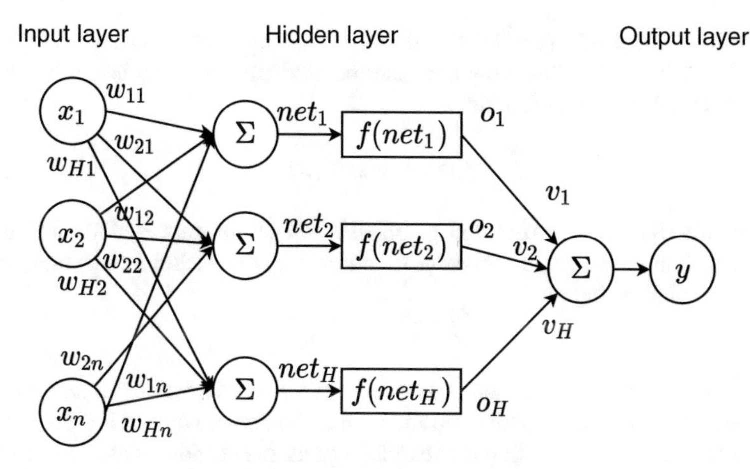

Fig. 4.7 The structure of a multi-layer perceptron with one hidden layer

regression problems. A simplest MLP has three layers: input, hidden, and output layers. Note that there are can be more than one hidden layer in an MLP. Given the structure in Fig. 4.7, the output y of the MLP can be expressed as:

$$\begin{aligned} y &= \sum_{i=1}^{H} v_i f(net_i), \\ net_i &= \sum_{j=1}^{n} w_{ij} x_j, \end{aligned} \tag{4.8}$$

where w_{ji} and v_i are weights connecting nodes between the input and hidden layers, $f(net_i)$ is the activation function. To be capable of dealing with the nonlinear relation between y and $\mathbf{x}$, the activation function of the hidden neurons usually adopts a nonlinear function. The existing activation functions can be divided into saturated and non-saturated functions, which depends on whether the function has a compact range. The sigmoid function and tanh function are two saturated functions, whose analytical expressions are shown below.

$$f(x) = \frac{1}{1+e^{-x}} \tag{4.9}$$

$$f(x) = \frac{e^x - e^{-x}}{e^x + e^{-x}} \tag{4.10}$$

The softmax function is an extension of the sigmoid function for multi-class neural networks (J-class), as shown below:

$$f_i(\mathbf{x}) = \frac{e^{x_i}}{\sum_{j=1}^{J} e^{x_j}}. \tag{4.11}$$

The rectified linear unit (ReLU) function is a non-saturated functions as shown in Eq. (4.12), which can effectively alleviate the gradient vanishing issue, in contrast to the saturated activation functions.

$$f(x) = \max(0, x) \tag{4.12}$$

Additionally, the leaky ReLU and parametric ReLU are improved version of the ReLU function. Recently, the maxout function is a general ReLU function, which can be described as below:

$$f_i(\mathbf{x}) = \max x_i. \tag{4.13}$$

It should be noted that the output neurons can also adopt a nonlinear activation function, however, a linear activation function in the output neurons has been shown to be beneficial for accelerating the learning speed. In addition, adding connections between the input and output neurons directly has also been found to be useful for speeding up the learning process.

4.2.3 Radial-Basis-Function Networks

Radial basis function networks (RBFNs) are another class of artificial neural networks, typically containing three layers (input, hidden, and output layers) (Moody and Darken, 1989). The activation functions in RBFNs are radial basis functions as described in the following equations:

$$\varphi_i(\mathbf{x}) = \varphi(\|\mathbf{x} - \mathbf{c}\|), \tag{4.14}$$

where $\mathbf{c}$ is the center of $\varphi_i(\mathbf{x})$ and $\|\mathbf{x} - \mathbf{c},\|$ is the distance between $\mathbf{x}$ and the center. The Gaussian function, reflected sigmoid, and multiquadric functions in Eqs. (4.15)–(4.17) have been used as distance metrics in RBFNs:

$$\varphi(\mathbf{x}) = \exp\left(-\frac{\|\mathbf{x} - \mathbf{c}\|^2}{2\sigma^2}\right), \tag{4.15}$$

$$\varphi(\mathbf{x}) = \frac{1}{1 + \exp(\frac{\|\mathbf{x} - \mathbf{c}\|^2}{\sigma^2})}, \tag{4.16}$$

$$\varphi(\mathbf{x}) = \sqrt{\|\mathbf{x} - \mathbf{c}\|^2 + \sigma^2}, \tag{4.17}$$

where σ controls the spread.

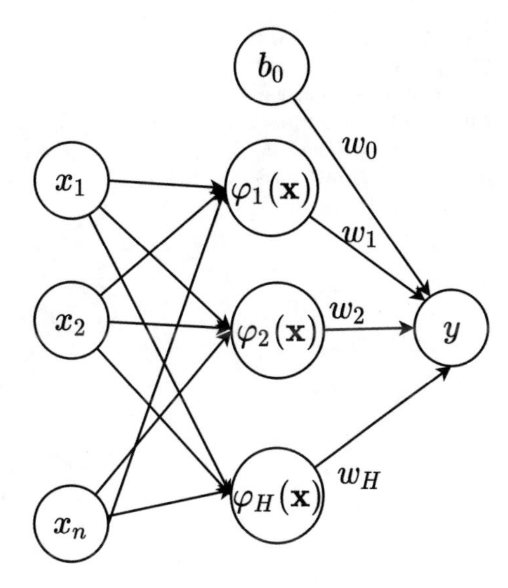

Fig. 4.8 An example structure of radial basis function networks

As shown in Fig.4.8, the hidden layer consists H radial basis functions (neurons, nodes), and a bias node b_0. Thus, the output y can be calculated as below.

$$y = w_0 b_0 + \sum_{i=1}^{H} w_i \varphi_i(\mathbf{x}), \tag{4.18}$$

where w_i are weights. To train an RBFN, parameters $\mathbf{c}$, σ, and w can be trained using supervised learning on d training data $\mathcal{D} = (\mathbf{x}_k^d, y_k^d)$. Alternatively, the centers $\mathbf{c}$ and spread σ can be pre-defined. Typically, the centers can be determined by clustering the training data into H groups and then use the cluster centers as the centers of the radial basis functions, and the spread according to the distance between the cluster centers. Them, the weights w can be estimated using the least square method (Moody and Darken, 1989).

4.2.4 Support Vector Machines

Support vector machines (SVMs) are classic statistical machine learning models that defines a $n - 1$ hyperplane to classify n-dimensional labelled data points $\mathbf{x}$ (Steinwart and Christmann, 2008). There might be many hyperplanes that can distinguish two classes, but the best hyperplane is the one with the largest margin.

Figure 4.9 shows an example of a linear SVM on a classification problem, where the square points are the positive class data ($y = 1$) and the dots are the negative class data ($y = -1$). The hyperplane can be expressed by:

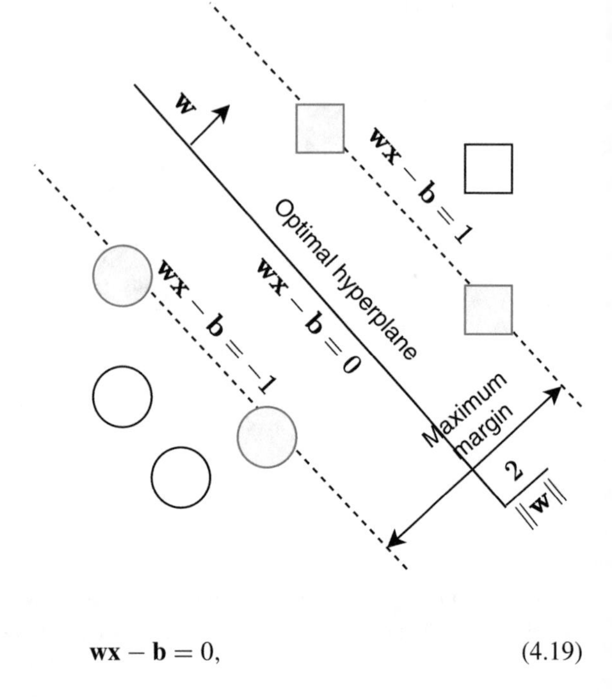

Fig. 4.9 An illustration of the hyperplane in support vector machines, where the square points are the positive class data ($y = 1$) and the dots are the negative class data ($y = -1$)

$$\mathbf{wx} - \mathbf{b} = 0, \tag{4.19}$$

where $\mathbf{w}$ and $\mathbf{b}$ are the normal vector and intercept of the hyperplane. In fact, we have two class boundaries $\mathbf{wx} - \mathbf{b} = 1$ and $\mathbf{wx} - \mathbf{b} = -1$, whose margin is $\frac{2}{\|\mathbf{w}\|}$. To calculate $\mathbf{w}$ and $\mathbf{b}$, the following optimization problem needs to be solved:

$$\begin{aligned} &\min \|\mathbf{w}\| \\ &s.t.\ y_i(\mathbf{w}\mathbf{x}_\mathrm{i} - \mathbf{b}) \geq 1, \\ &1 \leq i \leq d, \end{aligned} \tag{4.20}$$

where d is the number of training data points. To extend SVM to linearly inseparable classification problems, loss functions like hinge loss, logistic loss and exponential loss can be defined. Furthermore, the original linear SVM can deal with nonlinear classification problems by introducing kernel functions to map the training data onto a high-dimensional space (Jin and Wang, 2012).

4.2.5 Gaussian Processes

Gaussian process regression model is also known as kriging model (Matheron, 1963; Emmerich, 2005), which is an interpolation method governed by prior covariances. Assume the random variables follow a joint normal distribution, a Gaussian process regression model can be written as:

$$y = \mu(\mathbf{x}) + \varepsilon(\mathbf{x}), \tag{4.21}$$

where $\mu(\mathbf{x})$ is the global tend of the regression model and $\varepsilon(\mathbf{x})$ is the deviation. To estimate $\mu(\mathbf{x})$ and $\varepsilon(\mathbf{x})$ based on d training data $\mathcal{D} = (\mathbf{x}_k^d, y_k^d)$, the following posterior can be determined by the maximum likelihood estimation.

$$p(\theta|\mathbf{x}, y) = \frac{p(y|\mathbf{x}, \theta)p(\theta)}{p(y|\mathbf{x})}, \tag{4.22}$$

where θ is the hyperparameter to be optimized. Then, for a new $\mathbf{x}^*$, $\mu(\mathbf{x}^*)$ and $\varepsilon(\mathbf{x}^*)$ can be inferred as:

$$\mu(\mathbf{x}^*) = k(\mathbf{x}^*, \mathbf{x})(\mathcal{K} + \sigma_n^2 I)^{-1} y, \tag{4.23}$$

$$\varepsilon(\mathbf{x}^*) = k(\mathbf{x}^*, \mathbf{x}^*) - k(\mathbf{x}^*, \mathbf{x})(\mathcal{K} + \sigma_n^2 I)^{-1} k(\mathbf{x}, \mathbf{x}^*), \tag{4.24}$$

where function k is the covariance function or kernel function and $\mathcal{K}$ is the kernel matrix of d training data points. Many popular kernel functions like RBF, exponential, and rational quadratic kernels have been adopted in Gaussian process regression models.

4.2.6 Decision Trees

Decision trees have been used for representing decision making. Also, they can be used as prediction models for classification or regression tasks in supervised learning (Rokach and Maimon, 2008).

A decision tree has three types of nodes: root, decision, and leaf nodes. Fig. 4.10 gives an example decision tree to predict whether a person is fit or not, where the input includes age, dietary habit, and exercise habit. Root and decision nodes divide

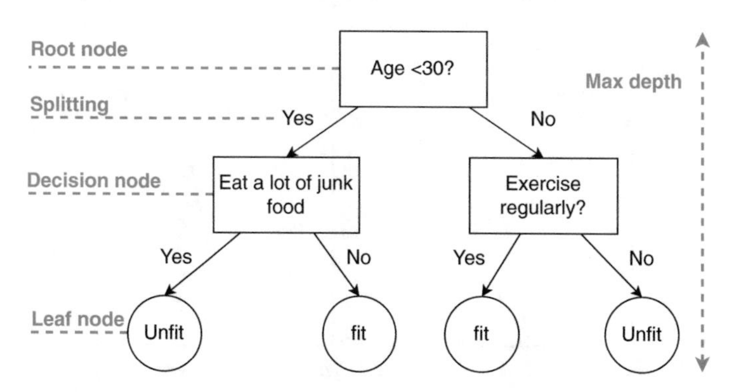

Fig. 4.10 Illustration of decision trees

data into several classes as leaf nodes. To grow a decision tree from the training data, root and decision nodes are splitted based on a metric (Gini impurity or information gain or variance reduction) until the stopping criterion is met (Breiman et al., 1984). The popular stopping criteria are maximum depth and MSE reduction threshold. ID3, C4.5, and classification and regression tree (CART) (Steinberg and Colla, 2009) are three widely used decision tree learning algorithms.

4.2.7 *Fuzzy Rule Systems*

Fuzzy rule systems are widely used for representing human understandable knowledge on the basis of fuzzy sets (Zadeh, 1965). Compared to crisp sets where an element is either belong to or not belong to a set, fuzzy sets allow an element to belong to the sets with a degree specified by a membership function. As shown in Fig. 4.11, for a set of "tall" person, a characteristic function can be defined to specify whether a person is tall or not, whereas in fuzzy sets, a person may belong to a fuzzy set of "tall" person with a degree between 0 and 1.0.

For the crisp set of "Tall", anyone who is taller than 1.8 m is defined to be in the set. As a result, a person who is 1.75 m tall is not belong to the set. By contrast, for a fuzzy set "Tall", a membership function is defined, and for persons who are 1.75 m, they belong to the fuzzy set "Tall" with a degree of 0.5, while for those taller than 1.8 m, the degree becomes 1.0.

Fuzzy sets have been used to define *linguistic terms* that are used in human reasoning. For example, 'hot', 'cold' for temperature, 'young', 'middle-aged' and 'old' for age. In this context, temperature and age are called *linguistic variables* (Zadeh, 1975). A linguistic variable consists of linguistic terms, which are defined by synthetic rules. The linguistic terms are fuzzy subsets, each defined by fuzzy membership function, which are the semantic rules. These fuzzy subsets consist of a fuzzy partition of the universe of discourse, all possible values the linguistic variable can take. An example of defining the linguistic variable variable 'age' is shown in Fig. 4.12.

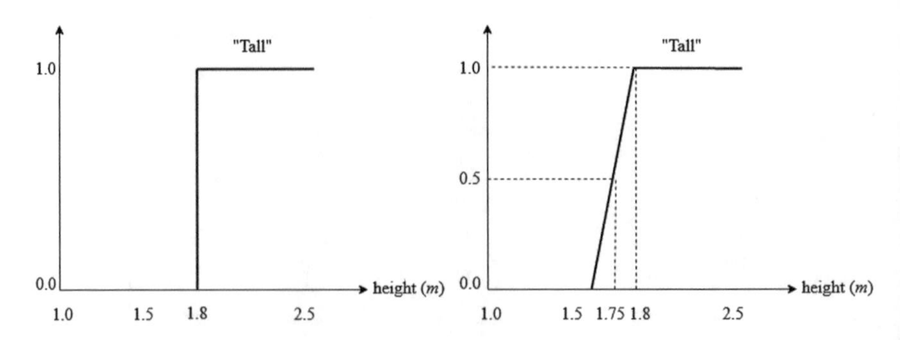

Fig. 4.11 Left: Characteristic function of a crisp set. Right: Membership function of a fuzzy set

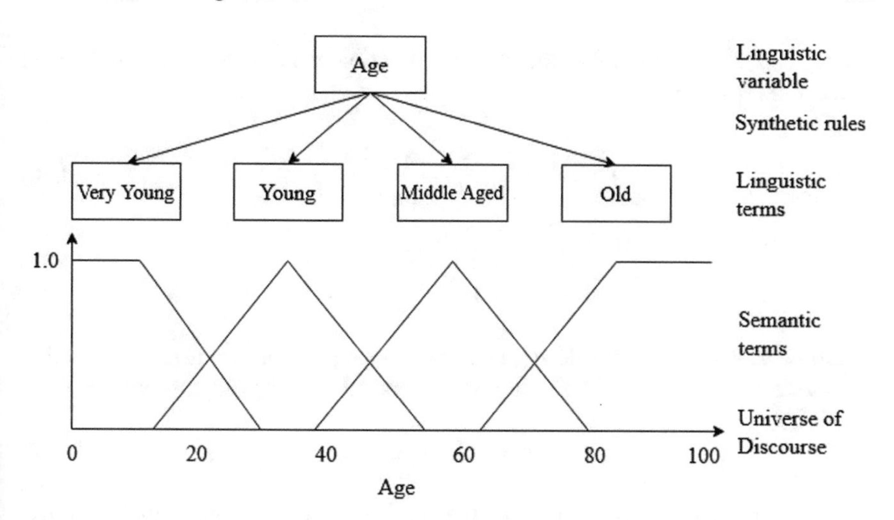

Fig. 4.12 Definition of the linguistic variable age

The linguistic variables can be used to represent human expert knowledge in terms of fuzzy IF-THEN rules. For instance,

$$\text{IF } age \text{ is } Young, \text{ THEN } risk \text{ is } Low. \tag{4.25}$$

In the above fuzzy rule, both age and risk are linguistic variables, *Young* and *Low* are linguistic terms defined by a fuzzy membership function.

Fuzzy rule systems have been very successful in represent human expert knowledge, in particular in control systems (Michels et al., 2006). Two main types of fuzzy rules have been investigated, one called Mamdani rules, in which both the rule premise and consequence use fuzzy variable:

$$\text{IF } x \text{ is } A, \ y \text{ is } B;$$

By contrast, Takagi-Sugeno-Kang (TSK) rules use crisp expression in the consequence part of the rule:

$$\text{IF } x \text{ is } A, \ y = f(x).$$

If $f(x) = c$ is a constant, the TSK rule is called zero-order; if $f(x) = a_0 + a_1 x$, it is called the first-order TSK rule, and if the consequent part is a dynamic system, i.e., $\dot{x} = f(x)$, it is then called a fuzzy dynamic system.

Fuzzy rule systems can be used as a general machine learning model for regression and classification. Compared to artificial neural networks, fuzzy systems are believed to be more understandable to human users than neural networks. For example, for a TSK fuzzy rule system consisting of n premises and N fuzzy rules:

$$R^j\text{: IF } x_1 \text{ is } A_1^j(x_1) \text{ AND } x_2 \text{ is } A_2^j(x_2) \text{ AND } \ldots\text{, AND } x_n \text{ is } A_n^j(x_n),$$
$$\text{THEN } y^j = a_0^j + a_1^j x_1 + \cdots + a_n^j x_n;$$

where $A_i^j(x_i)$ is the j-th fuzzy membership function of variable x_i. The final output of the TSK fuzzy rule is

$$y = \frac{\sum_{i=1}^{N} w^j y^j}{\sum_{i=1}^{N} w^j}, \tag{4.26}$$

$$w^j = \prod_{j=1}^{n} A_i^j(x_i), \tag{4.27}$$

where w^j is the weight of rule R^j, and $\prod(\cdot)$ is the probabilistic fuzzy intersection operator for fuzzy AND. If a Zadeh fuzzy intersection is applied, then we have

$$w^j = \min\{A_i^j(x_i)\}. \tag{4.28}$$

It has been shown that zero-order TSK fuzzy rule systems are mathematically equivalent to RBFNs, if Gaussian functions are used as the membership functions (Jang and Sun, 1993).

While the membership functions of early fuzzy systems are determined heuristically, data-driven fuzzy systems have become popular in the 1990s when learning algorithms such as the gradient based method has been applied to train the parameters in the fuzzy membership functions as well as in the consequence of the fuzzy rules, where are known as neuro-fuzzy systems. While data-driven fuzzy systems are more flexible and powerful, the interpretability of the rules may get lost. In this case, improvement of the interpretability of fuzzy rules will be necessary (Jin, 2000, 2003).

4.2.8 Ensembles

Single machine learning models are subject to the bias-variance tradeoff, i.e., the smaller the bias (i.e., the more accurate the model is on the training data), the bigger the variance the prediction on unseen data will be larger (i.e., the more likely the model may overfit), and it is impossible to simultaneously minimize both bias and variance.

It has been shown, however, that the bias-variance trade-off can be resolved if an ensemble of accurate and diverse base learners can be generated. Figure 4.13 provides an illustrative example of machine learning ensemble consisting of three base learners. Typically, the final output an ensemble can be plain or weighted averaging of, or the majority voting of all base learners.

The most important challenge in constructing machine learning ensembles is to build multiple base learners that are accurate and diverse. The most popular methods for generating base learners are as follows:

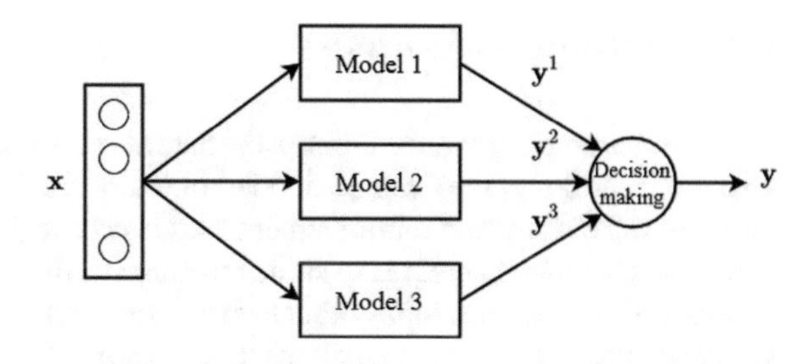

Fig. 4.13 An ensemble consisting of three base learners, and the decision-making component determines the final output typically by averaging or majority voting

- *Statistical methods.* One basic idea is to generate multiple data sets $D_1, D_2, \ldots, D_m$ by sampling the original data set D with replacement. As a result, m different models $M_1(\mathbf{x}), M_2(\mathbf{x}), \ldots, M_m(\mathbf{x})$ can be trained using the m data sets. The final output is usually the plain averaging, i.e., $M(\mathbf{x}) = \frac{1}{m}\sum_{i=1}^{m} M_i(\mathbf{x})$. This is usually known as bootstrap aggregating, or bagging. A variant of bagging is known as boosting, where the base learners are built sequentially, instead of in parallel as in bagging. One benefit of training each base learner sequentially is that one can emphasize the training instances on which the existing base learners perform poorly, so-called weak learners. The most popular bagging method is Adaptive Boosting, AdaBoost for short. Note, however, AdaBoost may be sensitive to noise and outliers.
- *Different initialization and learning algorithms.* Since gradient-based learning algorithms often depend on the initial model, different initialization will lead to different final models. Alternatively, one may achieve slightly different models by using different learning algorithms.
- *Negative correlation.* One idea to promote diversity is to explicitly promote negative correlation between the outputs of the base learners. For m base learners $M_i(\mathbf{x}), i = 1, 2, \ldots, m$, the loss function for promoting negative correlation can be expressed as following:

$$J = E + \lambda C, \tag{4.29}$$

$$C = (M_i(\mathbf{x}) - M(\mathbf{x})) \sum_{j \neq i} \left(M_j(\mathbf{x}) - M(\mathbf{x})\right), \tag{4.30}$$

  where E is the normal error function, C is to penalize correlation between the outputs of the base learning for a given input, and $0 \leq \lambda \leq 1$ is a hyperparameter.
- *Structural diversity.* Structural diversity of base learners can be obtained by means of using multiple models of the same type, e.g., neural networks having different number of layers or different number of hidden nodes, models of different types, e.g., a neural network, a support vector machine, and a radial-basis-function network, or models with different features. For example, boosting can be applied to combine multiple decision trees to generate random forests (Zhou, 2009; Liaw and Wiener, 2002).

4.3 Learning Algorithms

Machine learning algorithms differ in the data they use and the task they aim to solve. Generally speaking, they can be divided into three main categories, namely supervised, unsupervised, and reinforcement learning. Note, however, that these learning algorithms assume that there are sufficient amounts of training data and the data satisfies certain conditions. Many other learning approaches have been proposed, which combine different learning paradigms to address challenges in learning, in particular the lack of training data, the low quality of the data, and the dynamism in the data.

4.3.1 Supervised Learning

Supervised learning is the most popular learning paradigm where there are pairs of training data containing input and desired output. The purpose of supervised learning is to modify the parameters of the model so that its output learns the desired output for the given input. This is achieved by minimizing a loss function that calculates the difference between the desired output and the output of the model. Given d pairs of training data $(\mathbf{x}^{(i)}, y^{(i)}), i = 1, 2, \ldots, d$, where $\mathbf{x} = (x_1, x_2, \ldots, x_n)$, the following loss function can be defined:

$$\text{minimize} \sum_{i=1}^{d} \ell\left(h(\theta, \mathbf{x}^{(i)}), y^{(i)}\right), \tag{4.31}$$

where $h(\theta, \mathbf{x})$ is a generic machine learning model, θ is the parameter vector, and $\ell(\cdot)$ is the loss function. From the optimization point of view, a supervised learning algorithm aims to solve the minimization problem defined in Eq. (4.31), where the model parameters θ are the decision variables.

The most widely used loss function is the squared loss function:

$$\text{minimize} \sum_{i=1}^{d} \ell(\left(h(\theta, \mathbf{x}^{(i)}), y^{(i)}\right) = \text{minimize} \sum_{i=1}^{d} \left(h(\theta, \mathbf{x}^{(i)}) - y^{(i)}\right)^2. \tag{4.32}$$

In addition to the squared loss function, the absolute loss function can be used:

$$\text{minimize} \sum_{i=1}^{d} \ell\left(h(\theta, \mathbf{x}^{(i)}), y^{(i)}\right) = \text{minimize} \sum_{i=1}^{d} |h(\theta, \mathbf{x}^{(i)}) - y^{(i)}|. \tag{4.33}$$

To solve the minimization problem in (4.32), the gradient-based method described in Sect. 2.1.1 can be applied:

$$\Delta\theta = -\alpha \nabla_\theta \sum_{i=1}^{d} \ell\left(h(\theta, \mathbf{x}^{(i)}), y^{(i)}\right), \tag{4.34}$$

where $\nabla_\theta \ell(\cdot)$ is the differentiation of $\ell(\cdot)$ with respect to θ.

For example, the multi-layer perceptron with one hidden layer in Sect. 4.2.2 can be trained using the gradient based method, which is also known as error backpropagation (Rumelhart et al., 1985). Given d labelled data $\mathcal{D} = (\mathbf{x}_k^*, y_k^*)$, weights w_{ji} and v_i can be updated by minimizing the following loss function:

$$\ell(\cdot) = \frac{1}{2} \sum_{k=1}^{d} (y_k - y_k^*)^2, \tag{4.35}$$

where y_k is the output of the neural network input x_k^*. Then, the weights w_{ij} and v_i In equations, ξ and η are the learning rate.

$$\Delta v_i = -\xi \frac{\partial \ell(\cdot)}{\partial y_k} \frac{\partial y_k}{\partial v_i} = -\xi \sum_{k=1}^{d} (y_k - y_k^*) f(net_i) \tag{4.36}$$

$$\Delta w_{ij} = -\eta \frac{\partial \ell(\cdot)}{\partial y_k} \frac{\partial y_k}{\partial w_{ij}} = -\eta \sum_{k=1}^{d} (y_k - y_k^*) v_i f'(net_i) x_j^* \tag{4.37}$$

The above gradient based method is called batch gradient based method, which is very likely to converge to a local minimum. In practice, the stochastic gradient descent based method (SGD) is more often used, which updates the weights sample by sample. Thus, the above two equations can be rewritten as:

$$\Delta v_i(k) = -\xi \frac{\partial \ell(\cdot)}{\partial y_k} \frac{\partial y_k}{\partial v_i} = -\xi (y_k - y_k^*) f(net_i), k = 1, 2, \ldots, d. \tag{4.38}$$

$$\Delta w_{ij}(k) = -\eta \frac{\partial \ell(\cdot)}{\partial y_k} \frac{\partial y_k}{\partial w_{ij}} = -\eta (y_k - y_k^*) v_i f'(net_i) x_j^*, k = 1, 2, \ldots, d. \tag{4.39}$$

Although the SGD usually converges faster than the batch gradient based method, it may suffer from instability in learning when the data size is very large. In this case, the so called mini-batch SGD can be used, where a batch of $d' < d$ instances are randomly chosen from the training data at each epoch.

A large number of variants of the gradient based method have been proposed. For example, a momentum can be added to the SGD to accelerate stable convergence while reducing oscillation. Many ideas for adapting the learning rate and the hyperparameter for the momentum term have also been suggested (Ruder, 2016).

4.3.2 Unsupervised Learning

Unsupervised learning is a class of learning algorithms for dimension reduction and feature extraction on the basis of artificial neural networks. In contrast to supervised learning, unsupervised learning does not require an external teacher signal. Instead, the neural network receives a number of different input patterns and during the learning and discovers significant features in these patterns.

4.3.2.1 Hebbian Learning

Hebbian learning is probably the first unsupervised learning algorithm inspired from findings in computational neuroscience known as Hebbian Law. It states that for two connected neurons (neuron i and neuron j), if neuron i is near enough to excite neuron j and repeatedly participates in its activation, the synaptic connection between these two neurons is strengthened and neuron j becomes more sensitive to stimuli (input signal) from neuron i. If two neurons on either side of a connection are activated synchronously, then the weight of that connection is increased. On the contrary, the weight of that connection is decreased if two neurons on either side of a connection are activated asynchronously. Mathematically, the Hebbian rule can be expressed as follows:

$$\Delta w_{ij}(t) = \alpha y_j(t)\, x_i(t), \tag{4.40}$$

where α is the learning rate, and $x_i(t)$, $y_j(t)$ are the neuronal activity of the two connected neurons at time instant t, respectively. A forgetting factor can be added to make the Hebbian learning more stable:

$$\Delta w_{ij}(t) = \alpha y_j(t)\, x_i(t) - \phi y_j(t)\, w_{ij}(t), \tag{4.41}$$

where ϕ is the forgetting factor.

4.3.2.2 Competitive Learning

In competitive learning, the outputs of a neural network compete to be activated and the one that wins the competition is called the winner-takes-all neuron. The neural network consists of an input layer and a competition layer, and the number of input neurons equals the dimension of the data, while the number of the outputs equals the number of clusters each data is going to assigned to. The output neuron whose incoming weights have the shortest Euclidean distance from the input vector is the winner neuron and only the incoming weights of the winner node will be updated. As shown in Fig. 4.14, for n-dimensional data, if the number of clusters is set to M, and for any data point $\mathbf{x}$, the Euclidean distance between the weight vector of the j-th neuron, $\mathbf{w}_j = (w_{j1}, \cdots, w_{jn})$, $j = 1, \cdots, M$, can be calculated by:

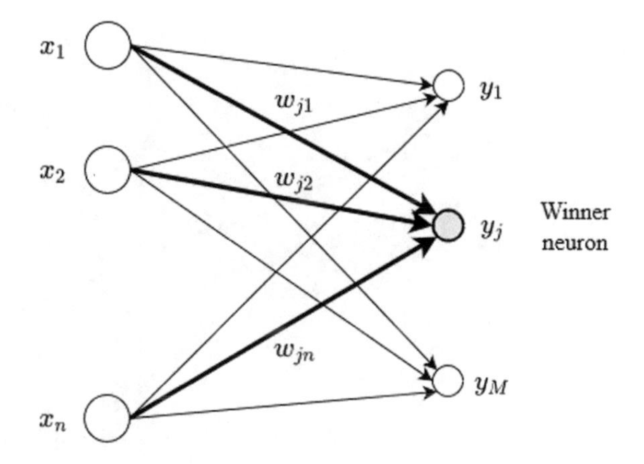

Fig. 4.14 Winner-takes-all competitive learning, where neuron y_j wins for input data $\mathbf{x}$ and then all weights connected to it will be updated

$$d_j = \sqrt{\sum_{i=1}^{n}(x_i - w_{ji})^2}, \ j = 1, \ldots, M \tag{4.42}$$

Then neuron j that has the minimum d_j is the winner and its weights are updated as follows:

$$\Delta w_{ji} = \eta(x_i - w_{ji}), \ i = 1, \ldots, d, \tag{4.43}$$

where η is the learning rate.

Note that in Eq. (4.42) the Euclidean distance is used. However, other distance metrics can also be used so that competitive learning can also be applicable to different data types.

4.3.2.3 Self-organizing Maps

In competitive learning, only the incoming weights of the winner neuron will be updated. By contrast, self-organizing maps (SOMs) include competition, cooperation, and adaptation. In this way, SOMs can transform given data into one- or two-dimensional space while preserving the topological structure, making it a powerful tool for visualizing high-dimensional data.

An SOM also consists of two layers, and mostly typically, the output neurons are ranked in a two-dimensional space, as shown in Fig. 4.15. Similar to competitive learning, the number of input neurons equals to the number of features, while the number of output neurons is user-defined. The input and output layers are fully connected.

Given an input data $\mathbf{x} = [x_1, x_2, \ldots, x_n]^T$ and a weight vector connecting to the j-th output neuron $\mathbf{w}_j = [w_{j1}, w_{j2}, \cdots, w_{jn}]^T, j = 1, 2, \ldots, L$, where L is the total output neurons. After initialization of the weights, the training of the SOM is composed of three main steps:

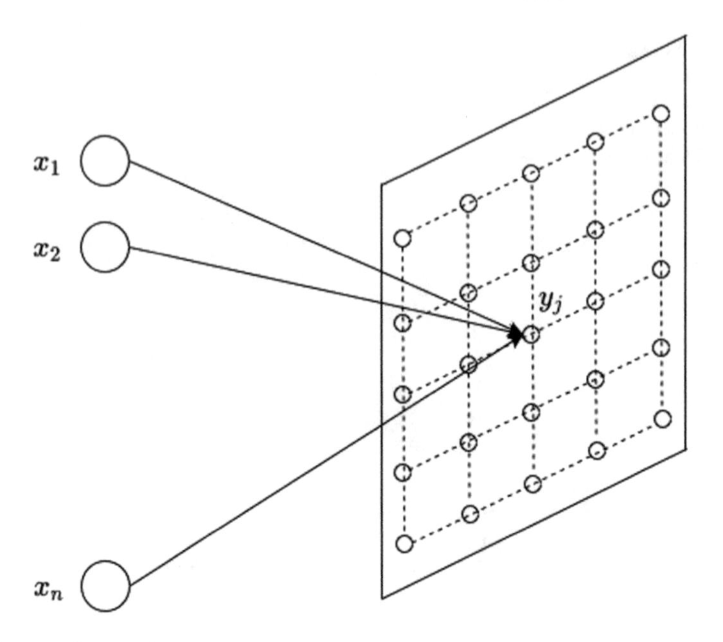

Fig. 4.15 A self-organizing map

- *Competition.* Determine the winner neuron according to the following criterion, which is similar to competitive learning:

$$\arg\min_j \parallel \mathbf{x} - \mathbf{w}_j \parallel, j = 1, 2, \ldots, L. \tag{4.44}$$

- *Cooperation.* The main difference between the SOM and competitive learning lies in the cooperation step, which specifies a spatial neighborhood of the winning neuron and all neurons in this neighborhood are considered to be excited and will participate in weight adaptation in the third step. Note that the topological neighborhood is symmetric to the winner neuron, and its size shrinks over the learning iterations, as shown in Fig. 4.16. In addition, the strength of weight update becomes weaker when the distance between an excited neuron and the winner neuron is larger, which is defined by a distance function to be defined in the next step.
- *Weight adaptation.* All excited neurons adjust their weights as follows:

$$\Delta\mathbf{w}_i = \eta h_{ji}(\mathbf{x} - \mathbf{w}_i), \tag{4.45}$$

where i is an index for any excited neuron, and h_{ji} is called a neighborhood function, which is typically a Gaussian:

$$h_{ji} = e^{-\frac{d_{ji}^2}{2\sigma^2}}, \tag{4.46}$$

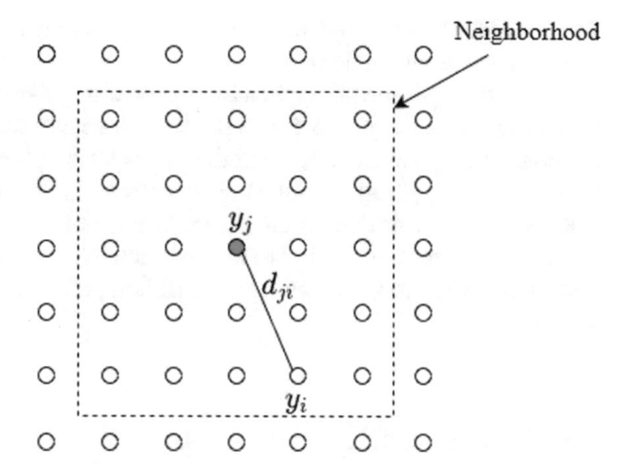

Fig. 4.16 Definition of a neighborhood and the distance function in a self-organizing map

where d_{ji} is the Euclidean distance between the location of excited neuron i and the winner neuron j, as illustrated in Fig. 4.16.

Note that the learning rate η should usually linearly decrease over the learning iterations.

4.3.3 Reinforcement Learning

Reinforcement learning (RL) (Sutton and Barto, 2016) differs from both supervised learning and unsupervised learning. RL is different from unsupervised learning in that its task is not to find out the structure in the input data. Similar to supervised learning, RL involves two spaces, with one (the input space) being the environmental state, and the other (output) being an action or a decision. Unlike in supervised learning, however, there is no desired action (teacher signal) for a given state, and consequently, no direct correction (error) can be made available. Instead of a teacher signal, there is a feedback from the environment, called reward, that reflects whether the action is good for achieving a particular goal. The RL learns how to behave successfully to achieve a goal by interacting with the external environment through and learning from experience.

The basic idea of reinforcement originated from both control theory, in particular optimal control and dynamic programming, and psychology, such as classical conditioning and operant conditioning, and later neuroscience such as dopamine system in the brain.

Mathematically, a RL task can be described by four elements: a set of states $\mathcal{S}$, a set of actions $\mathcal{A}$, a transition model $T(s, a, s') = P(s'|s, a)$ that defines the probability of transition from state s to s', $s, s' \in \mathcal{S}$ for given an action $a \in \mathcal{A}$, and the expected reward for taking action a in the given state s, $R(s, a) = \sum_{s'} T(s, a, s')r(s, a, s')$.

The challenge is we do not know about the environment, i.e., the transition function $T(s, a, s')$ is usually unknown.

There are two general approaches to solve the above problem, one is a model based approach that aims to learn a model and use it to derive the optimal policy, and the other is a model free approach that derives the optimal policy without learning the model, e.g., Q-learning. Other challenges in RL include that the reward may be delayed or stochastic and the environment may not be fully observable.

Reinforcement learning has found wide applications in control and game playing. It can also be as a tool for solving optimization problems such as neural architecture search.

4.3.4 Advanced Learning Algorithms

In addition the three basic categories of machine learning techniques we discussed above, many other learning techniques have been proposed to address different challenges, including active learning, semi-supervised learning, multi-task learning, and transfer learning. Most of these techniques aim to address the lack of labelled data, which is a key challenge in machine learning, since data collection, particularly collection of labelled data, is expensive. This challenge is also particularly relevant to data-driven optimization, where optimization must be carried out with only small data.

4.3.4.1 Active Learning

Active learning aims to prioritize the data that needs to be collected or labelled in training a supervised model, as shown in Fig. 4.17. The main purpose of active learning is to achieve a highly accurate model with as a small amount of data as possible. One idea is to query a data point where the model predicts most uncertainly, and therefore it is important to have a means to estimate the uncertainty of a prediction by the model at a given input. This can be achieved by query by committee (a normal ensemble of models), by using a probabilistic model such as a Gaussian process, or by calculating the distance to the existing labelled data. Another approach is uncertainty sampling, which can help identify unlabeled data points that are near a decision boundary in the current model. Three methods are commonly used, including least confidence (Culotta and McCallum, 2005), smallest margin (Scheffer et al., 2001), and maximum entropy-based sampling (Dagan and Engelson, 1995).

4.3.4.2 Semi-supervised Learning

Semi-supervised learning is another approach to addressing the issue of lack of labelled data. Different to active learning, semi-supervised learning does not add

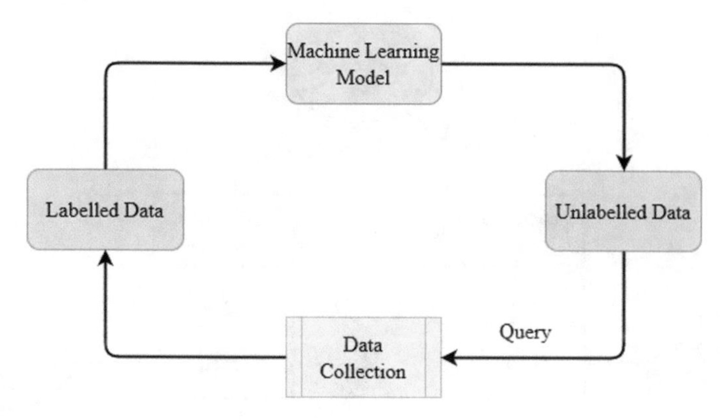

Fig. 4.17 The main components of active learning

new labelled data; it aims to use information from the unlablled data and generate synthetic labelled data from a subset of selected unlabelled data. The key issue becomes if, and which unlablled data can be reliably predicted by the existing model so that they can be added to the labelled data to improve the learning performance. A few main approaches to semi-supervised learning are listed below.

- *Co-training, tri-training and multi-training*. Assume there are two different, ideally independent views of the same data and there exists a small amount of labelled data and a large amount of unlabelled data for each view. Then, one separate classifier can be trained for each of the view and the most confident predictions on the unlabeled data can be used to train the other classifier (Blum and Mitchell, 1998). The assumption of existing two different views may hold only for some particular data such as image data. Interestingly, this assumption can be lifted and much work has demonstrated that training of a diverse and accurate ensemble consisting of two or more baser learners can also be used to select unlabeled data for augmentation of the labelled (Zhou and Li, 2005; Gu and Jin, 2017). A diagram of semi-supervised learning based on co-training is given in Fig. 4.18.
- *Continuity based approaches*. Another class of semi-supervised learning is based on the assumption that there exists some relationship in distributions between labelled and unlabelled data. These relations include continuity, neighborhoodness, connectivity, or manifold. Machine learning models such as Bayesian models, support vector machines and graph-based methods have been developed.

4.3.4.3 Multi-task Learning and Transfer Learning

Multi-task machine learning is a subfield of machine learning that aims to improve the learning performance in the presence of data paucity by simultaneously learning multiple related tasks. As shown in Fig. 4.19a, normally different learning tasks are

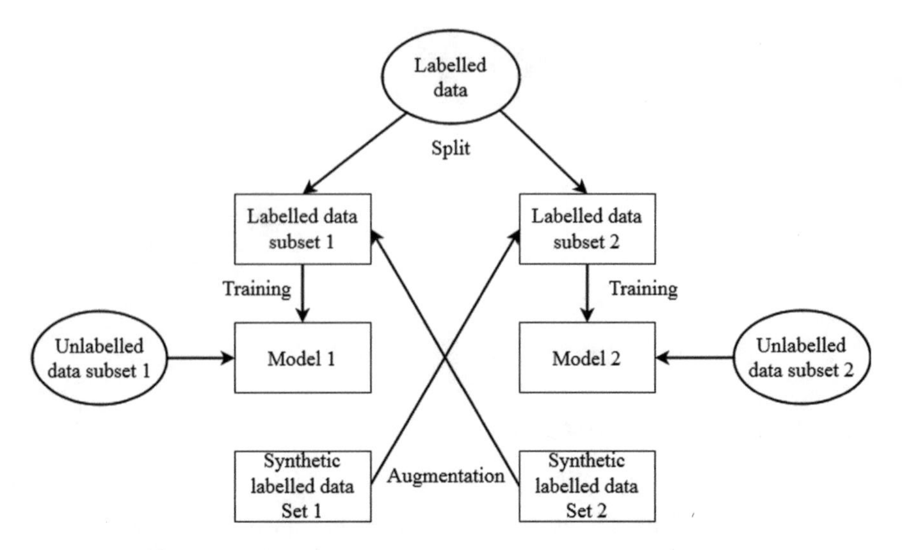

Fig. 4.18 Semi-supervised learning based on co-training using two diverse base learners

accomplished separately. The simplest form of multi-task learning is sketched in Fig. 4.19b, where the data for different tasks are shared in training the different models. Fig. 4.19c illustrates a more general form of multi-task learning, in which there are shared layers and task-specific layers.

Theoretically, learning multiple tasks simultaneously can improve the learning performance on all tasks. Nevertheless, one can make use of multi-task learning to improve the learning of a hard task with little training data. For example, in (Jin and Sendhoff, 1999), a catalytic task, which is an approximation of the hard task, is used to accelerate the training of the original task, as shown in Fig. 4.20.

Multi-task learning can be seen as an early form and more general type of machine learning that can reuse knowledge from other related tasks, known as transfer learning. In transfer learning, there are a source task and a target task. In supervised transfer learning, it is assumed that the source task has a large amount of labelled data, while the target task has only limited labelled data available. It is also assumed that knowledge from the source task is able to improve the learning performance of the source task, even if the source and target tasks have different domains, i.e., different feature space and/or different label space. The main ideas for knowledge transfer are summarized as follows (Zhuang et al., 2020):

- Instance transfer. In this case, knowledge is transferred by means of data. If the source and target tasks have the same domain, some data of the source task can be reused directly, which is similar to multi-task learning. If the source and target tasks have different domains, domain adaptation becomes essential in knowledge transfer via a latent space before synthetic data can be generated to augment the data of the target task.

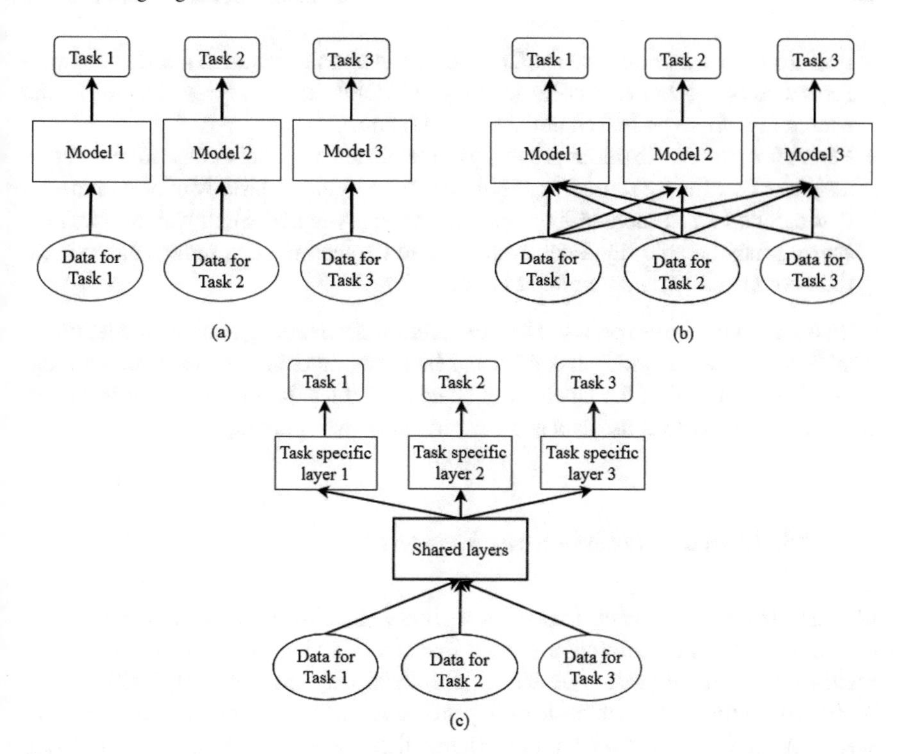

Fig. 4.19 Single-task learning versus multi-task learning

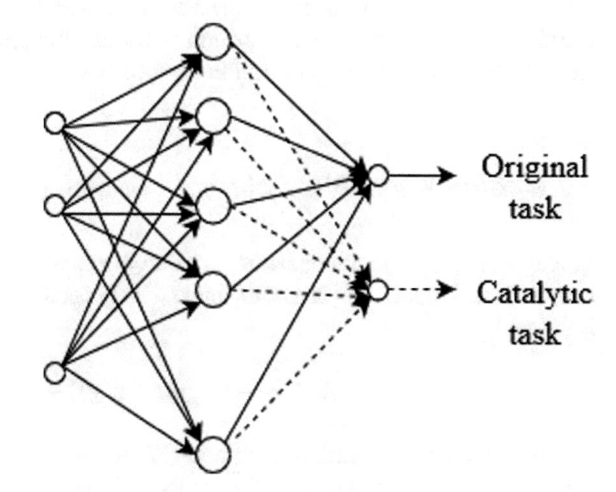

Fig. 4.20 Catalytic learning, where the output neuron with connections denoted by dashed lines learns an approximate task of the original one

- Feature-representation transfer. This approach aims to improve the learning performance of the target task by identifying useful feature representations in the source domain to be reused in the target domain.
- Parameter transfer. In this approach, parameters of the model trained for the source task can be directly reused in the model for the target task. Model parameters that can be reused include hyperparameters and weights. Such weight transfer / sharing may be able significantly accelerate the learning and reduce the training time compared to random model initialization.

In addition to semi-supervised learning, multi-task learning and transfer learning, weakly supervised learning has attracted increased attention in machine learning, which is very useful of the labels of the data are subject to uncertainty or mistakes, which is widely seen if the data is collected via crowd sourcing.

4.4 Multi-objective Machine Learning

Machine learning has been treated as a single-objective optimization problem in the field of machine learning, although machine learning problems are inherently multi-objective optimization problems (Jin, 2006; Jin and Sendhoff, 2008), which is true for almost all machine learning problems ranging from clustering, feature selection and feature extraction, to regularization, model selection, sparse learning, neural structure optimization, and interpretable machine learning. In the following sections, we are going to define the Pareto approach to multi-objective machine learning, and then provide a few examples for handling machine learning problems from the multi-objective optimization perspective.

4.4.1 Single- and Multi-objective Learning

Model selection is the fundamental issue in machine learning, and Akaike's Information Criterion (AIC) is one of the most widely used criteria. Mathematically, AIC is defined as follows:

$$AIC = -2\log(L(\theta|\mathbf{y}, g)) + 2K, \tag{4.47}$$

where $L(\theta|\mathbf{x}, g)$ is maximized likelihood of the function g, θ are the model parameters, $\mathbf{y}$ is the given data, and K is the number of effective parameters. In model selection, AIC should be minimized, meaning that the likelihood should be maximized and the number of free parameters should be minimized. Although these terms are aggregated to a scalar loss function maximization of the likelihood and minimization of the effective model parameters are actually two conflicting objectives that cannot be simultaneously optimized.

Let us consider the following neural network regularization problem, which can be seen as a special case of AIC:

$$\min J = E + \lambda\Omega, \tag{4.48}$$

where E is the error on the training data, Ω is the model complexity, and λ is a hyperparameter. Usually, Ω can be the sum of the squared weights of the neural network, which is meant to penalize overly complex models:

$$\Omega = \sum_{i=1}^{M} \theta_i^2, \tag{4.49}$$

where θ_i is a weight of the neural network, and M is the total number of weights in the neural network. By means of using a hyperparameter, the two objectives, minimization of the training error and minimization of the model complexity are converted into a single objective. However, it is non-trivial to determine λ, since an overly small λ may fail to address the overfitting problem, while an overly large λ may lead to underfitting.

From the multi-objective optimization point of view, the above regularization problem can be formulated as the following bi-objective optimization problem:

$$\min \{f_1, f_2\} \tag{4.50}$$

$$f_1 = E, \tag{4.51}$$

$$f_2 = \Omega. \tag{4.52}$$

The main differences between the single-objective formulation (4.49) and multi-objective formulation 4.50 of neural network regularization are that in the latter, no hyperparameter is needed, and multiple neural network models presenting the trade-off between the training error and model complexity will be obtained, as illustrated in Fig. 4.21.

As shown in (Jin and Sendhoff, 2008), the multi-objective approach to neural network regularization may provide some new insight into neural network training.

- First, it has been shown that simple network models are interpretable in that understandable symbolic rules can be extracted from them. For example, Fig. 4.22 plots two simplest neural networks obtained by evolutionary multi-objective structure optimization of neural networks on the Iris data set. The Iris data contains 150 samples in total, of which 120 are used for training. Each data has four attributes, namely Sepal-length (x_1), Sepal-width (x_2), Petal-length (x_3), and Petal-width (x_4), and three classes, Iris-Setosa (y_1), Iris-Versicolor (y_2), and Iris-Virginica (y_3). The simplest neural network uses the third attribute (x_3) only, as shown in Fig. 4.22 (a). From this network, the following rule can be extracted:

$$R_1: \text{ If } x_3 < 2.2, \text{ then Iris-Setosa.} \tag{4.53}$$

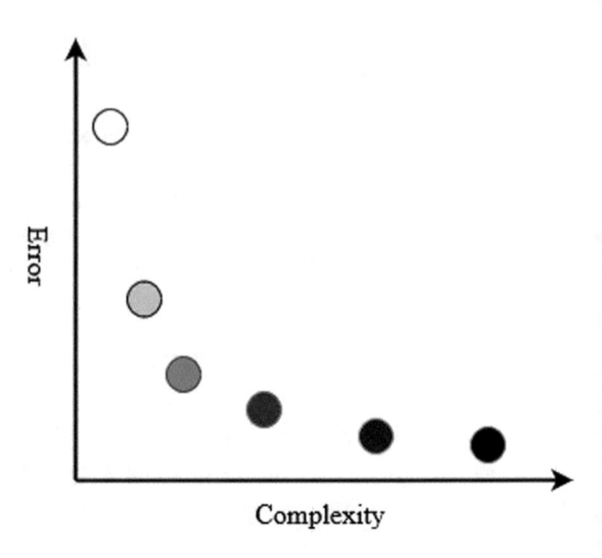

Fig. 4.21 Multi-objective neural network regularization, where multiple models that trade off between complexity and accuracy will be obtained

The second simplest neural network is plotted in 4.22 (b). This network selects two features, x_3 and x_4, however, the weights connecting to x_4 are much smaller than those to x_3. As a result, the following three rules are extracted:

$$R_1: \text{ If } x_3 < 2.2 \text{ and } x_4 < 1.0, \text{ then Iris-Setosa;} \quad (4.54)$$

$$R_2: \text{ If } x_3 > 2.2 \text{ and } x_4 < 1.4, \text{ then Iris-Versicolor;} \quad (4.55)$$

$$R_3: \text{ If } x_4 > 1.8, \text{ then Iris-Virginica.} \quad (4.56)$$

The extracted rules can be valid from the data plotted in Fig. 4.23. More detailed analysis can be found in (Jin and Sendhoff, 2008).

- Second, multi-objective approach to neural network regularization is helpful for model selection, and those models near the knee point on the Pareto front can generalize well.
- Third, multi-objective structure optimization provides a powerful means for identifying base learners having different structures, resulting in diverse and accurate ensemble models.

4.4.2 *Multi-objective Clustering, Feature Selection and Extraction*

Pareto-based multi-objective clustering has shown to be helpful for determining the number of clusters (Handl and Knowles, 2007). Here, the two objectives are cluster compactness, described by overall cluster deviation, and cluster connectivity, which is expressed by the degree to which neighboring data points are grouped in the same

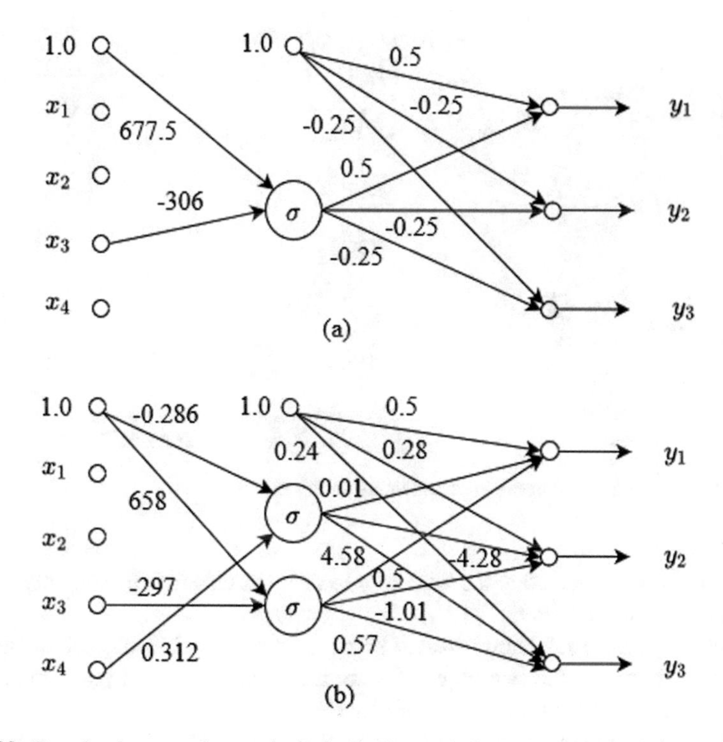

Fig. 4.22 Two simplest neural network obtained using evolutionary multi-objective structure optimization

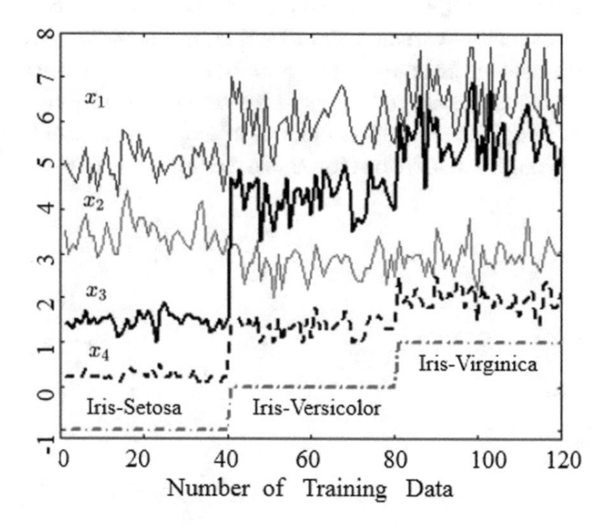

Fig. 4.23 Iris data and their class labels

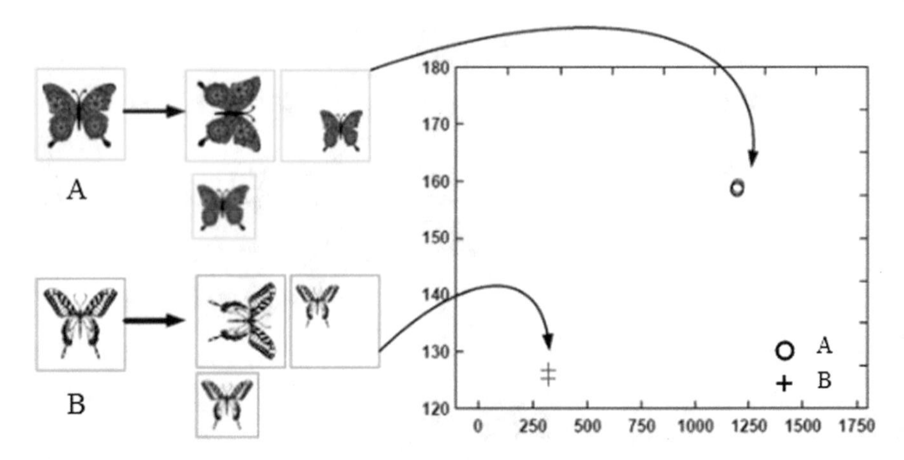

Fig. 4.24 Within-class and between-class feature variances

cluster. It has been found that the knee solution may be able to automatically identify the optimal cluster number.

Similarly, feature selection can also be handled by the multi-objective optimization approach, where the two objectives are the maximization of performance and the minimization of the number of features (Xue et al., 2012).

Finally, extraction of image features can also be formulated as a multi-objective optimization problem (Albukhanajer et al., 2015). The two objectives are the the minimization of the within-class feature variance and the maximization of the between-class feature variance. For example, in Fig. 4.24, two-dimensional features can be extracted from images using trace transform (Kadyrov and Petrou, 2001). For rotated, scaled or translated images of the original A, slightly different features $\mathbf{x}_1^A$, $\mathbf{x}_2^A$, and $\mathbf{x}_3^A$ may be extracted. Similarly, $\mathbf{x}_1^B$, $\mathbf{x}_2^B$, and $\mathbf{x}_3^B$ can be extracted from the original image B. Then, we can formulate the following two-objective optimization problem:

$$\min\{f_1, f_2\}; \tag{4.57}$$

$$f_1 = \mathbf{s}_w; \tag{4.58}$$

$$f_2 = \frac{1}{\mathbf{s}_b + \epsilon}; \tag{4.59}$$

where,

$$\mathbf{s}_w = \sum_{i=1}^{3} \parallel \mathbf{x}_i^A - \mu^A \parallel \tag{4.60}$$

$$\mathbf{s}_b = \parallel \mu^A - \mu^B \parallel \tag{4.61}$$

$$\mu^A = \frac{1}{3}\sum_{i=1}^{3} \mathbf{x}_i^A, \tag{4.62}$$

$$\mu^B = \frac{1}{3} \sum_{i=1}^{3} \mathbf{x}_i^B. \tag{4.63}$$

where > 0 is a small constant.

This way, multiple Pareto optimal features can be extracted from images, which can contribute to the enhancement of classification performance (Albukhanajer et al., 2017).

4.4.3 Multi-objective Ensemble Generation

Ensemble generation is concerned with two objectives, accuracy of the base learners and diversity between the base learners. Thus, it is very natural to consider using the multi-objective approach to ensemble generation (Gu et al., 2015a). Here, the following two issues are of main concern:

- *Objectives to be used.* Different objectives can be considered, such as accuracy versus diversity, different loss functions, accuracy versus complexity. In addition, three-objective formulations can also be adopted, such as accuracy, complexity and diversity.
- *Selection of base learners.* The multi-objective approach will result in multiple trade-off models, and only a subset of them should be selected for construction of the ensemble. One may select the most accurate models, or models whose accuracy is better than a given threshold, or those of a larger degree of diversity, and those near the knee point of the Pareto front (Smith and Jin, 2014).

4.5 Deep Learning Models

Deep learning is an area of machine learning, which has contributed to the great success of artificial intelligence since 2010. Conceptually, however, deep learning is a class of machine learning methods based on artificial neural networks with multiple hidden layers, and its success can be largely attributed to the availability of huge computational power and big data. Technically, many new models and more powerful learning algorithms have been proposed, making deep learning to be able to solve a wide range of scientific and technical problems such as complex game playing (e.g., AlphaGo), natural language processing, protein folding, autonomous driving, among many others.

In the following, a few widely used deep learning models are very briefly introduced.

4.5.1 Convolutional Neural Networks

The most popular deep learning model is probably the convolutional neural network (CNN) and its variants. CNN is a feedforward neural network, containing convolutional units, pooling layers, and fully-connected layers. The convolutional layers are specified by the kernel size, stride, and zero-padding, converting the input (e.g., an image) into an abstract representation. The pooling layers are meant to reduce the dimension of the representations, and typical pooling methods include averaging-pooling, max-pooling, and stochastic pooling. The convolution and pooling together can be seen as an automated feature extraction process, and the extracted features are then classified by a fully connected layer. Thus, the classification performance of CNNs depends more on the convolutional and pooling layers. An illustrative example of a CNN is plotted in Fig. 4.25.

Widely used variants of the CNN include AlexNet, VGGNet (Simonyan and Zisserman, 2014), ResNet (He et al., 2016), and U-Net.

4.5.2 Long Short-Term Memory Networks

Long short-term memory (LSTM) networks are a variant of recurrent neural networks, except that the neurons in the hidden layers are replaced by memory blocks. A common LSTM unit consists of a cell state, an input gate, a forget gate, and an output gate. The cell state represents the memory of the LSTM and undergoes changes via forgetting of old memory (the forget gate) and addition of new memory (the input gate). The input gate receives its activation from both output of the previous time and current input, which has a sigmoidal activation function. The forget gate computes a value between 0 and 1 using the sigmoid function from the input and the current state, determining which information should be forgotten by output a value close to 0. Finally, the output gate conditionally decides what to output from the memory. One LSTM cell is illustrated in Fig. 4.26.

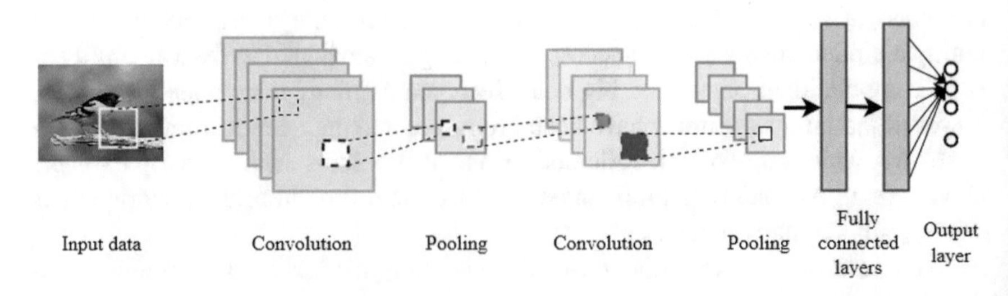

Fig. 4.25 An illustrative example of CNN containing two convolutional layers, two pooling layers and two fully connected layers

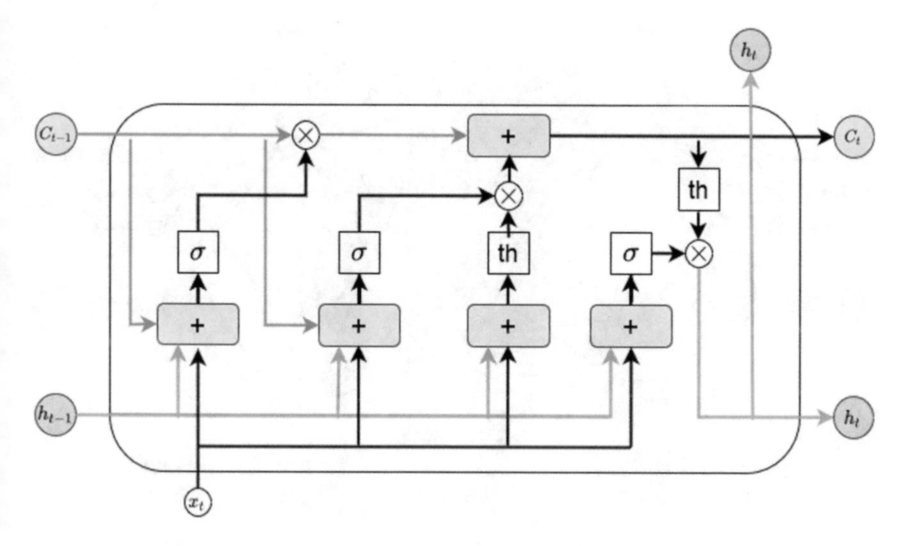

Fig. 4.26 A LSTM cell, where C_{t-1} is the memory from the previous cell, x_t is the input, h_{t-1} is the output of the previous block. σ is the sigmoid function, and th is the hyperbolic tangent function

As its name suggests, LSTMs are able to handle long-term dependencies and avoid the gradient vanishing or exploding problem. Since they have recurrent connections, LSTMs are well suited for analyzing time series, speech or video data containing temporal information.

4.5.3 *Autoassociative Neural Networks and Autoencoder*

In Sect. 4.1.2, we briefly introduced principal components analysis (PCA), a dimension reduction technique. However, PCA can mainly capture linear correlation and does not work well if there is nonlinear correlation between the attributes or features. Autoassociative neural networks (Baldi and Hornik, 1989; Kramer, 1991), were proposed as a nonlinear PCA, which are feedforward neural networks whose hidden layer is called *bottleneck layer* and the output layer aims to learn the input signal, as shown in Fig. 4.27.

By minimizing the following loss (error) function, the autoassociative neural network is able to realize nonlinear dimension reduction, if the number of neurons in the bottleneck layer (m) is set to be smaller than the dimension of the data (n):

$$E = \sum_{j=1}^{d} \sum_{i=1}^{n} (x_i^{(j)} - y_i^{(j)})^2, \tag{4.64}$$

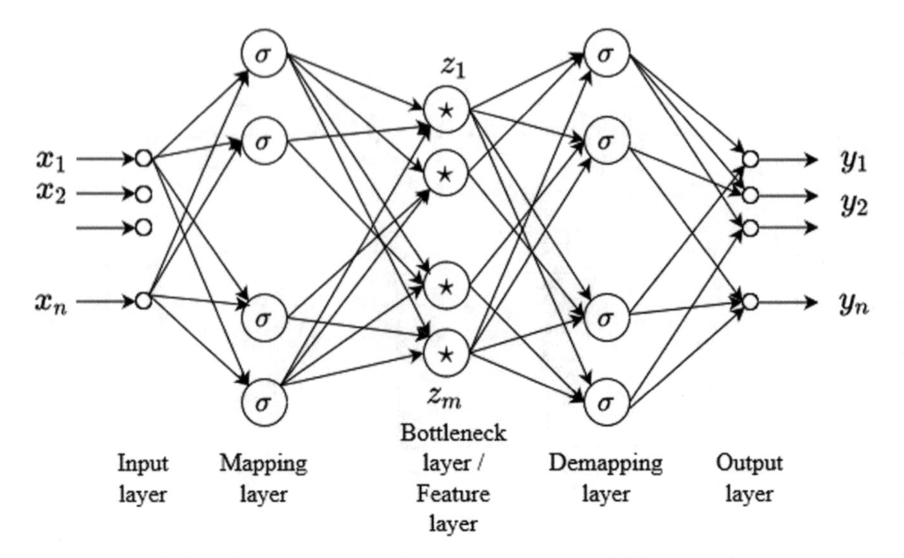

Fig. 4.27 An autoassociative neural network for nonlinear principal component analysis

where d is the number of data pairs used in training the autoassociative neural network, and $\mathbf{x}$, $\mathbf{y}$ are n-dimensional vectors, and

$$z_j = \sum_{k=1}^{T^{(1)}} v_{jk}^{(1)} \sigma \left(\sum_{i=1}^{n} w_{ik}^{(1)} x_i + b_j^{(1)} \right), \ j = 1, 2, \ldots, m, \tag{4.65}$$

$$y_l = \sum_{p=1}^{T^{(2)}} v_{lp}^{(2)} \sigma \left(\sum_{p=1}^{m} w_{jp}^{(2)} z_p + b_l^{(2)} \right), \ l = 1, 2, \ldots, n, \tag{4.66}$$

where $T^{(1)}$ and $T^{(2)}$ are the number of hidden nodes of the mapping and demapping layers, respectively, $\mathbf{b}^{(1)}$ and $\mathbf{b}^{(2)}$ are the threshold vector of the layers, and $\sigma(\cdot)$ is the sigmoid function.

Note that the autoassociative neural networks can be equivalent to PCA if the sigmoid function is changed to be a linear function. Recently, autoassociative neural networks have more widely been known as autoencoders (Baldi, 2012; Bank et al., 2020), in particular deep autoencoders that contains a large number of encoding (mapping) and decoding (demapping) layers. In addition, several variants of autocoders have been proposed, including denoising autoencoders, sparse autoencoders and contractive autoencoders (Bank et al., 2020).

4.5.4 Generative Adversarial Networks

Machine learning models can be generally divided into two categories, one called discriminative models, such as regression models, feedforward neural networks, support vector machines and decision tree, and the other called generative models, such as mixture of Gaussians, Bayesian neural networks, and Boltzmann machines. Statistically, given a set of data instances $\mathbf{X}$ and a set of labels $\mathbf{Y}$, discriminative models capture the conditional probability $p(Y|X)$, while generative models capture the joint probability $p(X, Y)$ in supervised learning, or $p(X)$ only in unsupervised learning.

The generative adversarial network (GAN) is one of the most influential deep learning models and the most successful deep generative model. It consists of a generator and a discriminator (Goodfellow et al., 2014), where the generator continuously generates fake samples trying to fool the discriminator, whereas the discriminator endeavors to distinguish between real and fake samples. The generator and discriminator are trained against each other until both are improved, i.e., the samples generated by the generator can no longer be distinguished by the discriminator. Ideally, the generator learns the probability distribution of the real data and consequently, data generated by the generator can be seen as the real data. A diagram of the GAN is given in Fig. 4.28.

Given a set of real data (training data), $\{\mathbf{x}^{(1)}, \mathbf{x}^{(2)}, \ldots, \mathbf{x}^{(d)}\}$, a random noise $\mathbf{z}$, $\{\mathbf{z}^{(1)}, \mathbf{z}^{(2)}, \ldots, \mathbf{z}^{(d)}\}$ in a latent space of the original data, the loss function for training the GAN is defined as follows (Goodfellow et al., 2014)

$$\min_G \max_D V(D, G), \tag{4.67}$$

where the discriminator is trained to maximize $V(G, D)$, while the generator learns to minimize the discriminator's reward, or to maximize the discriminator's loss. Specifically, the loss function:

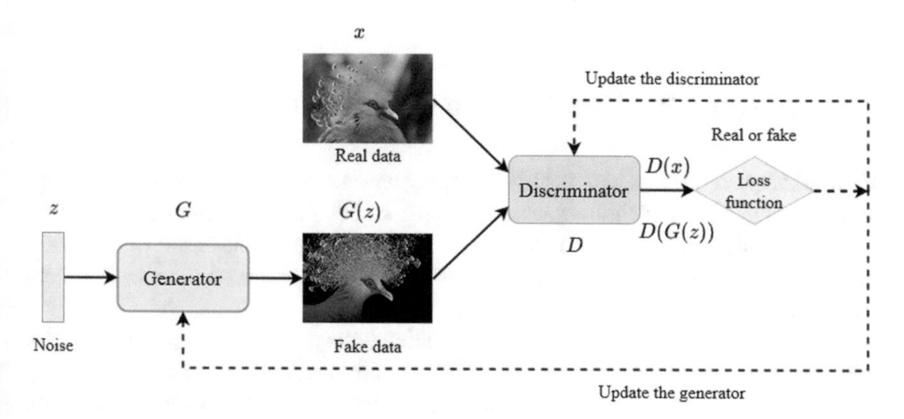

Fig. 4.28 A diagram of a generative adversarial network

$$V(G, D) = E_{\mathbf{x}\sim p(\mathbf{x})}[\log D(\mathbf{x})] + E_{\mathbf{z}\sim q(\mathbf{z})}[\log(1 - D(G(\mathbf{z})))]. \quad (4.68)$$

Thus, the parameters in the discriminator θ_D can be updated by ascending the gradient:

$$\nabla_{\theta_D} \frac{1}{d} \sum_{i=1}^{d} \left[\log D\left(\mathbf{x}^{(i)}\right) + E_{\mathbf{z}\sim q(\mathbf{z})} \log\left(1 - D\left(G\left(\mathbf{z}^{(d)}\right)\right)\right)\right]. \quad (4.69)$$

Similarly, the parameters of the generator θ_G are updated by descending the gradient:

$$\nabla_{\theta_G} \frac{1}{d} \sum_{i=1}^{d} \log\left[\left(1 - D\left(G\left(\mathbf{z}^{(d)}\right)\right)\right)\right]. \quad (4.70)$$

Successfully training a GAN is non-trivial. The gradient vanishing issue may occur if the discriminator is too good, making it fail to provide sufficient information for the generator to make progress. It has been shown that the loss function based on the Wasserstein distance (Wasserstein GAN) may be able to prevent vanishing gradients even the discriminator is trained to be optimal. Alternatively, one can use the modified minimax loss function as suggested in (Goodfellow et al., 2014). Sometimes, the generator may constantly generate similar data samples that over-optimize for a particular discriminator, whilst the discriminator gets stuck in a local optimum, which is known as mode collapse in GAN. To alleviate this problem, a loss function that takes into account not only the current discriminator's classification errors, but also the outputs of future discriminator may be helpful.

Many GAN variants, such as deep convolutional GAN (DCGAN), conditional GAN, cycle-consistent adversarial networks (CycleGANs), InfoGAN, and styleGAN (Jabbar et al., 2020), among many others, have been proposed either to address the weaknesses of the original GAN, or for some particular applications, such as data augmentation for evolutionary optimization (He et al., 2020a), image blending and image in-painting, face aging, and text-to-image synthesis.

4.6 Synergies Between Evolution and Learning

Evolutionary algorithms and machine learning have both common and distinct properties. On the one hand, both machine learning and evolutionary optimization solve an optimization problem, the former with usually the gradient-based or statistical methods, while the latter uses nature inspired population-based stochastic search methods.

Since optimization and learning are complementary, many interesting synergies will be generated by integrating evolutionary optimization and machine learning. On the one hand, machine learning plays an important role in assisting optimization by means of knowledge acquisition and reuse, approximation of objective functions or constraint functions, and selection the most important decision variables, just

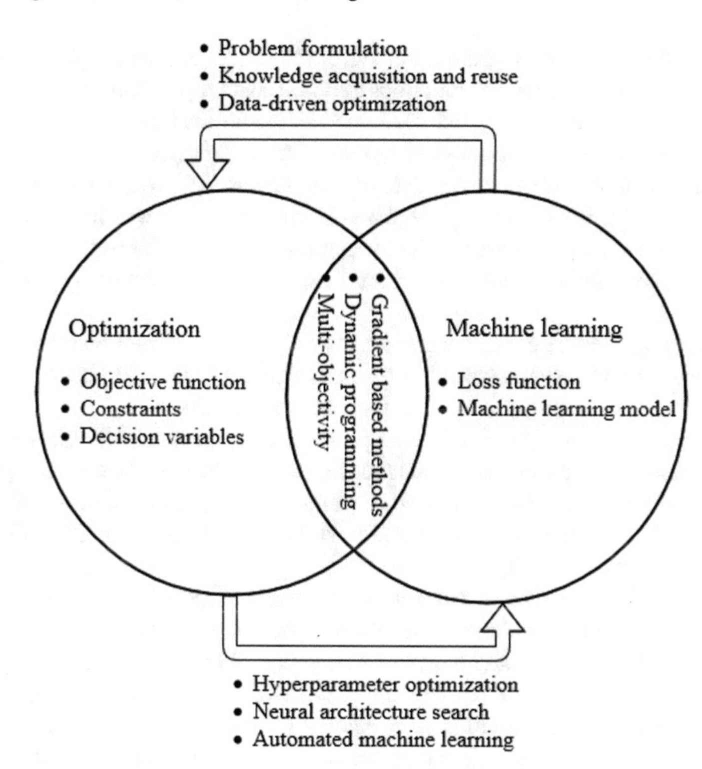

Fig. 4.29 Relationships and synergies between machine learning and optimization

to name a few. On the other hand, optimization has also attracted increasing attention in machine learning, including hyperparameter optimization, neural architecture search, and automated machine learning. Fig. 4.29 summarizes the relationship and interactions between optimization and learning.

In the following, we elaborate a few synergies between evolutionary optimization and machine learning, which are categorized into evolutionary learning and learning for evolutionary optimization. Indeed, data-driven surrogate-assisted evolutionary optimization is a perfect example of synergies between optimization and learning.

4.6.1 Evolutionary Learning

Evolutionary algorithms can play a significant role in learning or enhancing the performance of learning systems. By evolutionary learning, we may identify at least the following three scenarios:

- *Symbolic regression*. Symbolic regression can be seen as a direct learning system in which both the structure and parameters of the machine learning models are

optimized by genetic programming. Here, the model is a tree representation in terms of basic mathematical functions and arithmetic operations, as discussed in Sect. 3.5, while the loss function is defined as the fitness function.
- *Learning Classifier Systems*. Learning classifier systems (LCSs) can be seen as an evolving symbolic rule system that learns on the basis of evolutionary algorithms (Bull, 2015). LCSs can be categorized into Michigan style, in which each individual represents one rule, and the whole population together is the rule based system; and Pittsburgh style, where each rule represents a rule system consisting of a set of rule.
- *Evolutionary Feature Selection and Feature Extraction*. Evolutionary algorithms have been shown to be powerful in feature selection (Xue et al., 2016; Gu et al., 2018), and feature extraction (Albukhanajer et al., 2015).
- *Evolutionary multi-objective learning*. As already discussed in Sect. 4.4, all machine learning problems are inherently multi-objective. Due to the particular strength of EAs, multi-objective machine learning has been most effectively solved with evolutionary algorithms (Jin, 2006; Jin and Sendhoff, 2008; Gu et al., 2015b; Zhu and Jin, 2020).
- *Neural architecture search and automated machine learning*. In deep learning, evolutionary neural architecture search has been shown to be effective in automatically determine the architecture of deep neural networks (Liu et al., 2020b) and federated architecture search (Zhu et al., 2021). A step further is to use EAs to automatically select training data, select or extract features, and optimize the machine learning model, automating the overall machine learning process.

4.6.2 Learning for Evolutionary Optimization

Learning helps improve the performance of evolutionary optimization. Indeed, data-driven evolutionary optimization, the main topic of the book, is all about using machine learning to assist evolutionary optimization. Below we very briefly summarize the main applications of machine learning to evolutionary optimization.

- *Problem formulation*. Machine learning can help identify the most important decision variables and objective functions using techniques such as correlation analysis and dimension reduction.
- *Diversity enhancement*. Clustering techniques can often been used to group the population of an evolutionary algorithm into subpopulations for promote diversity.
- *Problem decomposition*. Large-scale optimization problems can be decomposed into smaller ones by detecting the correlations between the decision variables.
- *Solving dynamic optimization problems*. Prediction techniques have shown to be powerful in predicting the location of the new optimum based on history data to enhance the search efficiency.

- *Modeling objective or constraint functions*. Using machine learning models, also called surrogates to approximate objective and/or constraint functions is the main mechanism behind data-driven evolutionary optimization.
- *Knowledge acquisition and knowledge transfer*. Many evolutionary algorithms cannot explicitly acquire knowledge about the problem. By contrast, machine learning techniques are very useful to detect the problem structure and acquire useful domain knowledge during the search process, and reuse it. In addition, machine learning techniques are effective in transferring knowledge from one objective to another, and from previous optimization tasks to the current one.

4.7 Summary

Data-driven optimization relies heavily on a wide range of basic and advanced machine learning techniques to efficiently formulate the optimization problem, build models for the objective functions and constraints, address the issue of data paucity and to acquire and reuse knowledge for enhancement of the optimization efficiency. In this chapter, we provide a brief introduction to the fundamental machine learning models and algorithms, many of which will be used in addressing various challenges in data-driven evolutionary optimization. In addition, we indicate that integrating machine learning and evolutionary computation can reveal new insight and produce many interesting synergies.

References

Albukhanajer, W. A., Briffa, J. A., & Jin, Y. (2015). Evolutionary multi-objective image feature extraction in the presence of noise. *IEEE Transactions on Cybernetics, 45*(9), 1757–1768.

Albukhanajer, W. A., Briffa, J. A., & Jin, Y. (2017). Classifier ensembles for image identification using multi-objective pareto features. *Neurocomputing, 238,* 316–327.

Baldi, P. (2012). Autoencoders, unsupervised learning, and deep architectures. In Guyon, I., Dror, G., Lemaire, V., Taylor, G., & Silver, D., (Eds.), *Proceedings of ICML Workshop on Unsupervised and Transfer Learning, Proceedings of Machine Learning Research, 27,* 37–49.

Baldi, P., & Hornik, K. (1989). Neural networks and principal component analysis: Learning from examples without local minima. *Neural Networks, 2,* 53–58.

Bank, D., Koenigstein, N., & Giryes, R. (2020). Autoencoders. arxiv.org/abs/2003.05991.

Blum, A., & Mitchell, T. (1998). Combining labeled and unlabeled data with co-training. In *Proceedings of the 11th Annual Conference on Computational Learning Theory, ACM (1998)*, pages 92–100.

Box, G. E. (1979). All models are wrong, but some are useful. *Robustness in Statistics, 202*(1979), 549.

Breiman, L., Friedman, J., Stone, C. J., & Olshen, R. A. (1984). *Classification and regression trees.* CRC Press.

Bull, L. (2015). A brief history of learning classifier systems: from cs-1 to xcs and its variants. *Evolutionary Intelligence, 8,* 55–70.

Culotta, A. & McCallum, A. (2005). Reducing labeling effort for structured prediction tasks. In *Proceedings of the National Conference on Artificial Intelligence(AAAI)*, (pp. 746–751). AAAI Press.

Dagan, I., & Engelson, S. P. (1995). Committee-based sampling for training probabilistic classifiers. In *In Proceedings of the Twelfth International Conference on Machine Learning*, (pp. 150–157). Morgan Kaufmann.

Draper, N. R., & Smith, H. (1998). *Applied regression analysis*, (Vol. 326). Wiley.

Emmerich, M. (2005). *Single-and multi-objective evolutionary design optimization assisted by Gaussian random field metamodels*. Dortmund: University of Dortmund.

Gardner, M. W., & Dorling, S. (1998). Artificial neural networks (the multilayer perceptron)—A review of applications in the atmospheric sciences. *Atmospheric environment, 32*(14–15), 2627–2636.

Goodfellow, I. J., J. Pouget-Abadie, M. M., Xu, B., Warde-Farley, D., Ozairy, S., Courville, A., & Bengio, Y. (2014). Generative adversarial nets. In *Proceedings of the International Conference on Neural Information Processing Systems (NIPS 2014)*, (pp. 2672–2680).

Gu, S., Cheng, R., & Jin, Y. (2015a). Multi-objective ensemble generation. *WIREs Data Mining and Knowledge Discovery, 5*(5), 234–245.

Gu, S., Cheng, R., & Jin, Y. (2015b). Multi-objective ensemble generation. *WIREs Data Mining and Knowledge Discovery, 5*(5), 234–245.

Gu, S., Cheng, R., & Jin, Y. (2018). Feature selection for high dimensional classification using a competitive swarm optimizer. *Soft Computing, 22*(3), 811–822.

Gu, S., & Jin, Y. (2017). Multi-train: A semi-supervised heterogeneous ensemble classifiers. *Neurocomputing, 249,* 202–211.

Handl, J., & Knowles, J. (2007). An evolutionary approach to multiobjective clustering. *IEEE Transactions on Evolutionary Computation, 11*(1), 56–76.

He, et al. (2016). "Deep residual learning for image recognition." *Proceedings of the IEEE conference on computer vision and pattern recognition.*

He, C., Huang, S., Cheng, R., Tan, K. C., & Jin, Y. (2020a). Evolutionary multi-objective optimization driven by generative adversarial networks. *IEEE Transactions on Cybernetics*.

Jabbar, A., Li, X., & Omar, B. (2020). A survey on generative adversarial networks: Variants, applications, and training. arXiv:2006.05132.

Jang, J.-S., & Sun, C.-T. (1993). Functional equivalence between radial basis functions and fuzzy inference systems. *IEEE Transactions on Neural Networks, 4*(1), 156–159.

Jin, Y. (2003). *Advanced Fuzzy Systems Design and Applications*. Physica. Springer.

Jin, Y. & Sendhoff, B. (2008). Pareto-based multiobjective machine learning: An overview and case studies. *IEEE Transactions on Systems, Man, and Cybernetics, Part C (Applications and Reviews)*, 38(3):397–415.

Jin, C. & Wang, L. (2012). Dimensionality dependent pac-bayes margin bound. In *NIPS*, volume 12, pages 1034–1042. Citeseer.

Jin, Y. (2000). Fuzzy modeling of high-dimensional systems: Complexity reduction and interpretability improvement. *IEEE Transactions on Fuzzy Systems, 8*(2), 212–221.

Jin, Y. (Ed.). (2006). *Multi-Objective Machine Learning*. Heidelberg: Springer.

Jin, Y., & Sendhoff, B. (1999). Knowledge incorporation into neural networks from fuzzy rules. *Neural Processing Letters, 10*(3), 231–242.

Kadyrov, A., & Petrou, M. (2001). The trace transform and its applications. *IEEE Transactions on Pattern Analysis and Machine Intelligence, 23*(8), 811–828.

Kramer, M. (1991). Nonlinear principal component analysis using autoassociative neural networks. *AIChE Journal, 37*(2), 233–243.

Kriegel, H.-P., Kröger, P., & Zimek, A. (1999). Clustering high-dimensional data: A survey on subspace clustering, pattern-based clustering, and correlation clustering. *ACM Transactions on Knowledge Discovery from Data, 3*(1), Article No. 1.

Liaw, A., & Wiener, M. (2002). Classification and regression by random forest. *R News, 2*(3), 18–22.

Liu, Y., Sun, Y., Xue, B., Zhang, M., Yen, G. G., & Tan, K. C. (2020b). A survey on evolutionary neural architecture search. arXiv:2008.10937.

Matheron, G. (1963). Principles of geostatistics. *Economic Geology, 58*(8), 1246–1266.

Michels, K., Klawonn, F., Nürnberger, A., & Kruse, R. (2006). *Fuzzy Control: Fundamentals, Stability and Design of Fuzzy Controllers*. Berlin/Heidelberg: Springer.

Moody, J., & Darken, C. J. (1989). Fast learning in networks of locally-tuned processing units. *Neural Computation, 1*(2), 281–294.

Rokach, L., & Maimon, O. Z. (2008). *Data mining with decision trees: theory and applications*, (vol. 69). World Scientific.

Ruder, S. (2016). An overview of gradient descent optimization algorithms. page arXiv:1609.04747.

Rumelhart, D. E., Hinton, G. E., & Williams, R. J. (1985). Learning internal representations by error propagation. Technical report, California Univ San Diego La Jolla Inst for Cognitive Science.

Scheffer, T., Decomain, C., & Wrobe, S. (2001). Active hidden markov models for information extraction. In *Proceedings of the International Conference on Advances in Intelligent Data Analysis*, (pp. 309–318). Springer-Verlag.

Smith, C., & Jin, Y. (2014). Evolutionary multi-objective generation of recurrent neural network ensembles for time series prediction. *Neurocomputing, 143*, 302–311.

Simonyan, K., & Zisserman, A. (2014). "Very deep convolutional networks for large-scale image recognition." arXiv preprint arXiv:1409.1556.

Steinberg, D., & Colla, P. (2009). CART: classification and regression trees. *The Top Ten Algorithms in Data Mining, 9*, 179.

Steinwart, I., & Christmann, A. (2008). *Support vector machines*. Springer Science & Business Media.

Sutton, R. S. & Barto, A. G. (2016). *Reinforcement Learning: An Introduction*. MIT Press.

Wang, S., & Yao, X. (2012). Multiclass imbalance problems: Analysis and potential solutions. *IEEE Transactions on Systems, Man, and Cybernetics, Part B: Cybernetics, 42*(4), 1119–1130.

Xue, B., Zhang, M., & Browne, W. (2012). Particle swarm optimization for feature selection in classification: A multi-objective approach. *IEEE Transactions on Cybernetics, 43*(6), 1656–1671.

Xue, B., Zhang, M., Browne, W. N., & Yao, X. (2016). A survey on evolutionary computation approaches to feature selection. *IEEE Transactions on Evolutionary Computation, 20*(4), 606–626.

Zadeh, L. A. (1965). Fuzzy sets. *Information and Control, 8*(3), 338–353.

Zadeh, L. A. (1975). The concept of a linguistic variable and its application to approximate reasoning-I. *Information Sciences, 8*(3), 199–249.

Zhou, Z.-H. (2009). Ensemble learning. *Encyclopedia of Biometrics, 1*, 270–273.

Zhou, Z.-H., & Li, M. (2005). Tri-training: Exploiting unlabeled data using three classifiers. *IEEE Transactions on knowledge and Data Engineering, 17*(11), 1529–1541.

Zhu, H., Zhang, H., & Jin, Y. (2021). From federated learning to federated neural architecture search: A survey. *Complex & Intelligent Systems*.

Zhuang, F., Qi, Z., Duan, K., Xi, D., Zhu, Y., Zhu, H., et al. (2020). A comprehensive survey on transfer learning. *Proceedings of IEEE, 109*(1), 43–76.

Zhu, H., & Jin, Y. (2020). Multi-objective evolutionary federated learning. *IEEE Transactions on Neural Networks and Learning Systems, 31*(4), 1310–1322.

Chapter 5
Data-Driven Surrogate-Assisted Evolutionary Optimization

5.1 Introduction

Most evolutionary algorithms (EAs) assume that analytical objective functions are available for quality evaluations, and the quality evaluations are computationally cheap so that tens of thousands of evaluations can be done to find optimums of the problem. These assumptions may not hold for many real-world optimization problems. The first category of such problems belongs to a large class of engineering optimization problems where quality evaluations rely on time-consuming numerical analysis such as computational fluid dynamics simulations and finite element analysis. These include aerodynamic design optimization problems such as design of turbine engines, high-lift wings, fuselage of aircraft, and race cars. Other examples include optimization of complex industrial processes, drug design, and material design. Most of these numerical simulations and numerical analysis are computationally very intensive, with each evaluation taking from minutes to hours, and even days. A second category of problems involves human judgement and assessment that are hard to be modeled mathematically, such as clothing design, art and music design, and other aesthetic design of tools or products. Finally, there are real-world optimization problems which can be optimized based on data only, which are collected from daily life or production process.

Accordingly, problems that are optimized based on data collected from historical records, numerical simulations, or physical experiments are called data-driven optimization problems (Jin et al. 2018; Wang et al. 2016). The interdisciplinary research area of data-driven evolutionary optimization involves techniques in data science, machine learning, and evolutionary algorithms, as shown in Fig. 5.1. In an evolutionary data-driven optimization framework, data will be collected at first, which will be pre-processed to enhance the data quality, since the data may be subject to noise or errors. Then, surrogate models, which are machine learning models, are built from the data to approximate the real objective functions and / or constraint functions.

Y. Jin et al., *Data-Driven Evolutionary Optimization*,
Studies in Computational Intelligence 975,
https://doi.org/10.1007/978-3-030-74640-7_5

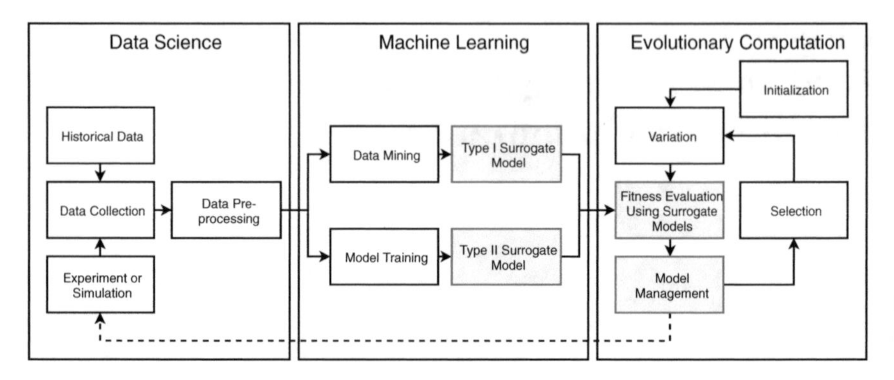

Fig. 5.1 A diagram of data-driven evolutionary optimization

Given the approximated objective or constraint functions, evolutionary algorithms can then be applied to perform optimization.

According to the types of surrogate models Jin et al. (2018), data can be divided into direct and indirect data.

- **Direct data** are samples with decision variables, corresponding objective and / or constraint values, which are from simulation or experiments. Surrogate models approximating the objective or constraint functions (termed type I surrogate models) can be built directly from such types of data.
- **Indirect data** are not samples with decision variables, corresponding objective and / or constraint values, but the objective and constraint calculations require such data. Indirect data can be employed to further train surrogate models (termed type II surrogate models).

Both type I and II surrogate models can be generic or specific models. Existing machine learning models can be employed as generic models, which do not require much pre-knowledge of the problem. If there are useful information available, specific models in the problem domain, such as simulations with the various fidelity control Wang et al. (2016), may be more accurate than generic models.

Different from conventional EAs, the data-driven evolutionary optimization process employs surrogate models for function evaluations and model management strategies to efficiently control the models and data Jin (2005). To control the models, the model type, hyper-parameters, or ensemble structure of surrogate models can be changed over generations. To control the data, new data can be actively sampled by the EA. However, in some cases, the new data cannot be controlled by the EA, or there is no new data at all. Therefore, data-driven EAs can be divided into offline and online algorithms, according to whether new data can be be collected during the optimization process Wang et al. (2016).

It should be noted that data-driven evolutionary optimization, in particular online data-driven evolutionary optimization, is also known as surrogate-assisted evolutionary optimization (Jin 2005, 2011), or evolutionary optimization with approximate fitness functions Jin (2005). Some offline data-driven optimization, e.g., in Wang et al. (2016), no explicit surrogates are involved.

5.2 Offline and Online Data-Driven Optimization

5.2.1 Offline Data-Driven Optimization

In the offline data-driven evolutionary optimization process, no new data can be actively sampled or continuously generated, which makes the optimization highly related to the surrogate models solely based on the offline data. In other words, both the quality and quantity greatly affect the surrogate models and optimization performance of offline data-driven EAs, because the accuracy of surrogate models cannot be validated Wang et al. (2018).

- **Non-ideal quality**: Generally, the data in the real world can have non-ideal quality, such as noisy, imbalanced, even incomplete data. It is very hard to build surrogate models with high accuracy from those non-ideal data. Thus, the optimization direction is easy to be misled using such poor surrogate models.
- **Extreme quantity**: Either big or small data poses challenges to offline data-driven EAs. In the case of big data, the computational cost of surrogate construction is high. In the case of small data, the accuracy of surrogate model cannot be guaranteed due to data paucity.

To deal with those challenges, how to efficiently use the limited data to build robust and accurate surrogate models is the key issue of offline data-driven EAs. The following techniques can be helpful, some of them have been employed in offline data-driven evolutionary optimization.

1. **Data pre-processing**: When the offline data is non-ideal, data pre-processing is a very straightforward way to enhance the quality of data and surrogate models. The data in the blast furnace problem in Chugh et al. (2017) is very noisy. A local regression smoothing method Cleveland and Loader (1996) is used to denoise before building then Kriging models.
2. **Data mining**: To deal with the high computational cost in big data-driven EAs, data redundancy Wu et al. (2017) should be reduced by using existing data mining techniques for extracting patterns. A clustering algorithm has been applied to generate multi-fidelity surrogate models in the trauma system design problem Wang et al. (2016).
3. **Synthetic data generation**: In some extreme cases, especially for high-dimensional problems, the data amount is too small to build a surrogate model. Synthetic data can be generated to enrich the original data, which is the situation of the fused magnesium furnace optimization problem Guo et al. (2016). In the algorithm Guo et al. (2016), a low-order polynomial model is employed to generate synthetic data.
4. **Semi-supervised learning**: An alternative way of handling insufficient data is semi-supervised learning Zhu (2005), where the implicit information in the unlabeled data is used in the model training process. Thus, the required size of offline data can be reduced (Sun et al. 2018; Huang et al. 2021).

5. **Model selection**: Although there are various types of surrogate models, their advantages are different. Since offline data-driven EAs have no chance to validate or update their surrogate models, the choice of surrogate models (i.e. model selection) is important to the algorithm. In Habib et al. (2019), validation error is adopted as a criterion to select one type of surrogate models.
6. **Ensemble learning**: To improve the robustness of surrogate models, according to the ensemble learning theory Zhou (2009), a large number of homogeneous/heterogeneous models can be combined as an ensemble. A selective ensemble learning algorithm Wang et al. (2018) has been used in offline data-driven single-objective optimization.
7. **Multi-form optimization**: In addition to data and modelling techniques, multiple problem formulations, such as multi-fidelity fitness evaluations Wang et al. (2018d), can be adopted to further assist the search Gupta et al. (2018).
8. **Transfer learning**: Further, the processes of solving similar problems share similar searching path, where common knowledge can be reused to save computational costs Gupta et al. (2018). Transfer optimization and multi-task optimization (Gupta et al. 2016; Ding et al. 2017; Min et al. 2017) might be helpful to offline data-driven EAs.

5.2.2 *Online Data-Driven Optimization*

Different from offline data-driven optimization, online data-driven optimization has new data during the evolutionary search, which can make use of the online data to enhance the quality of surrogate models. Thus, how to utilize the online data is key to the performance of online algorithms. Note that, the offline data affects the performance, the technique to deal with the offline data in the previous sub-section can be applied to online algorithms as well.

In fact, there are two scenarios for online data-driven optimization: whether the online data can be controlled by the algorithm or not Wang et al. (2016). In the cases with streaming data, online data-driven EAs cannot control the data generation, so the main hardness is how to capture the information from the dynamic data.

When online data-driven EAs can actively sample the online data, the performance can be greatly improved if proper model management strategies are applied. In this case, both the frequency and the choice of online data sampling need to be adjusted during the optimization process. The aim of online data sampling is to economically improve global and local accuracy of surrogate models, thus the optimization performance can be enhanced (Jin et al. 2000, 2002; Branke and Schmidt 2005). The further details of existing online surrogate management methods are shown in Sect. 5.3.

5.3 Online Surrogate Management Methods

In online data-driven surrogate-assisted evolutionary optimization, we assume that some additional data can be actively collected during the optimization to update the surrogate and guide the search. The first question is of course why it is necessary to collect additional data during the optimization. This is mainly because the surrogate that is trained before optimization may introduce optimums that are not the real optimums of the original objective function, termed *false optimums* Jin et al. (2002). Fig. 5.2 illustrates as example, where the surrogate $\hat{f}(x)$ introduces a false minimum (x^{Δ}). Thus, one will end up with a solution (x^{Δ}) that has very poor quality and is very far from the true minimum (x^*), if an evolutionary algorithm is run on the surrogate $\hat{f}(x)$ instead of the true objective function $f(x)$.

Thus, surrogate model management, a process of actively querying new data points and updating the surrogate, is essential for surrogate-assisted evolutionary optimization. In surrogate management, recursively identifying a solution, or solutions to be queried, either by performing numerical simulations or physical experiments is of paramount importance. Different ideas for selecting solutions to be samples have been proposed, typically by selecting the promising solutions as predicted by the surrogate, or those whose predicted fitness is most uncertain, or representative solutions of a population. Methods for surrogate management can be categorized into population based, generation based, and individual based.

5.3.1 Population-Based Model Management

One early idea for model management is based on populations, where multiple populations co-evolve at the same time, each using a model of various fidelity. In the simplest case, two populations are involved, one using a surrogate model as the fitness function, and the other using the true fitness function, i.e., time-consuming

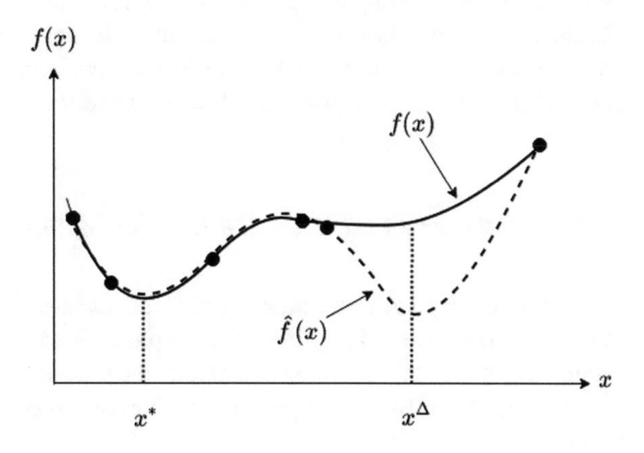

Fig. 5.2 Illustration of a false minimum introduced by a surrogate model

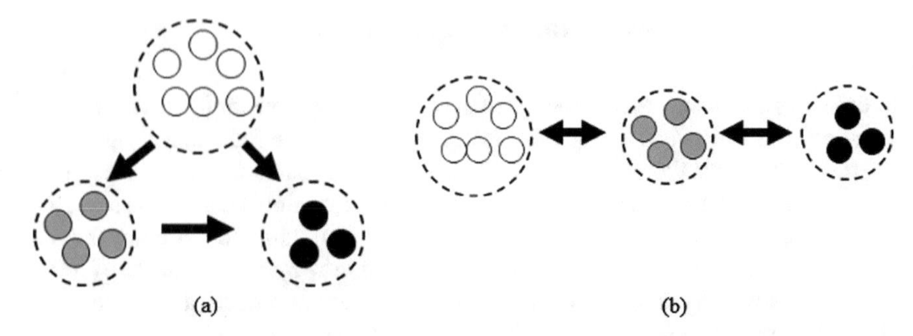

Fig. 5.3 Population based model management, where solid dots mean a high-fidelity evaluation method, e.g., a 3D Navier-Stokes solver, shaded dotes are medium-fidelity evaluation method such as a 2D Navier-Stokes solver, and a circle is a low-fidelity surrogate model. **a** Individuals evaluated using a lower fidelity method are sent to populations using a higher fidelity evaluation method; **b** Populations exchange individuals

numerical simulations or expensive physical experiments. Taking aerodynamic optimization as an example, we my have three methods for evaluating the pressure loss of a designed turbine blade, a 3D Navier-Stokes solver , a 2D Navier-Stokes solver, or a surrogate model, e.g., a neural network. Of these three fitness evaluation methods, there is a trade-off between accuracy and computational complexity, i.e., the more accurate the evaluation method is, the computationally more intensive it is to make the evaluations. In this case, three populations can be employed for evolutionary optimization, each using a different evaluation method, as show in Fig. 5.3. There are two slightly different ways of exchanging individuals between the populations. In the first approach, as illustrated in Fig. 5.3a, individuals of a population using a lower fidelity evaluation method are sent to populations using a higher fidelity method. In the second approach, by contrast, individuals evolved based on a medium fidelity method are exchanged with those using a lower or higher fidelity evaluation method. The hypothesis here is, by co-evolving the three populations using models of various levels of fidelity, the optimization can be accelerated compared to the case where only the highest fidelity evaluation method is adopted.

In the population based methods, the surrogates will be updated once fitness evaluations using the highest fidelity method are performed. Models methods such as co-kriging that is able to use multi-fidelity data for training may also be helpful.

5.3.2 Generation-Based Model Management

As its name suggests, generation based model management strategies aim to manage the surrogates generation-wise. The simplest idea is, a surrogate is trained using some data collected before optimization starts. Then optimization is performed on the surrogate until it converges. Then, the converged solutions will be evaluated

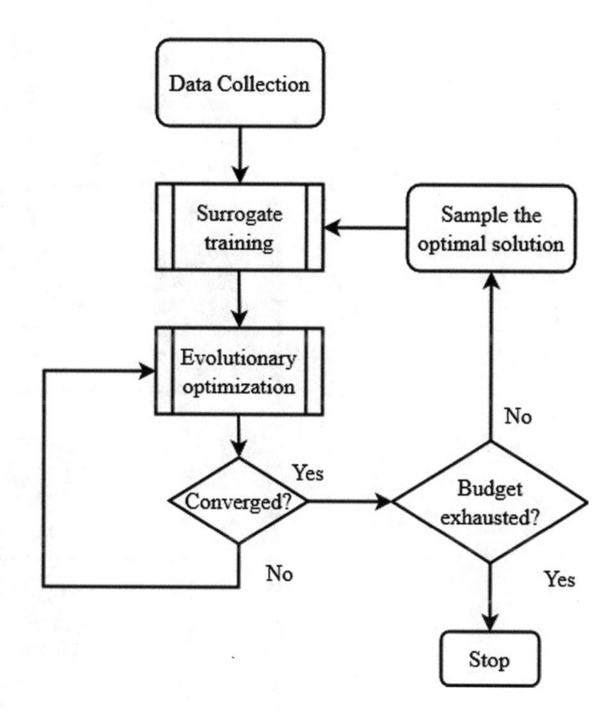

Fig. 5.4 Generation based surrogate model management

using the true fitness evaluation method. Then, the surrogate is updated, followed by another round of search on the surrogate. A flowchart of generation based surrogate management is provided in Fig. 5.4.

It should be noted that there is a risk in the above simplistic generation based model management framework that the model gets stuck in a local solution. Thus, it is recommended some random solutions are also samples in addition to the optimum found by the search on the surrogate.

One straightforward variant of the above approach is to switch the evaluation method in every a given number of generations. For example, optimization is performed using the expensive fitness evaluation method for five generations, and then the surrogate is updated using the data generated in these generations and optimization is run on the surrogate for 20 generations. This process repeats until the computational budget is exhausted. An adaptive generation based surrogate management method has been proposed in Jin et al. (2002), which aims to adjust the frequency of the generations in which the true fitness evaluation is used. Here, a cycle specifying a fixed number of generations is defined. Within a cycle, an initial frequency of using the expensive fitness evaluation, e.g., the computational fluid dynamic simulation (CFD), is predefined. For example, a cycle contains ten generations, of which the CFD is used in five generations and the surrogate is used in the next five generations. At the end of each cycle, the fidelity of the surrogate is estimated based on the average approximation error of the model for these solutions evaluated in the five generations. According to the change of the average error, the frequency of using the expensive

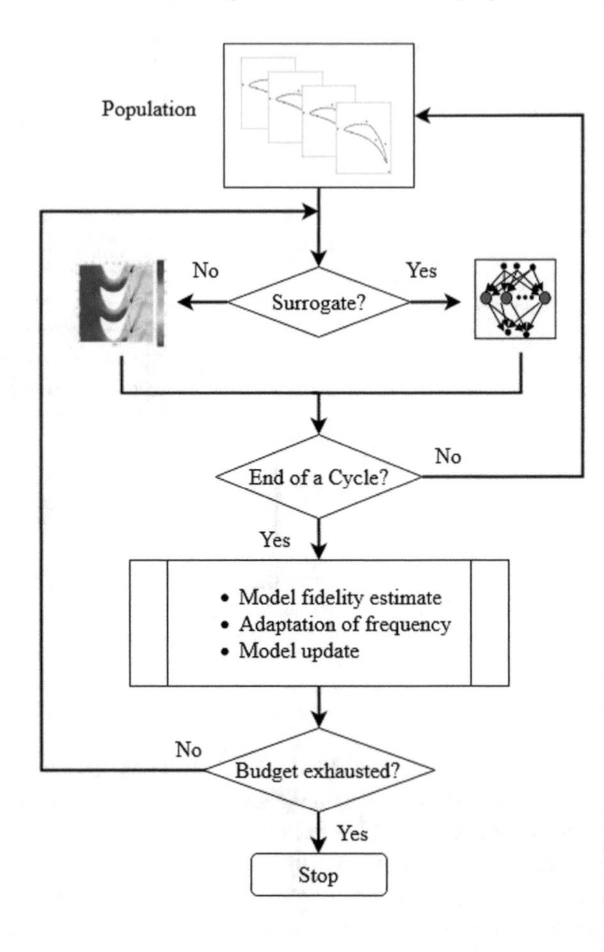

Fig. 5.5 An illustration of adaptive generation based surrogate model management

fitness evaluation is adapted. That is, if the average error decreases, the frequency of using the expensive fitness can be decreased, and accordingly, the frequency of using the expensive fitness is increased if the average error increases. Note that in the adaptation, the minimum frequency is one, i.e., the original fitness function is used in at least one generation of each cycle. An illustration of the adaptive generation based surrogate management is given in Fig. 5.5.

5.3.3 *Individual-Based Model Management*

Individual based model management is probably the most flexible and most widely used surrogate management approach. Different ideas for individual based model management have been proposed proposed. They can be roughly divided into pre-selection method, random strategy, best strategy, clustering based method, and uncertainty based method.

- *Pre-selection*. Pre-selection is a sifting method that uses surrogates to reduce randomness and speed up the evolution. For example, for a genetic algorithm of a population size of N, instead of generating N offspring, one can generate $2N$ offspring. Then, a trained surrogate can be employed to select the better N individuals, which will then be re-evaluated using the expensive fitness evaluation method. Similarly, for a (μ, λ) evolution strategy, instead of generating λ offspring from μ parents, one can generate $\lambda' >$ offspring, and then select the better λ offspring individuals to be reevaluated using the expensive fitness function.
- *Random strategy*. This strategy does not expand the population size as done in pre-selection. Instead, one randomly selects a specified number of individuals from the offspring population to be evaluated using the expensive fitness function. For a population size of N, $N' < N$ individuals can be chosen.
- *Best strategy*. In contrast to the random strategy, one selects the best N' individuals from N offspring to be evaluated using the original fitness function. There is an empirical comparison of the random strategy and best strategy Jin et al. (2002). It has been found that the best strategy is a more effective model management method, since the population will converge to the right optimum when roughly 50% of the better individuals in the population are evaluated using the real fitness function. By contrast, at least 80% of the individuals must be evaluated using the real fitness function for the population to converge to the correct optimum, if the random strategy is used.
- *Clustering method*. The clustering method is meant to reduce the number of individuals to be evaluated using the expensive real fitness function by grouping the population into a number of clusters and then only the individual closest to each cluster center is evaluated using the real fitness function. The rest individuals in each cluster are estimated based on the distance to the cluster center, or using a normal surrogate. Alternatively, the best strategy can be applied to each cluster so that only a subset of each cluster of individuals will be evaluated using the expensive fitness function Gräning et al. (2005).
- *Uncertainty based strategy*. While it makes good sense to select the best individuals predicted by the surrogate to be evaluated using the real fitness function, it is also reasonable from both optimization and model update point of views to select individuals that are far from the existing training samples. For model update, those points that are far from existing may be effective in enhancing the model quality, while for search, sampling solutions in these unexplored areas will promote exploration. One idea to estimate the uncertainty of the predicted fitness of an individual is to calculate the inverted distance between the solution and all training data Branke and Schmidt (2005):

$$\delta_i = \frac{1}{\sum_{j=1}^{m} \frac{1}{\|\mathbf{x}_i - \mathbf{x}^{(j)}\|}}, \tag{5.1}$$

where δ_i is the uncertainty degree of the i-th individual, $\mathbf{x}^{(j)}$ is the j-th training data, m is the total number of samples, and $\|\cdot\|$ is usually the Euclidean distance.

Different to the idea of selecting the most uncertain solution, one can try to maximize the overall uncertainty decrease of the population in selecting individuals for evaluation. It is also possible to combine the quality of individual to its uncertainty degree:

$$r_j = \sum_{i \in \mathcal{Q} \setminus \{j\}} p_i \delta_i, \tag{5.2}$$

where p_i is the probability of individual i is selected, typically proportional to its predicted fitness value.

5.3.4 Trust Region Method for Memetic Algorithms

Memetic algorithms that combine evolutionary search with local search can also be accelerated by surrogates. In surrogate-assisted memetic algorithms, typically the local search is assisted by a surrogate. For this reason, the trust region based method, which is a black-box model based search method developed in traditional optimization, is often adopted in surrogate-assisted memetic algorithms. The trust region method, proposed by Powell in Powell (1970), is one of iterative optimization methods and plays a significant role in the area of nonlinear optimization. Assume that $\mathbf{x}$ is a solution of the optimization problem, in the trust region method, a model will be constructed near this solution and an optimal step $\mathbf{s}$ will be obtained by solving the following problem:

$$\begin{aligned} &\min q(\mathbf{s}) = f(\mathbf{x}) + g^T \mathbf{s} + \frac{1}{2}\mathbf{s}^T \mathbf{H}\mathbf{s} \\ &\text{s.t. } \|\mathbf{s}\| \leq \delta, \end{aligned} \tag{5.3}$$

where $g = \nabla f(\mathbf{x})$, $\mathbf{H}$ is an $n \times n$ symmetric approximation matrix of $\nabla^2 f(\mathbf{x})$, and $\delta > 0$ is the trust region radius. Once an optimal step $\mathbf{s}_b$ in this iteration is chosen, the objective value of position $\mathbf{x} + \mathbf{s}_b$ will be evaluated. The ratio ρ between the reduction of the actual function value $f(\mathbf{x}) - f(\mathbf{x} + \mathbf{s})$ and that of the predicted function value $q(\mathbf{0}) - q(\mathbf{s})$, i.e, $\rho = \frac{f(\mathbf{x}) - f(\mathbf{x}+\mathbf{s})}{q(\mathbf{0}) - q(\mathbf{s})}$, plays an important role on deciding whether the step $\mathbf{s}$ should be acceptable or not and how the trust region radius would be adjusted for the next iteration. If ρ is close to 1, it shows that the model is quite good and the region can be enlarged in size. Conversely, if ρ is too small, the region needs to be decreased in size. Specifically, given a threshold η,

1. If $\rho < \eta$, then $\mathbf{s}$ will be rejected, $\mathbf{x}(t+1) = \mathbf{x}(t)$ and $\delta(t+1) = \gamma\delta(t)$
2. If $\rho \geq \eta$, then $\mathbf{s}$ will be accepted, $\mathbf{x}(t+1) = \mathbf{x} + \mathbf{s}$, and $\delta(t+1) \geq \delta(t)$

where γ is a pre-defined parameter.

The pseudocode of the trust region method is given in Algorithm 1. In Algorithm 1, η, γ_i and γ_d are parameters pre-defined. $0 < \eta, \gamma_d < 1$ and $\gamma_i \geq 1$.

Algorithm 1 The pseudocode of the trust region method

Input: $\mathbf{x}$: a solution of the optimization problem;
Output: $\mathbf{x}$
1: $k = 0$;
2: **while** the stopping criterion is not met **do**
3: Construct a model $q(\mathbf{s})$ using Eq. (5.3);
4: Find the optimal step $\mathbf{s}_b$;
5: **if** $\rho \geq \eta$ **then**
6: $\mathbf{x} = \mathbf{x} + \mathbf{s}_b$ and $\delta = \gamma_i \delta$;
7: **else**
8: $\delta = \gamma_d \delta$
9: **end if**
10: $k = k + 1$;
11: **end while**

In Algorithm 1, a model will be firstly constructed in a given trust region δ. Then the model will be solved, which is a seriously significant step in the trust region method. A number of optimization methods have been proposed for efficiently solving this problem. However, when the dimension of the optimization problem gets larger and larger, some methods turn out to be less efficient.

5.4 Bayesian Model Management

Bayesian optimization (BO) is a class of machine-learning-based optimization method for solving the black-box or computationally expensive problems. Generally, Bayesian optimization builds a probability representation of the objective function and tries to find a valuable position for exact objective evaluation so that the optimal solution can be found in a very limited computational budget. The probability representation of BO is called surrogate model which is used to approximate the objective function. The acquisition function utilizes the posterior to direct sampling for expensive objective evaluation. Algorithm 2 gives the pseudocode of the general Bayesian optimization. A number of positions will be sampled at first and used to train a surrogate model after they are evaluated. Then the activated function is optimized to search for a valuable solution, which will be evaluated using the exact objective evaluation and further be used to update the surrogate model.

A popular surrogate model for Bayesian optimization is Gaussian process model, which has been described in Chap. 4 and thus will not be repeated here. The acquisition function, which considers the trade-off between exploration and exploitation, is used to determine which data point in the decision space is to be evaluated using the exact expensive objective. Formally, the position $\mathbf{x}_t = \arg\max_{\mathbf{x}} \mathcal{A}(\mathbf{x}|\mathcal{D}_{1:t-1})$ will be sampled, where $\mathcal{A}$ is the acquisition function and $\mathcal{D}_{1:t-1} = \{(\mathbf{x}_1, y_1), (\mathbf{x}_2, y_2), \ldots, (\mathbf{x}_{t-1}, y_{t-1})\}$ is the data set storing all data sampled and evaluated so far. Popular acquisition functions, also known as infill criteria, include the probability of improvement (PI), expected improvement (EI), and upper confidence bound (UCB).

Algorithm 2 The pseudocode of BO

1: Sampling N_I data in the decision space and evaluating them using the black-box or expensive problem;
2: Keep the best position of N_I data, denoted as $\mathbf{x}_b$;
3: Train a surrogate model M;
4: **while** $N_I \leq N_{FE}$ **do**
5: Optimize the activate function $\mathcal{A}$ and get its optimal position;
6: Evaluate this position using the black-box or expensive problem;
7: Update the best position $\mathbf{x}_b$ found so far;
8: Update the surrogate model;
9: **end while**
10: Output the optimal solution $\mathbf{x}_b$;

5.4.1 Acquisition Functions

5.4.1.1 Probability of Improvement

The probability of improvement is defined by

$$PI(\mathbf{x}) = \psi\left(\frac{f(\mathbf{x}_b) - \mu(\mathbf{x}) - \xi}{\sigma(\mathbf{x})}\right) \tag{5.4}$$

where $\mu(\mathbf{x})$ and $\sigma(\mathbf{x})$ are the mean and variance of the model at position $\mathbf{x}$, f is the objective function to be optimized with the estimated minimum at $\mathbf{x}_b$, ξ is a parameter controlling the degree of exploration and $\psi(\cdot)$ denotes the cumulative distribution function of a standard Gaussian distribution.

5.4.1.2 Expected Improvement

The expected improvement is defined by

$$\begin{aligned} EI(\mathbf{x}) = & \ (f(\mathbf{x}_b) - \mu(\mathbf{x}) - \xi)\,\psi\left(\frac{f(\mathbf{x}_b) - \mu(\mathbf{x}) - \xi}{\sigma(\mathbf{x})}\right) \\ & + \sigma(\mathbf{x})\phi\left(\frac{f(\mathbf{x}_b) - \mu(\mathbf{x}) - \xi}{\sigma(\mathbf{x})}\right) \end{aligned} \tag{5.5}$$

where $\phi(\cdot)$ denotes the density function of a standard Gaussian distribution.

5.4.1.3 Lower Confidence Bound

Lower confidence bound is defined by

$$LCB(\mathbf{x}) = \mu(\mathbf{x}) - \beta\sigma(\mathbf{x}) \tag{5.6}$$

where β is a parameter controlling the degree of exploration.

5.4.2 Evolutionary Bayesian Optimization

Bayesian optimization has been very popular not only in the area of black-box optimization, but also machine learning, mainly thanks to its mathematically solid ideas for model management, i.e., the acquisition functions. However, the success of Bayesian optimization has mainly limited to low-dimensional systems and single-objective optimization, due to the following challenges. First, the time complexity of Gaussian processes is cubic in the number of training samples, which makes it unrealistic to be used a surrogate for high-dimensional systems. Second, the optimization of the acquisition function using a mathematical programming method is nontrivial, since it is piece-wise linear and highly multi-modal. An illustrative example is provided in Fig. 5.6, where the upper panel plots the original function (solid line) to be approximated, nine training samples (denoted by diamonds) for training, the obtained Gaussian process model (the dashed line denotes the expected prediction, and shaded region indicates the uncertainty). In addition, the lower panel plots the obtained acquisition function, which we can see is highly multi-modal. The solution that maximizes the acquisition function is denoted by the triangle, which will be sampled and added to the training set.

Finally, Bayesian optimization was developed for single-objective optimization, which makes it not directly applicable to multi-objective optimization, in particular when the number of objectives becomes high.

For the above reasons, meta-heuristics have widely been used to solve the acquisition function in single- or multi-objective Bayesian optimization. A generic diagram for using an evolutionary algorithm for solving the acquisition function in Bayesian optimization is shown in Fig. 5.7, which is called evolutionary Bayesian optimization.

Note that for multi-objective optimization, different ideas have been developed either to convert a multi-objective optimization problem into a single-objective one so that the traditional acquisition function can be applied, or non-dominated sorting according to the acquisition function value of different objective will be performed to select the solutions to be samples. More detailed will be discussed in Chap. 7.

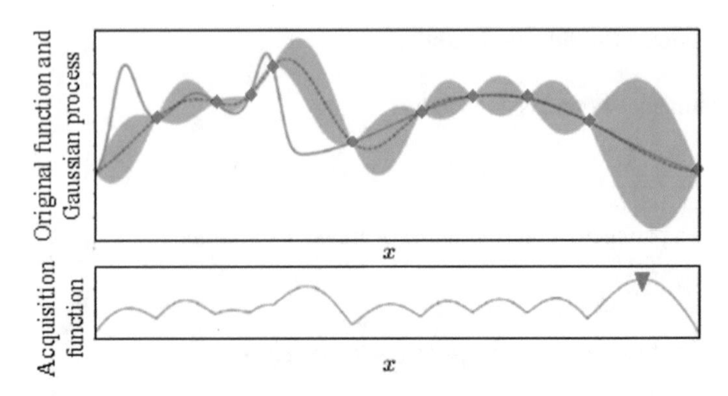

Fig. 5.6 An illustration of Bayesian optimization, where the upper panel shows a constructed Gaussian process model, and the lower panel is the acquisition function for selecting the next data to be queried (sampled) and added to the training data

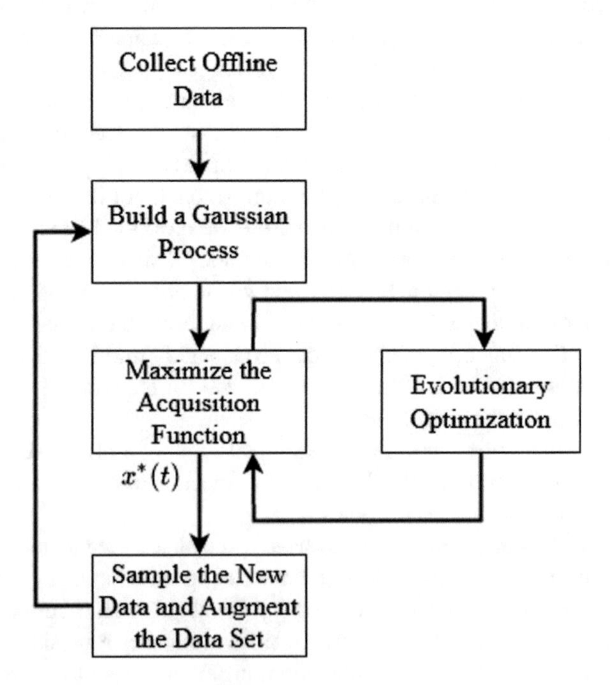

Fig. 5.7 A diagram of evolutionary Bayesian optimization

5.4.3 Bayesian Evolutionary Optimization

Not much attention has been devoted to distinguish the difference between the use of an evolutionary algorithm to optimize the acquisition function in Bayesian optimization, and the use of an acquisition function for individual-based model management in evolutionary optimization. In the latter, the optimization is performed on the surrogate and no optimization of the acquisition function is conducted. Instead, the acquisition function value for each individual in the offspring population is calculated and then those (one or more) having the maximum acquisition function values will be chosen for sampling (fitness evaluation using the real objective function). A diagram of data-driven evolutionary optimization using an acquisition function for model management is provided in Fig. 5.8.

A few differences can be identified between evolutionary Bayesian optimization (EBO) and Bayesian evolutionary optimization (BEO). In EBO, the search of the optimum is basically a single point search, and population based search is applied to the optimization of the acquisition function. Once the optimum of the acquisition function is found, i.e., when the population converges, the optimal solution is sampled and added to the training data before the Gaussian process is updated. Then, the acquisition function is also updated and another round of optimization of the acquisition function is performed anew. This process repeats until the the computational budget is exhausted.

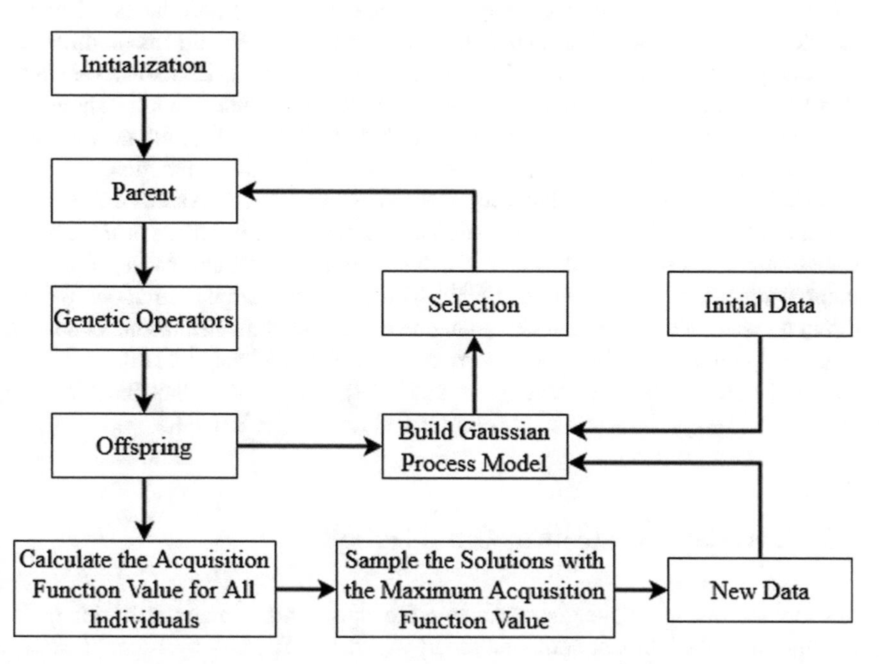

Fig. 5.8 A diagram of Bayesian evolutionary optimization, where an acquisition function is used for model management in surrogate-assisted evolutionary optimization

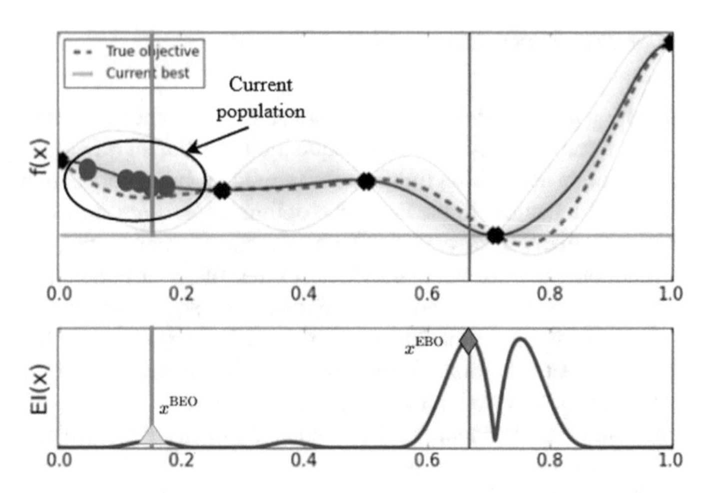

Fig. 5.9 Comparison of the selected solutions to be sampled in BEO and EBO. In BEO, the solution that obtains the maximum acquisition function value is x^{BEO} denoted by a triangle, while the solution of the EBO approach that maximizes the acquisition function will obtain x^{EBO} denoted by a diamond

By contrast, in BEO, the evolutionary algorithm is run on the Gaussian process and the expected value is used for fitness evaluations. After a number of generations, usually before the population converges, the acquisition function value is calculated for each solution in the offspring population, and the one having the maximum value of the acquisition function is sampled and added to the training data. Then, the Gaussian process model is updated and evolutionary search continues. Fig. 5.9 shows the difference in the solutions selected in EBO and BEO. For EBO, the evolutionary algorithm is supposed to find the maximum of the acquisition function, which is denoted by x^{EBO}. For EBO, however, the acquisition function value of the population is calculated before the population converges to the optimum of the current Gaussian process model and as a result, the offspring individual that has the maximum acquisition function value is x^{BEO}. Thus, the new samples solutions will be different, leading to a different Gaussian process model in the next round of search. Empirical comparisons have been carried out on the efficiency of BEO and EBO for multi-objective optimization Qin et al. (2019), which shows that BEO is more effective, although the scope of the empirical study may be limited.

5.5 Bayesian Constrained Optimization

In many real-world engineering problems, the constraints of an optimization problem are often black-box or computationally expensive. Therefore, surrogate models are proposed for replacing the constrained functions so as to save the computational expense in searching for the feasible optimal solutions of the optimization problems. Generally, there are three ways to establish surrogate models for constrained optimization problems.

1. A surrogate model is built for a constrained function.
2. A surrogate model is built for all constrained functions.
3. A classification model is built for identifying the feasible solutions from the infeasible ones.

Ideally, the surrogate models are expected to be close as much as possible to the real constrained functions, however, like the models for the objective functions, it is normally difficult to fit the original constrained functions well. If a feasible solution is approximated to be an infeasible one, the real optimal feasible solution may not be able to be found. Conversely, if an infeasible solution is estimated to be a feasible one, the search direction may be misled. Therefore, it is important to select solutions to be evaluated using the exact constrained functions, which will be used to update the surrogate model to assist approximating the feasibility of a solution correctly as much as possible. Normally, the accuracy of the approximation on the constraints can be improved by evaluating the solution with the maximum approximation uncertainty. However, the main objective of the optimization on a constrained problem is to find the optimal feasible solution. Therefore, both the trade-off between the exploration and exploitation of the algorithm and the improvement of the approximation accuracy on the constrained functions should be considered in the optimization of the expensive constrained problems. Generally, the infill criteria for expensive constrained problems can be classified into two categories. One category takes into account both the objective and constraints in the infill criterion, and the other one considers the objective and constraints independently to find solutions for exact objective and constraint evaluations. In the following, we will give a detailed description on these two categories.

5.5.1 Acquisition Function for Constrained Optimization

Using penalty functions to transform a constrained optimization problem into an unconstrained one is a common method for solving constrained problems. Therefore, in the optimization of the expensive constrained problems, surrogate models can be built for the unconstrained problems and the infill criteria proposed for unconstrained optimization can be adopted directly for solving expensive constrained optimization problems.

Nevertheless, there are also some methods that are dedicated to addressing constraints in the optimization process. As the Gaussian process model can provide both the approximated mean value and the approximation uncertainty, some methods have been proposed to use the Gaussian process to approximate the objective and constraints separately, and variants of the infill criterion are proposed to consider the objective and constraints simultaneously for selecting promising solutions to be evaluated using the exact objective and constraints. In the Gaussian process model assisted constrained optimization, each constraint of a solution is approximated by the Gaussian process model and thus have a probability of feasibility. Therefore, it is called the probabilistic approach. In the probabilistic approach, the constrained optimization problem can be formulated and solved as an unconstrained problem by maximizing the product of the expected improvement and the probability of feasibility:

$$\max_{\mathbf{x}} EI(\mathbf{x}) \times P(\hat{g}(\mathbf{x}) \leq 0) \tag{5.7}$$

where

$$I(\mathbf{x}) = \max\{f^* - \hat{f}(\mathbf{x}), 0\}, \tag{5.8}$$

$$EI(\mathbf{x}) = (f^* - \hat{f}(\mathbf{x}))\psi\left(\frac{f^* - \hat{f}(\mathbf{x})}{s_f(\mathbf{x})}\right) + s_f(\mathbf{x})\phi\left(\frac{f^* - \hat{f}(\mathbf{x})}{s_f(\mathbf{x})}\right) \tag{5.9}$$

and

$$P(\hat{g}(\mathbf{x}) \leq 0) = \psi\left(\frac{g^* - \hat{g}(\mathbf{x})}{s_g(\mathbf{x})}\right) \tag{5.10}$$

In Eqs. (5.8)–(5.10), $EI(\mathbf{x})$ is the expected improvement of solution $\mathbf{x}$ on its objective, $P(\hat{g}(\mathbf{x}) \leq 0)$ represents the probability of feasibility. f^* and g^* represent the best fitness value and the best constraint violation found so far, respectively. $\hat{f}(\mathbf{x})$ and $\hat{g}(\mathbf{x})$ are approximated values on fitness and constraints, and $s_f(\mathbf{x})$ and $s_g(\mathbf{x})$ are the squared root of the predicted variance of the objective and constraint, respectively. $\psi(\cdot)$ is the standard normal cumulative function and $\phi(\cdot)$ is the standard probability density function. Note that if there are a number of constraints, the probability of feasibility is the product of the individual probability. Therefore, the mathematical model of the the probability approach will be

$$\max_{\mathbf{x}} EI(\mathbf{x}) \times \prod_{i=1}^{l} P(\hat{g}_i(\mathbf{x}) \leq 0) \tag{5.11}$$

where l is the number of constraints.

From Eq. (5.11), we can see that the value of constrained EI will be close to 0 if the probability of feasibility of any constraint is very low. Thus, in this situation, Eq. (5.11) cannot be used as the infill criterion to choose solutions for exact expensive evaluations. Jiao et al. Jiao et al. (2019) proposed a modified probability approach, in which two different infill criteria are utilized, respectively, for two cases to select solutions to be evaluated using the expensive objective and constraint functions. One infill criterion, whose mathematical model is given in the following, is used for the situation in which no feasible solution is available.

$$\max EI(\mathbf{x}) = \int_{0}^{g*} \prod_{i=1}^{l} \psi\left(\frac{z - \hat{g}_i(\mathbf{x})}{s_{g_i}(\mathbf{x})}\right) d_z - g^* \times \prod_{i=1}^{l} \psi\left(\frac{-\hat{g}_i(\mathbf{x})}{s_{g_i}(\mathbf{x})}\right) \tag{5.12}$$

The other infill criterion is utilized when feasible solutions are available, and its mathematical expression is:

$$EI(\mathbf{x}) = \left[(f^* - \hat{f}(\mathbf{x}))\psi\left(\frac{f^* - \hat{f}(\mathbf{x})}{s_f(\mathbf{x})}\right) + s_f(\mathbf{x})\phi\left(\frac{f^* - \hat{f}(\mathbf{x})}{s_f(\mathbf{x})}\right)\right] \times \prod_{i=1}^{l} \psi\left(\frac{-\hat{g}_i(\mathbf{x})}{s_{g_i}(\mathbf{x})}\right) \tag{5.13}$$

5.5.2 *Two-Stage Acquisition Functions*

Another type of infill criterion for expensive constrained optimization is provided by COBRA (Constrained Optimization By Radial basis function interpolation) and Extended ConstrLMSRBF Regis (2014), both of which follow a two-phase approach where the first phase is used to find a feasible point while the second phase aims to improve the feasible point. In COBRA, a feasible point will be found to be evaluated using the expensive functions by optimizing the following objective function in the first phase.

$$\begin{aligned} \min \quad & \sum_{i=1}^{l} \max\{\hat{g}_i(\mathbf{x}), 0\} \\ \text{s. t.} \quad & \mathbf{a} \leq \mathbf{x} \leq \mathbf{b} \\ & \hat{g}_i(\mathbf{x}) + \epsilon_i \leq 0, i = 1, 2, \ldots, l \\ & \|\mathbf{x} - \mathbf{x}_j\| \geq \rho, j = 1, 2, \ldots, n \end{aligned} \tag{5.14}$$

where ϵ_i represents the margin of solution i, which is used to force the solution to move away from the RBF constraint boundaries to increase the chance of falling into a feasible region. ρ is the threshold of the distance required for the solution to be distant from its previous position, $\rho = \gamma\ell([\mathbf{a}, \mathbf{b}])$, where $0 < \gamma < 1$ is called a distance requirement factor and $\ell([\mathbf{a}, \mathbf{b}])$ is the length of the smallest side of the box $[\mathbf{a}, \mathbf{b}] \subseteq R^D$.

In the second phase, a better feasible point will be searched by the optimization of the following mathematical model, which is assisted by RBF surrogate models of the objective and constraint functions.

$$\begin{aligned} \min \quad & \hat{f}(\mathbf{x}) \\ \text{s. t.} \quad & \mathbf{a} \leq \mathbf{x} \leq \mathbf{b} \\ & \hat{g}_i(\mathbf{x}) + \epsilon \leq 0, i = 1, 2, \ldots, l \\ & \|\mathbf{x} - \mathbf{x}_j\| \geq \rho, j = 1, 2, \ldots, n \end{aligned} \tag{5.15}$$

The optimal solution of Eq. (5.15) will be evaluated using the expensive objective and constrained functions. The margin will then be adjusted. Note that in COBRA, if an initial feasible solution is given or is obtained by the space-filling design, the COBRA method only run the second phase.

The extended ConstrLMSRBF is an extension of ConstrLMSRBF Regis (2014), which is a heuristic surrogate-based approach and sassumes that there is a feasible solution among the initial points. Similar to COBRA, the extended ConstrLMSRBF also has two-phase structure, in which the first phase is used to find a feasible point and the second phase is expected to find better feasible solution. In the first phase of the extended ConstrLMSRBF, a set of random candidate points will be generated by adding normal perturbations to some or all of the components of the current best position, if the current best point is infeasible. Then the solution with the minimum number of predicted constraint violations among these candidate points will be evaluated using the expensive objective and constrained function for updating the best solution found so far. Note that if there are a number of solutions with the same minimum number of predicted constraint violations, then the solution with the smallest maximum predicted constraint violation will be evaluated using the expensive function. The iteration will continue until a feasible solution is found, and the second phase, which is the same to the second phase of COBRA, will be conducted.

5.6 Surrogate-Assisted Robust Optimization

5.6.1 Bi-objective Formulation of Robust Optimization

Search for robust optimal solutions using explicit averaging requires additional fitness evaluations. This will give rise to big problems if one single fitness evaluation is time consuming. One solution to this problem is to introduce fitness estimation based on surrogate for both single- and multi-objective approach to robust optimization. The bi-objective formulation of robust optimization based on a surrogate model can be expressed by:

$$\widehat{f}_{\text{exp}}(\mathbf{x}^0) = \frac{1}{d}\sum_{i=1}^{d} \widehat{f}(\mathbf{x}_i), \tag{5.16}$$

$$\widehat{f}_{\text{var}}(\mathbf{x}^0) = \frac{1}{d}\sum_{i=1}^{d} \left[\widehat{f}(\mathbf{x}_i) - \widehat{f}_{\text{exp}}(\mathbf{x}^0)\right], \tag{5.17}$$

where $\mathbf{x}^0$ is the solution whose robustness is to be estimated, $\widehat{f}(\cdot)$ is the approximated fitness function, $\widehat{f}_{\text{exp}}(\mathbf{x}^0)$ is the estimated expected fitness value, $\mathbf{x}_i = \mathbf{x}^0 + \delta_i$, δ_i is the perturbation, and d is the number of samples. The samples can be generated using the Latin hypercube sampling.

5.6.2 *Surrogate Construction*

Different surrogate techniques can be applied Paenke et al. (2006):

- *A single surrogate.* In this case, one approximation model per individual is built, i.e., the models' fitting points are the same as the individuals' locations in the decision space, and the surrogate model will be used for all sample points for estimating the expected fitness and fitness variance. Depending on the model used, a single surrogate may not work well within the range of expected disturbances δ.
- *A nearest model.* This is also a single surrogate model to be built around each individual, however, the nearest model will be used to estimate the fitness of a sample. Note that the nearest model may not be always the one associated to the individual.
- *An ensemble.* In this approach, the function value of a sample ($\mathbf{x}_s$) point will be estimated with a weighted combination (ensemble) of models that correspond to the k nearest fitting points. The expected fitness value can be estimated by:

$$\begin{aligned} \widehat{f}_{\mathrm{ENS}}(\mathbf{x}_s) &= \frac{1}{\sum_{1 \le i \le k} w_i} \sum_{1 \le i \le k} w_i \, \widehat{f}_i(\mathbf{x}_s) \\ w_i &= \frac{1}{\|\mathbf{x}_{\mathrm{fp}} - \mathbf{x}_s\|_2}, \end{aligned} \tag{5.18}$$

  where w_i is a weight and k the ensemble size, $\mathbf{x}_{\mathrm{fp}}$ is the fitting point of the corresponding surrogate.
- *Multiple models.* In this approach, a separate model is constructed around each sample, which is used to estimate the sample's fitness values.

5.7 Performance Indicators for Surrogates

It is not straightforward to measure the quality of the surrogate model, as the purpose of a surrogate is not to accurately approximate the original fitness function, but to efficiently guide the search instead.

As illustrated in Fig. 5.10, the surrogate model $\widehat{f}(x)$ does not fit the original objective function $f(x)$ well. However, if the optimization is run on the surrogate model, the minimum of $\widehat{f}(x)$ is very close to that of $f(x)$. From this perspective, $\widehat{f}(x)$ is a perfect surrogate model. Therefore, the performance indicator for surrogate models may differ from that in machine learning, where the prediction or classification accuracy on unseen data is the most important.

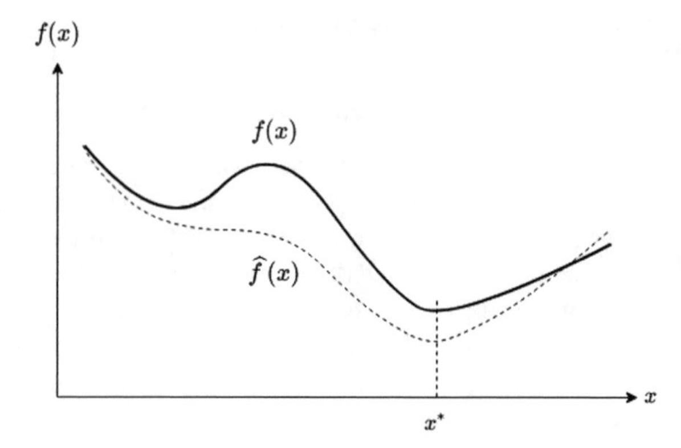

Fig. 5.10 An illustration of a surrogate having relatively poor approximation quality can perfectly guide the search to the true optimum (x^*) of the original problem

5.7.1 Accuracy

The most widely used measure for the quality of surrogate models is the mean squared error between the individual's original fitness function $f(\mathbf{x})$ and the output of the surrogate model $\widehat{f}(\mathbf{x})$

$$E = \frac{1}{d}\sum_{j=1}^{d}\left(\widehat{f}(\mathbf{x}_j) - f(\mathbf{x}_j)\right)^2 . \tag{5.19}$$

Here, the mean squared error is averaged over d different individuals taken into account of the estimation of the quality measure, e.g., the $n = \lambda$ offspring individuals in individual based model management, or the $n = \eta\,\lambda$ controlled individuals in one control cycle. Note that measures should be taken to alleviate overfitting.

Generally speaking, a surrogate model of good approximation accuracy can make sure that the fitness of the individuals can be reasonable correctly estimated so that the good individuals will be selected for the next generation. However, from the evolutionary optimization point of view, the actual fitness value of the individuals are not critical, if the ranking of the individuals is correct to ensure correct selection. In the following, a number of performance indicators with respect to correct selection rather than the accuracy of the surrogate will be presented. Note that the definitions of some measures depends on the environmental selection strategy.

5.7.2 Selection-based Performance Indicator

In the following discussions, we take (μ, λ)-selection with $\lambda \geq 2\mu$ as an example, which is continuous optimization. In principle, however, it is be extended to other selection strategies.

The first measure is based on the number of individuals that are correctly selected using the approximate model:

$$\rho^{(\text{sel})} = \frac{\xi - \langle \xi \rangle}{\mu - \langle \xi \rangle}, \tag{5.20}$$

where ξ $(0 \leq \xi \leq \mu)$ is the number of correctly selected individuals, i.e., the number of individuals that would have also been selected if the original fitness function had been used for fitness evaluation. The expectation

$$\begin{aligned} \langle \xi \rangle &= \sum_{m=0}^{\mu} m \frac{\binom{\mu}{m}\binom{\lambda-\mu}{\mu-m}}{\binom{\lambda}{\mu}} \\ &= \frac{\mu^2}{\lambda}. \end{aligned} \tag{5.21}$$

of ξ in case of random selections is used for normalization. It can be seen that if all μ parent individuals are correctly selected, the measure reaches its maximum of $\rho^{(\text{sel})} = 1$, and that negative values indicate that the selection based on the surrogate model is worse than a random selection.

The measure $\rho^{(\text{sel})}$ only evaluates the absolute number of correctly selected individuals. However, in case $\rho^{(\text{sel})} < 1$, the measure does not indicate whether the $(\mu + 1)$-th or the worst offspring individual will be selected, significantly influencing the the evolutionary optimization process.

In the following, the measure $\rho^{(\text{sel})}$ is extended to include the rank of the selected individuals, which is calculated according to the original fitness function. A surrogate model is considered to be good, if the rank of the selected individuals based on the surrogate model is above-average according to the rank based on the original fitness function.

The definition of the extended measure $\widehat{\rho}^{(\text{sel})}$ is as follows. The surrogate model gets a score of $\lambda - m$, if the m-th best individual based on the original fitness function is selected. Thus, the quality of the surrogate model can be indicated by summing up the scores of the selected individuals, which is denoted by π. Note that π reaches its maximum, if all μ individuals are selected correctly:

$$\begin{aligned} \pi^{(\text{max.})} &= \sum_{m=1}^{\mu} (\lambda - m) \\ &= \mu \left(\lambda - \frac{\mu + 1}{2} \right). \end{aligned} \tag{5.22}$$

Similar to Eq. (5.20), the measure $\widehat{\rho}^{(\text{sel})}$ is defined by transforming π linearly, using the maximum $\pi^{(\text{max.})}$ and the expectation $\langle \pi \rangle = \frac{\mu\lambda}{2}$ in case of a purely random selection:

$$\widehat{\rho}^{(\text{sel})} = \frac{\pi - \langle \pi \rangle}{\pi^{(\text{max.})} - \langle \pi \rangle}. \tag{5.23}$$

5.7.3 *Rank Correlation*

The quality of a surrogate may also be measured by the rank correlation. The rank correlation can be defined given by

$$\rho^{(\text{rank})} = 1 - \frac{6 \sum_{l=0}^{\lambda} d_l^2}{\lambda(\lambda^2 - 1)}, \tag{5.24}$$

is a measure for the monotonic relation between the ranks of two variables. Here, d_i is the difference between the ranks of the i-th offspring individual based on the original fitness function $f(\mathbf{x})$ and on the surrogate model $(\mathbf{x})$. The range of $\rho^{(\text{rank})}$ is between $[-1; 1]$. The higher the value of $\rho^{(\text{rank})}$, the stronger the monotonic relationship with a positive slope between the ranks of the two variables. In contrast to $\rho^{(\text{sel})}$, the rank correlation not only takes into account the ranking of the selected individuals, but also the ranks of all individuals. This makes it possible to provide a good estimation of the ability of the surrogate model to distinguish between good and bad individuals, which is the basis for correct selection in surrogate-assisted evolutionary optimization.

5.7.4 *Fitness Correlation*

Fitness correlation, that is, the correlation between the the fitness predicted by the surrogate model and the true fitness value, provides another way to measure if the surrogate model is able to ensure correct selection, although not necessarily to accurately predict the fitness values:

$$\rho^{(\text{corr})} = \frac{\frac{1}{d} \sum_{j=1}^{d} \left(\widehat{f}_j(\mathbf{x}) - \tilde{f}(\mathbf{x}) \right) \left(f_j(\mathbf{x}) - \bar{f}(\mathbf{x}) \right)}{\sigma^f \sigma^{\widehat{f}}}. \tag{5.25}$$

Here, $\tilde{f}(\mathbf{x})$ and $\bar{f}(\mathbf{x})$ are the mean values of the estimated and true fitness, and σ^f and σ^f are the standard deviations of the surrogate model output and original fitness function, respectively.

The properties of the correlation is related to both the rank based measures introduced above and the mean squared error. It is not a measure for the difference between the surrogate and original fitness function, but evaluates a monotonic relationship between them. In addition, the range of this measure is known and therefore $\rho^{(\text{corr})}$ is easier to evaluate than mean squared error. In addition, $\rho^{(\text{corr})}$ is differentiable, allowing a gradient-based method to adapt the surrogate model.

5.8 Summary

This chapter provides the basic definitions in data-driven evolutionary optimization, and introduces the preliminary ideas for surrogate model management in single-objective optimization. In the next chapters, we will extend these basic surrogate management methods for addressing various challenges.

References

Branke, J., & Schmidt, C. (2005). Faster convergence by means of fitness estimation. *Soft Computing*, *9*(1), 13–20.

Chugh, T., Chakraborti, N., Sindhya, K., & Jin., Y., (2017). A data-driven surrogate-assisted evolutionary algorithm applied to a many-objective blast furnace optimization problem. *Materials and Manufacturing Processes*, *32*, 1172–1178.

Cleveland, W. and Loader, C. (1996). *Smoothing by Local Regression: Principles and Methods*, (pp. 10–49). Physica-Verlag HD.

Ding, J., Yang, C., Jin, Y., & Chai, T. (2017). Generalized multi-tasking for evolutionary optimization of expensive problems. *IEEE Transactions on Evolutionary Computation*, *23*, 44–58.

Gräning, L., Jin, Y., & Sendhoff, B. (2005). Efficient evolutionary optimization using individual-based evolution control and neural networks: A comparative study. In *European Symposium on Artificial Neural Networks (ESANN'2005)*, (pp. 273–278).

Guo, D., Chai, T., Ding, J., and Jin, Y. (2016). Small data driven evolutionary multi-objective optimization of fused magnesium furnaces. In *IEEE Symposium Series on Computational Intelligence*, (pp. 1–8). Athens, Greece: IEEE.

Gupta, A., Ong, Y.-S., & Feng, L. (2016). Multifactorial evolution: toward evolutionary multitasking. *IEEE Transactions on Evolutionary Computation*, *20*(3), 343–357.

Gupta, A., Ong, Y.-S., & Feng, L. (2018). Insights on transfer optimization: Because experience is the best teacher. *IEEE Transactions on Emerging Topics in Computational Intelligence*, *2*(1), 51–64.

Habib, A., Singh, H. K., Chugh, T., Ray, T., & Miettinen, K. (2019). A multiple surrogate assisted decomposition-based evolutionary algorithm for expensive multi/many-objective optimization. *IEEE Transactions on Evolutionary Computation*, *23*(6), 1000–1014.

Huang, P., Wang, H., & Jin, Y. (2021). Offline data-driven evolutionary optimization based on tri-training. *Swarm and Evolutionary Computation*, *60*, 100800.

Jiao, R., Zeng, S., Li, C., Jiang, Y., & Jin, Y. (2019). A complete expected improvement criterion for Gaussian process assisted highly constrained expensive optimization. *Information Sciences*, *471*, 80–96.

Jin, Y., Olhofer, M., & Sendhoff, B. (2000). On evolutionary optimization with approximate fitness functions. In *Proceedings of the Genetic and Evolutionary Computation Conference*, (pp. 786–793). Morgan Kaufmann Publishers Inc.

Jin, Y. (2005). A comprehensive survey of fitness approximation in evolutionary computation. *Soft Computing*, *9*(1), 3–12.

Jin, Y. (2011). Surrogate-assisted evolutionary computation: Recent advances and future challenges. *Swarm and Evolutionary Computation*, *1*(2), 61–70.

Jin, Y., Olhofer, M., & Sendhoff, B. (2002). A framework for evolutionary optimization with approximate fitness functions. *IEEE Transactions on Evolutionary Computation*, *6*(5), 481–494.

Jin, Y., Wang, H., Chugh, T., Guo, D., & Miettinen, K. (2018). Data-driven evolutionary optimization: An overview and case studies. *IEEE Transactions on Evolutionary Computation*, *23*(3), 442–458.

Min, A. T. W., Ong, Y.-S., Gupta, A., & Goh, C.-K. (2017). Multi-problem surrogates: Transfer evolutionary multiobjective optimization of computationally expensive problems. *IEEE Transactions on Evolutionary Computation*. To appear.

Paenke, I., Branke, J., & Jin, Y. (2006). Efficient search for robust solutions by means of evolutionary algorithms and fitness approximation. *IEEE Transactions on Evolutionary Computation, 10*(4), 405–420.

Powell, M. J. (1970). A new algorithm for unconstrained optimization. In *Nonlinear programming*, (pp. 31–65). Elsevier.

Qin, S., Sun, C., Jin, Y., & Zhang, G. (2019). Bayesian approaches to surrogate-assisted evolutionary multi-objective optimization: A comparative study. In *IEEE Symposium Series on Computational Intelligence*.

Regis, R. G. (2014). Constrained optimization by radial basis function interpolation for high-dimensional expensive black-box problems with infeasible initial points. *Engineering Optimization, 46*(2), 218–243.

Sun, C., Jin, Y., & Tan, Y. (2018). Semi-supervised learning assisted particle swarm optimization of computationally expensive problems. In *Proceedings of the Genetic and Evolutionary Computation Conference*, (pp. 45–52). ACM.

Wang, H., Jin, Y., & Doherty, J. (2018d). A generic test suite for evolutionary multi-fidelity optimization. *IEEE Transactions on Evolutionary Computation*. to appear.

Wang, H., Jin, Y., & Janson, J. O. (2016a). Data-driven surrogate-assisted multi-objective evolutionary optimization of a trauma system. *IEEE Transactions on Evolutionary Computation, 20*(6), 939–952.

Wang, H., Jin, Y., Sun, C., & Doherty, J. (2018e). Offline data-driven evolutionary optimization using selective surrogate ensembles. *IEEE Transactions on Evolutionary Computation, 23*(2), 203–216.

Wu, X., Zhu, X., Wu, G.-Q., & Ding, W. (2014). Data mining with big data. *IEEE Transactions on Knowledge and Data Engineering, 26*(1), 97–107.

Zhou, Z.-H. (2009). *Ensemble learning. Encyclopedia of biometrics, 1*, 270–273.

Zhu, X. J. (2005). Semi-supervised learning literature survey. Technical report, University of Wisconsin-Madison Department of Computer Sciences.

Chapter 6
Multi-surrogate-Assisted Single-objective Optimization

6.1 Introduction

In surrogate-assisted evolutionary optimization, the importance of selecting an appropriate surrogate model cannot be underestimated. Unfortunately, there is no clear, simple rules for selecting the right surrogate in most cases, since each type of surrogate models has its strengths and weaknesses. For example, Gaussian processes are powerful since they can provide both a predicted fitness and an estimate of the uncertainty of the prediction. However, Gaussian processes suffer from extremely high computational complexity when the number of training samples is large and incapability of learn incrementally.

To address the challenges in selecting the appropriate surrogate, multiple surrogates that are meant to play different roles in assisting search can be adopted. The surrogates can also be classified to global and local models. A local surrogate model is usually trained on the data of a sub-region of the decision space and used to assist searching for local optimal solution of this sub-region. While a global surrogate model is usually trained on the data that spread on the whole decision space, which is normally used to help the search for the global optimal solution. Also, we can classify the surrogate models into single and multiple according to the number of surrogate models used in each generation of the optimization. Furthermore, there are two ways for multiple surrogate models to be utilized. In the approach, one solution will be approximated simultaneously by different surrogate models and choose one strategy to determine the final approximated value. By contrast, different surrogate models are used to approximate different solutions in the optimization.

A straightforward idea to improve the quality of surrogates is to adopt an ensemble as the surrogate, instead of using a single model. For example, in Jin and Sendhoff (2004b), an neural network ensemble is used for estimating the fitness of individuals in clusters, where only the individual closest to each cluster center is evaluated using the real fitness function. To maximize the performance of the ensemble, an weighted sum of the outputs of the base learners can be used:

Y. Jin et al., *Data-Driven Evolutionary Optimization*,
Studies in Computational Intelligence 975,
https://doi.org/10.1007/978-3-030-74640-7_6

$$y^{\text{ens}} = \sum_{k=1}^{N} w^k y^k, \tag{6.1}$$

where y^k is the output of the k-th base learner, w^k is the weight of the k-th base learner, and N is the ensemble size.

It is non-trivial to optimize the weights of the ensemble. Since the expected approximation error of the ensemble is given by:

$$E^{\text{ens}} = \sum_{i=1}^{N} \sum_{j=1}^{N} w^i w^j C_{ij}, \tag{6.2}$$

where C_{ij} is the error correlation matrix between the i-th and j-th base learner networks of the ensemble:

$$C_{ij} = E\left[(y^i - y_d^i)(y^j - y_d^j)\right], \tag{6.3}$$

where y_d^i is the desired output of the i-th neural network. An optimal set of weights that minimize the expected prediction error of the ensemble can be obtained by:

$$w^k = \frac{\sum_{j=1}^{N}(C_{kj})^{-1}}{\sum_{i=1}^{N}\sum_{j=1}^{N}(C_{ij})^{-1}}, \tag{6.4}$$

where $1 \leq i, j, k \leq N$.

Nevertheless, it is non-trivial to reliably estimate the error correlation matrix since the prediction errors of the base learners in an ensemble are often strongly correlated. One measure to remedy this is to explicitly reduce the correlation between the outputs of the ensemble. Empirical methods for optimizing the weights using a meta-heuristics can also employed, although this may increase the computational complexity.

The benefits of using an ensemble include improved fitness predictions and the possibility of discarding very uncertain fitness predictions, in case the base learners strongly disagree with themselves.

In the following subsections, a number of representative methods for using multiple surrogates will be introduced.

6.2 Local and Global Surrogates Assisted Optimization

In surrogate-assisted optimization algorithms, the approximation errors of the surrogate model may usually mislead the optimization algorithm, can however, also speed up the search. In this section, we will introduce a generalized surrogate assisted

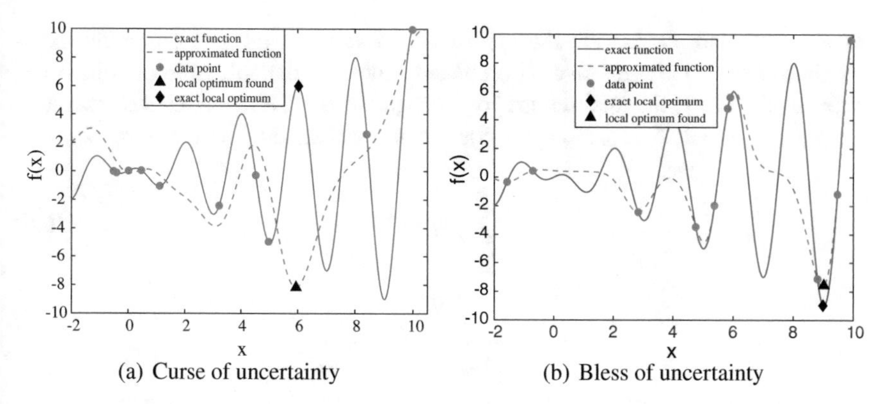

(a) Curse of uncertainty (b) Bless of uncertainty

Fig. 6.1 An example to show the curse and bless of uncertainty in the surrogate-assisted single-objective optimization

memetic algorithm for single-objective (denoted as GS-SOMA) and multi-objective (named as GS-MOMA) optimization problems proposed by Lim et al. (2010). The generalized surrogate assisted memetic algorithm aims to mitigate the 'curse of uncertainty', meaning to reduce the negative consequences caused by the approximation error of the surrogate model, and even to benefit from the 'bless of uncertainty'. The assumption here is that a rugged fitness landscape may be smoothed by the surrogate, which is helpful in speeding up the search for the global optimal solution. Fig. 6.1 gives a simple example to illustrate the meaning of 'curse of uncertainty' and 'bless of uncertainty', respectively. In Fig. 6.1a, the optimal solution of the surrogate model is not the true optimal solution of the original problem, thus the search for the optimal solution of the surrogate model will make the algorithm get stuck or converge to a false optimum, which is the 'curse of uncertainty'. On the other hand, from Fig. 6.1b, we can see that the optimal solution of the surrogate model locates at a promising position of the original problem, thus being able to speed up the search for the global optimal solution of the expensive optimization problem, which is called the 'bless of uncertainty'. In this section, the generalized surrogate models for single- and multi-objective optimization will be introduced.

6.2.1 *Ensemble Surrogate Model*

The surrogate ensemble has been shown to be more likely to generate reliable fitness prediction than a single surrogate model. Thus, in the proposed GSM framework, a surrogate ensemble is adopted to guide the global search:

$$\hat{f}_{ens}(\mathbf{x}) = \sum_{i=1}^{N_m} w_i \hat{f}_i(\mathbf{x}) \tag{6.5}$$

where $\hat{f}_i(\mathbf{x})$ and $\hat{f}_{ens}(\mathbf{x})$ are the approximated value of the i-th ensemble member (base learner) and the overall predicted fitness of the ensemble of solution $\mathbf{x}$, respectively. N_m is the total number of surrogate models utilized for the ensemble. $0 \leq w_i \leq 1, i = 1, 2, \ldots, N_m$ is the coefficient associated with the i-th model.

$$\sum_{i=1}^{N_m} w_i = 1 \tag{6.6}$$

$$w_i = \frac{\sum_{j=1, j \neq i}^{N_m} \epsilon_j}{(N_m - 1) \sum_{j=1}^{N_m} \epsilon_j} \tag{6.7}$$

where ϵ_j is the approximation error of the j-th surrogate model.

6.2.2 Multi-surrogate for Single-objective Memetic Optimization

Algorithm 1 gives the pseudocode of the generalized surrogate single-objective memetic algorithm. The method can be divided into two parts. In the first parts (lines 3-6), the population will be evolved for a number of generations, in which each solution will be evaluated using the exact fitness function. The second parts is a local search phase, in which two types of surrogate models, i.e., M_1 and M_2, will be trained for each solution using its N_n nearest solutions in Arc (lines 10-12). After that, the optimal solutions $\mathbf{x}^1_{opt}$ and $\mathbf{x}^2_{opt}$ of M_1 and M_2, respectively, will be found using an optimization algorithm. The two optimums found on $\mathbf{x}^1_{opt}$ and $\mathbf{x}^2_{opt}$ will be compared and the better one is considered as the optimal solution found by the local search. Note that all optimal solutions found in the local search will be evaluated using the exact fitness function and stored in Art for updating the surrogate models.

6.2.3 Multi-surrogate for Multi-objective Memetic Optimization

In the generalized surrogate multi-objective memetic algorithm, the framework is the same as that of GS-SOMA. However, different to the single-objective optimization, in the multi-objective optimization, another archive is required to save the non-dominated solutions and a selection pool is built for mating selection. Similar to GS-SOMA, the population will be evolved for a number of generations G_{db} in the first phase, in which each solution will be evaluated using the real objective functions and saved to Arc. The non-dominated solutions will be saved in Arc_{ND}. In the following phase, a new population will be generated using the multi-objective

Algorithm 1 Pseudo code of GS-SOMA.

Require: N_m: number of surrogate models used for ensemble;
N_n: number of nearest neighbors of each solution $\mathbf{x}$;
Arc: an archive to save solutions that have been evaluated using the expensive real objective function;
G_{db}: the maximum generation for building the database.
Ensure: $\mathbf{x}^*$ and $f(\mathbf{x}^*)$
1: $g = 0$;
2: **while** the stopping criterion is not met **do**
3: **if** the number of generation $g < G_{db}$ **then**
4: Generate ($g = 0$) or evolve ($g > 0$) a population, evaluate each solution using the real objective function, and save them to Arc;
5: Determine the best position found so far $\mathbf{x}^*$;
6: $g = g + 1$;
7: **else**
8: Evolve to a new population pop;
9: **for** each solution $\mathbf{x}$ in pop **do**
10: Select N_n nearest solutions from Arc;
11: Train N_m surrogate models and aggregate them to one model, denoted as M_1;
12: Train a low-order PR model, denoted as M_2;
13: Search for optimal solutions $\mathbf{x}^1_{opt}$ and $\mathbf{x}^2_{opt}$ of M_1 and M_2, respectively.
14: **if** $f(\mathbf{x}^1_{opt}) < f(\mathbf{x}^2_{opt})$ **then**
15: $\mathbf{x} = \mathbf{x}^1_{opt}$;
16: **else**
17: $\mathbf{x} = \mathbf{x}^2_{opt}$;
18: **end if**
19: Save all solutions that have been evaluated using the real objective function in the local search to Arc;
20: **end for**
21: **end if**
22: **end while**

evolutionary operators on the selection pool P_s, and will be evaluated using the expensive objective functions. After that, for each solution $\mathbf{x}$, a random weight vector $\mathbf{v}_i$ will be generated for converting the multi-objective problem into a single-objective one. As done in GS-SOMA, N_n solutions in Arc will be found and used to train two surrogate models, i.e., M_1 and M_2, for the aggregated single-objective function f_{aggr}, in which M_1 is an ensemble consisting of a number of base learners of f_{aggr}, and M_2 is a low-order PR model of f_{aggr}, respectively. The optimal solutions of M_1 and M_2 will be searched by a local search algorithm. Save all solutions that have been evaluated using the real objective functions to Arc and use these solutions to update Arc_{ND} as well. Finally, the selection pool will be formed by the current population pop, the solutions evaluated using the real objective function in the local search, which are stored in set P_l, and all solutions in archive Arc_{ND}.

Algorithm 3 gives the detail about when solution $\mathbf{x}$ will be replaced and when the archive Arc_{ND} will be updated. From Algorithm 3, we can see that there are six cases to update the archive Arc_{ND}.

Algorithm 2 Pseudo code of GS-MOMA.

Require: N_m: number of surrogate models used for ensemble;
N_n: number of nearest neighbors of each solution $\mathbf{x}$;
Arc: an archive to save solutions that have been evaluated using the real expensive problem;
Arc_{ND}: an archive to save the non-dominated solutions;
P_s: selection pool;
G_{db}: the maximum generation for building the database.
Ensure: $\mathbf{x}^*$ and $f(\mathbf{x}^*)$
1: $g = 0$;
2: **while** the stopping criterion is not met **do**
3: **if** the number of generation $g < G_{db}$ **then**
4: Generate ($g = 0$) or evolve ($g > 0$) a population, evaluate each solution using the real objective functions, and save them to Arc;
5: Save all non-dominated solutions in Arc_{ND};
6: $g = g + 1$;
7: **else**
8: Evolve to a new population pop;
9: **for** each solution $\mathbf{x}$ in pop **do**
10: Generate a random weight vector $\mathbf{v} = (v_1, v_2, \ldots, v_m)$, $\sum_{i=1}^{m} v_i = 1$, where m is the number of objectives.
11: Select N_n nearest solutions from Arc;
12: Train N_m surrogate models for $f_{aggr} = \sum_{i=1}^{m} v_i f_i(\mathbf{x})$, and ensemble them to one model, denoted as M_1;
13: Train a low-order PR model of $f_{aggr} = \sum_{i=1}^{m} v_i f_i(\mathbf{x})$, denoted as M_2;
14: Search for optimal solutions $\mathbf{x}_{opt}^1$ and $\mathbf{x}_{opt}^2$ of M_1 and M_2, respectively.
15: Update the position of $\mathbf{x}$ and Arc_{ND} (Refer to Algorithm 3);
16: **end for**
17: Form the selection pool, $P_s = P_l \cup pop \cup Arc_{ND}$
18: **end if**
19: **end while**

6.2.4 Trust Region Method Assisted Local Search

From Algorithms 1 and 2, we can see that local search is utilized to find the optimal solution of the surrogate model. In both GS-SOMA and GS-MOMA, the trust-region-regulated search strategy is adopted to search on the surrogate models. For each solution $\mathbf{x}$ in GS-SOMA and GS-MOMA populations, the local search is conducted by optimize the following problem:

$$\begin{aligned} \min \quad & \hat{f}(\mathbf{x}_{cb} + \mathbf{s}) \\ \text{s.t.} \quad & \|\mathbf{s}\| \leq \Omega, \end{aligned} \tag{6.8}$$

where $\hat{f}(\mathbf{x})$ is the approximation function corresponding to the objective function. $\mathbf{x}_{cb}$, $\mathbf{s}$ and Ω are the current best solution found so far, an arbitrary step and the trust-region radius, respectively. The sequential quadratic programming (SQP) is utilized to search for the minimum of Eq. (6.8). The initial trust-region radius Ω is initialized based on the minimum and maximum values of the N_n solutions for

Algorithm 3 Pseudo code of Replace&Archive procedure.

Require: $\mathbf{x}_{opt}^1$: the optimal solution of model M_1;
$\mathbf{x}_{opt}^2$: the optimal solution of model M_2;
Arc_{ND}: the archive to save the non-dominated solutions;
Ensure: Updated Arc_{ND}
1: **if** $\mathbf{x}_{opt}^1 \preceq \mathbf{x}$ **then**
2: $\mathbf{x} = \mathbf{x}_{opt}^1$;
3: **if** $\mathbf{x}_{opt}^2 \preceq \mathbf{x}_{opt}^1$ **then**
4: $\mathbf{x} = \mathbf{x}_{opt}^2$;
5: **else if** $\mathbf{x}_{opt}^2 \sim \mathbf{x}_{opt}^1$ **then**
6: Archive $\mathbf{x}_{opt}^2$ in Arc_{ND};
7: **end if**
8: **else if** $\mathbf{x}_{opt}^2 \preceq \mathbf{x}$ **then**
9: $\mathbf{x} = \mathbf{x}_{opt}^2$;
10: **if** $\mathbf{x}_{opt}^2 \sim \mathbf{x}_{opt}^1$ **then**
11: Archive $\mathbf{x}_{opt}^1$ in Arc_{ND};
12: **end if**
13: **else if** $(\mathbf{x}_{opt}^1 \sim \mathbf{x}) \wedge (\mathbf{x}_{opt}^2 == \mathbf{x})$ **then**
14: Archive $\mathbf{x}_{opt}^1$ in Arc_{NS}
15: **else if** $(\mathbf{x}_{opt}^2 \sim \mathbf{x}) \wedge (\mathbf{x}_{opt}^1 == \mathbf{x})$ **then**
16: Archive $\mathbf{x}_{opt}^2$ in Arc_{NS}
17: **else if** $(\mathbf{x}_{opt}^1 \sim \mathbf{x}) \wedge (\mathbf{x}_{opt}^2 \sim \mathbf{x})$ **then**
18: **if** $(\mathbf{x}_{opt}^1 \preceq \mathbf{x}_{opt}^2) \| (\mathbf{x}_{opt}^1 == \mathbf{x}_{opt}^2)$ **then**
19: Archive $\mathbf{x}_{opt}^1$ in Arc_{ND};
20: **else if** $\mathbf{x}_{opt}^2 \preceq \mathbf{x}_{opt}^1$ **then**
21: Archive $\mathbf{x}_{opt}^2$ in Arc_{ND};
22: **else**
23: Archive $\mathbf{x}_{opt}^1$ and $\mathbf{x}_{opt}^2$ in Arc_{ND};
24: **end if**
25: **end if**

training the surrogate models. During the local search, Ω is updated based on the following measure ρ (figure of merit):

$$\rho = \frac{f(\mathbf{x}_{cb}) - f(\mathbf{x}_{opt})}{\hat{f}(\mathbf{x}_{cb}) - \hat{f}(\mathbf{x}_{opt})} \tag{6.9}$$

where $\mathbf{x}_{opt}$ is the local optimal solution.

The Ω is then updated as follows:

$$\Omega = \begin{cases} C_1\Omega, & \text{if } \rho < C_2 \\ \Omega, & \text{if } C_2 \leq \rho \leq C_3 \\ C_4\Omega, & \text{if } \rho > C_3 \end{cases} \tag{6.10}$$

In Equ. (6.10), C_1, C_2, C_3 and C_4 are constants. Typically, $C_1 \in (0, 1)$ and $C_4 \geq 1$ to make the scheme work efficiently. Clearly, we can see from Eq. (6.10), the trust-region radius will be reduced if the accuracy of the surrogate model measured by ρ is low, otherwise it will be increased.

Note that the initial guess of the optimum at each iteration is

$$\mathbf{x}_{cb} = \begin{cases} \mathbf{x}_{opt}, & \text{if } \rho > 0 \\ \mathbf{x}_{cb}, & \text{otherwise} \end{cases} \tag{6.11}$$

6.2.5 Experimental Results

An empirical study on ten single-objective benchmark problems, whose characteristics can be found in Lim et al. (2010), is conducted to evaluate the performance of the GSM framework. The parameter settings used in the experiments are shown in Table 6.1. The results obtained by GS-SOMA and GS-MOMA and their variants are compared using the t-test with 95% confidence level. The '+', '−' and '≈' symbols represent that the proposed method performs statistically significantly better, worse, and equivalent, respectively, compared to their variants. Table 6.2 gives the statistical results on ten 30-dimensional single-objective benchmark problems obtained by GA, SS-SOMA-GP, SS-SOMA-PR, SS-SOMA-RBF and GS-SOMA. SS-SOMA-GP, SS-SOMA-PR and SS-SOMA-RBF are variants of GS-SOMA, in which only GP, PR and RBF is used to be the surrogate model M_1, respectively. From Table 6.2, we can see that GS-SOMA obtains better results than GA on all ten benchmark problems, which clearly shows that the surrogate model is truly beneficial for the evolutionary algorithms to find a good solution. Compared to SS-SOMA-GP, SS-SOMA-PR and SS-SOMA-RBF, we can find that the proposed GS-SOMA is out performed only on one, three and one out of ten test instances, respectively, showing that the surrogate ensemble is more reliable than one surrogate model for assisting the algorithms to find an optimal solution.

Table 6.1 Parameter setting in GS-SOMA

Population size (N_{pop})	100
Crossover probability	0.9
Mutation probability	0.1
Maximum number of exact evaluations	8000
Number of trust region iteration	3
Database building phase (G_{db})	20
Number of independent runs	20

Table 6.2 Statistical results (shown in terms of the mean ± standard deviation averaged over 20 independent runs) obtained by GS-SOMA on ten single-objective benchmark problems

Prob.	GA	SS-SOMA-GP	SS-SOMA-PR	SS-SOMA-RBF	GS-SOMA
F1	1.24e+01±9.50e−01 (+)	6.43e+00±9.73e−01 (+)	**1.39e+00±1.93e−01** (−)	4.91e+00±7.57e−01 (+)	3.58e+00±5.09e−01
F2	4.58e+01±8.61e+00 (+)	1.79e+01±8.58e+00 (+)	**1.18e−02±2.78e-02** (≈)	7.49e−01±8.98e-02 (+)	**2.20e−03±4.60e−03**
F3	4.10e+02±1.01e+02 (+)	**2.99e+01±7.73e-01** (−)	6.73e+01±2.55e+01 (+)	4.90e+01±2.92e+01 (≈)	4.63e+01±2.92e+01
F4	−5.46e+01±3.01e+01 (+)	−1.19e+02±1.87e+01 (≈)	−1.19e+02±1.23e+01 (≈)	**−1.65e+02±1.86e+01** (−)	−1.26e+02±1.60e+01
F5	1.26e+02±2.85e+00 (+)	1.19e+02±4.29e+00 (≈)	**5.67e+01±3.79e+00** (−)	1.21e+02±2.61e+00 (+)	1.19e+02±3.05e+00
F6	−9.57e+01±9.43e+00 (+)	−1.02e+02±2.99e+00 (+)	−1.06e+02±2.45e+00 (+)	−1.03e+02±2.43e+00 (+)	**−1.12e+02±1.05e+00**
F7	7.29e+02±5.92e+01 (+)	6.81e+02±7.23e+01 (+)	6.42e+02±5.80e+01 (+)	**6.27e+02±7.93e+01** (≈)	**6.07e+02±3.06e+01**
F8	4.83e+02±6.30e+01 (+)	4.52e+02±9.66e+01 (+)	3.94e+02±4.41e+01 (+)	**3.79e+02±3.30e+01** (≈)	**3.25e+02±1.17e+02**
F9	1.02e+03±2.35e+01 (+)	9.42e+02±1.71e+01 (≈)	**9.32e+02±8.26e+00** (−)	9.81e+02±1.43e+01 (+)	**9.42e+02±1.75e+01**
F10	1.51e+03±5.52e+01 (+)	1.26e+03±1.88e+02 (+)	**1.07e+03±1.07e+02** (≈)	1.12e+03±1.16e+02 (+)	**1.01e+03±7.85e+01**

The best solutions are highlighted

6.3 Two-Layer Surrogate-Assisted Particle Swarm Optimization

In this section, we present a different approach to using multiple surrogate ensembles, called a two-layer surrogates-assisted particle swarm optimization (TLSAPSO) algorithm proposed by Sun et al. (2015), in which the global and local surrogate models are trained in two stages and utilized cooperatively to approximate the fitness of a solution. A global radial-basis-function (RBF) surrogate model is trained in the first stage, which is expected to smooth out the local optima of the objective function and guide the population to the region where the global optimum is potentially located. Meanwhile, a local RBF surrogate model is built for each individual using the data in the neighborhood of this individual, which aims to approximate the local fitness landscape as accurately as possible. Note that in TLSAPSO, no local search is employed. Algorithm 4 gives the pseudo code of TLSAPSO. From Algorithm 4, we can see that the initialization process of TLSAPSO is the same to that of the canonical PSO, i.e., all individuals in the initial population will be evaluated using the exact fitness function and used to determine the personal best positions and the best position of the population found so far. After that, before the stopping criterion is met, a global surrogate model GM will be trained using the data in Arc and used to approximate the fitness values of all updated individuals in the population (lines 7–10 in Algorithm 4). Then, each individual will be further approximated on the fitness value by a local surrogate model and updated on the approximated value (lines

11–13 in Algorithm 4). Afterwards, the personal best position of each individual will be updated only when the individual gets a better approximated value than its personal best position and is evaluated using the real objective function. Therefore, it can be understood that probably no individual can be evaluated using the real objective function. Note that all individuals of the current population will be evaluated using the real fitness function if none of them satisfies the condition for updating its personal best position to avoid endless loop. In the following, we will give a detailed description of the main components of TLSAPSO.

Algorithm 4 Pseudo code of TLSAPSO.

1: Set the archive Arc to be a null set, i.e., $Arc = \emptyset$;
2: Initialize velocities and positions of a population pop;
3: Evaluate the fitness of each individual in pop using the real fitness function, and save them to the archive Arc;
4: Determine the personal best position of each individual $\mathbf{pbest}_i$, $i = 1, 2, \ldots, N$;
5: Determine the best position found so far of the population **gbest**;
6: **while** the stopping criterion is not met **do**
7: Train a global surrogate model GM;
8: Update the velocity and position of each individual i using Eqs. (3.38) and (3.39), respectively;
9: **for** each individual i **do**
10: Approximate the fitness individual i using the global surrogate model GM, denoted as $\hat{f}^{GM}(\mathbf{x}_i)$;
11: Train a local surrogate model for individual i, denoted as LM_i;
12: Approximate the fitness of individual i using LM_i, denoted as $\hat{f}^{LM_i}(\mathbf{x}_i)$;
13: $\hat{f}(\mathbf{x}_i) = \min\{\hat{f}^{GM}(\mathbf{x}_i), \hat{f}^{LM_i}(\mathbf{x}_i)\}$
14: **if** $\hat{f}(\mathbf{x}_i) < f(\mathbf{pbest}_i)$ **then**
15: Evaluate individual i using the exact expensive function;
16: Update the personal best position of individual i if its exact fitness value is better than its personal best position;
17: Save it to the archive Arc;
18: **end if**
19: **end for**
20: **if** no individual is evaluated using the exact fitness function **then**
21: Evaluate all individuals in the current population pop using the real fitness function;
22: Update the personal best position of each individual;
23: Save to the archive Arc;
24: **end if**
25: Update the global best position found so far **gbest**;
26: **end while**
27: Output **gbest** and its fitness value;

6.3.1 Global Surrogate Model

Generally, a global surrogate model will be trained using the data distributed in the whole decision space. However, with the population evolves, the individuals will not disperse in the whole decision space. Thus, in TLSAPSO, only part of the data in the archive Arc will be employed for training the global surrogate to reduce the computation time on the one hand, and to ensure the global nature of the surrogate on the other hand. Fig. 6.2 gives an example to show how to determine the region in which the data are chosen for training a global RBF surrogate model. In Fig. 6.2, the samples, denoted in circles in black, are distributed in the decision space while the individuals in the current population, denoted by empty circles, locate in a sub-region of the whole decision space. Therefore, it is not necessary to use all data to train a surrogate because the model is used to approximate the fitness values of the individuals in the current population. Thus, in TLSAPSO, the data in the subspace where the population locates will be used to train the global surrogate model. Let

$$maxd_j = \max\{x_{ij}(t), i = 1, 2, \cdots, N\}, \tag{6.12}$$

$$mind_j = \min\{x_{ij}(t), i = 1, 2, \cdots, N\}, \tag{6.13}$$

where $maxd_j$ and $mind_j$ represents the maximum and minimum values of the current population on j-th dimension of the decision space at the generation. N is the population size. Then we define a subspace:

$$ub_j^{sp}(t+1) = \min\{maxd_j + \alpha(maxd_j - mind_j), u_j\}, \tag{6.14}$$

$$lb_j^{sp}(t+1) = \min\{mind_j + \alpha(maxd_j - mind_j), l_j\}, \tag{6.15}$$

where u_j and l_j are the upper and lower bound of the decision space on dimension j, respectively. ub_j^{sp} and lb_j^{sp} are the maximum and minimum values of the subspace on j-th dimension. α is a spread coefficient in the range of [0, 1] to allow the data that are outside the space occupied by the current population to be used to train the surrogate model, as shown in Fig. 6.2.

6.3.2 Local Surrogate Model

As discussed in Sect. 6.2, the global surrogate model may not be able to approximate the fitness of a solution accurately because a poor data distribution, or a limited number of training data Jin (2005). However, if an adequate number of samples are available around a solution, it is desirable to train a local surrogate model that can estimate the fitness of this solution more accurately. Therefore, we propose to train a local surrogate model for each solution using the data in its neighborhood, if there is a sufficient amount of data. The size of a particle's neighborhood is adaptively set as follows:

$$ns_j = \beta(ub_j - lb_j) \tag{6.16}$$

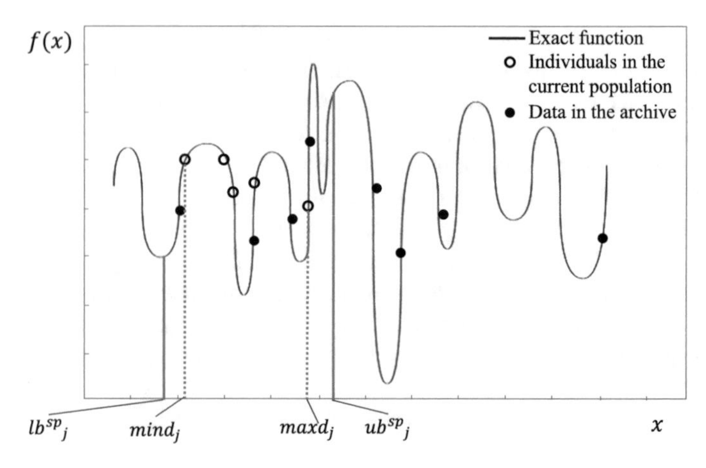

Fig. 6.2 Determination of the region of data samples for training the global surrogate model

Note that a local model can be trained only when an adequate number of training data around the individual is available, and local models may be different with each other because the data used to train the local models are usually not the same.

6.3.3 *Fitness Estimation*

Till now, each solution can be approximated using the global surrogate model or a local surrogate model if adequate samples are available around this solution. The next question is which approximated fitness value should be adopted as the final fitness value of the solution. Fig. 6.3 gives a simple example showing the idea to determine the fitness. From Fig. 6.3, we can see that a global surrogate model, denoted by a red dot dash line, is built in the sub-space $[lb^{sp}, ub^{sp}]$. The blue dot line represents the local surrogate model for solution x_1. As there is inadequate number of samples in the neighborhood of solution x_2, we can see that no local surrogate model can be trained for solution x_2.

For a solution that does not have a sufficient amount of training data in its neighborhood, no local surrogate model can be trained to approximate its fitness. In such a situation, the fitness value of this particle will be approximated by the global surrogate model only. For example, in Fig. 6.3, the approximated fitness value of solution x_2, $\hat{f}(\mathbf{x}_2)$ will be set to $\hat{f}^g(\mathbf{x}_2)$. In case both global and local approximations are all available, the lower one will be chosen to be the final approximated value of the solution, i.e., $\hat{f}(\mathbf{x}) = \min\{\hat{f}^g(\mathbf{x}), \hat{f}^l(\mathbf{x})\}$. In Fig. 6.3, $\hat{f}(\mathbf{x}_1) = \hat{f}^g(\mathbf{x}_1)$.

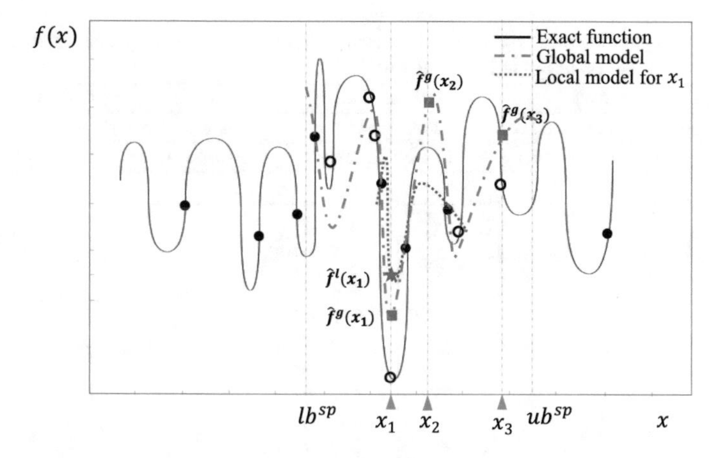

Fig. 6.3 Fitness estimation of a solution based on global and local surrogates

6.3.4 *Surrogate Management*

To improve the contribution of surrogate models, a number of solutions will be evaluated using the exact expensive function and used to update the global and local surrogate models. In particle swarm optimization, the personal best position plays a central role as well as the global best position to ensure that the population converges to the optimum. Therefore, in the surrogate-assisted particle swarm optimization method, the fitness values of all personal best positions are ensured to be exactly evaluated using the real expensive objective function, to make sure the the global best position is always evaluated by the real fitness function. To be specific, if $\hat{f}(\mathbf{x}_i) < f(\mathbf{pbest}_i)$, the solution $\mathbf{x}_i$ will be evaluated using the real objective function, and used to update the personal best position. All solutions that have been evaluated using the real objective function will be saved to the archive Arc for updating the global and local surrogate models.

It is possible that no solution in the current population is better than its personal best position. In this case, all individuals of the current population will be evaluated using the exact fitness function to avoid premature convergence.

6.3.5 *Experimental Results and Discussions*

To evaluate the performance of TLSAPSO and verify the hypothesis that the global surrogate model is able to smooth out the local optima while the local ones are expected to accurately approximate the local fitness landscape, experiments are conducted on two test problems, one unimodal [F1 in Suganthan et al. (2005)] and the other multimodel [F6 in Suganthan et al. (2005)]. The search dimension of both prob-

Table 6.3 Comparative results (shown as the mean optimal value of 10 independent runs±standard deviation) on F1 and F6 in Suganthan et al. (2005)

Prob.	Opt.	CPSO	CPSO-L	CPSO-G	TLSAPSO
F1	−4.50e+02	2.71e+03±2.17e+03	2.79e+03±2.20e+03	−2.57e+02±2.51e+02	**−4.50e+02±3.90e-03**
F6	3.90e+02	2.64e+08±2.59e+08	1.64e+08±2.23e+08	4.65e+06±8.81e+06	**1.57e+03±1.75e+03**

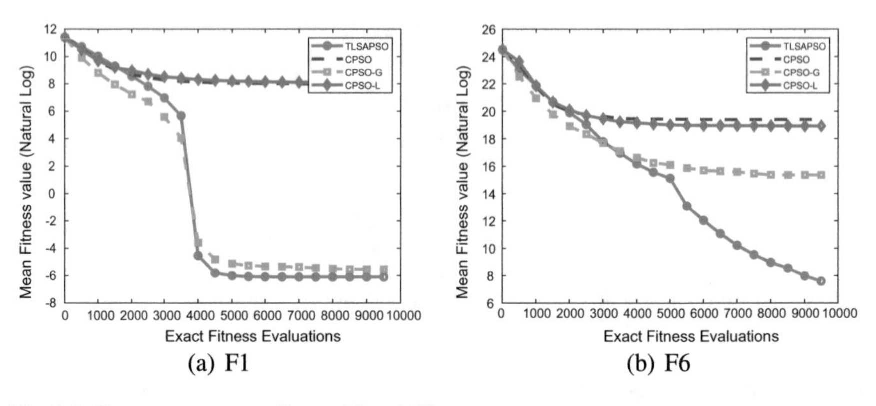

Fig. 6.4 The convergence profiles on F1 and F6

lems is set to $D = 30$. The parameters of the TLSAPSO used in the experiments are set as follows: the size of the swarm is 60, both the cognitive and social parameters are both set to 2.05. The maximum number of fitness evaluation is set to 10000. All algorithms perform 10 independent runs on each problem. Table 6.3 shows the comparative results and Fig. 6.4 gives the convergence profile of four different algorithms on F1 and F6 test problems. In Table 6.3 and Fig. 6.4, CPSO represents the PSO with a constriction factor without using a surrogate model, CPSO-L and CPSO-G are CPSO with local surrogate models only and CPSO with global surrogate model only, respectively. From the results given in Table 6.3 and Fig. 6.4, the following conclusions can be drawn. First, local surrogate models only are not able to speed up the PSO-based search, neither on unimodal nor multimodal optimization problems. Second, a global surrogate model is able to accelerate the PSO search on both unimodal and multimodal problems, however, the search may get stick in a local optimum on multimodal problems. Finally, a combination of a global surrogate and local surrogates can take advantage of the benefits of both the global and local models, and work well on both unimodal and multimodal optimization problems.

6.4 Committee Surrogate Assisted Particle Swarm Optimization

In the above sections, we show that a combination of multiple surrogates are helpful to enhance the search performance in data-driven surrogate-assisted optimization. In this section, we will show that using multiple surrogate models such as an ensemble in data-driven EAs can not only improve the prediction quality, but also provide the uncertainty information Guo et al. (2018) that is very useful for surrogate management. Specifically, we present a committee-based surrogate-assisted PSO (termed CAL-SAPSO) Wang et al. (2017a) for single-objective optimization. In CAL-SAPSO, multiple surrogate models are built for both global and local surrogate ensemble models, and three different infill sampling criteria are designed for improving the accuracy of the local and global models. A PSO is employed to find the infilling samples for each criterion.

6.4.1 Committee of Surrogate Models

In CAL-SAPSO, three heterogeneous surrogate models (PR, RBFN and GP model) are employed to constitute a committee of surrogates in order to maximize the model diversity. The settings of those three surrogate models ($\hat{f}_1$, $\hat{f}_2$, and $\hat{f}_3$) are shown below:

- $\hat{f}_1$: a second-order PR model (see more details in Sect. 4).
- $\hat{f}_2$: an RBFN model with $2n + 1$ Gaussian radial basis functions in the hidden layer, where n is the number of decision variables (see more details in Sect. 4).
- $\hat{f}_3$: a GP model Olea (2000) with a Gaussian kernel, where the Hooke & Jeeves method Kowalik and Osborne (1968) is used for its hyperparameter optimization (see more details in Sect. 4).

The above three surrogate models are combined for fitness prediction according to the following equation:

$$\hat{f}_{ens}(\mathbf{x}) = w_1 \hat{f}_1 + w_2 \hat{f}_2 + w_3 \hat{f}_3, \tag{6.17}$$

where w_1, w_2, and w_3 are the ensemble weights of the three surrogate models. The weighted aggregation method in Goel et al. (2007) is adopted, where w_i depends on the RMSEs (e_1, e_2, and e_3) of three models:

$$w_i = 0.5 - \frac{e_i}{2(e_1 + e_2 + e_3)}. \tag{6.18}$$

As GS-SOMA in Sect. 6.2, a global and a local surrogate ensemble model are built to balance exploration and exploitation in CAL-SAPSO.

- The global surrogate ensemble $\hat{f}^g_{ens}$ is built in the whole decision space, which is trained on $5n$ diverse data points from the whole data $\mathcal{D}$.
- The local surrogate ensemble $\hat{f}^l_{ens}$ is built in the local area trained with the top 10 % best data points.

6.4.2 Infill Sampling Criteria

As discussed in Sect. 5, the samples with the best predicted fitness and the largest uncertainty are promising samples. Usually, it is easy to sample the best predicted solution, but the definition of the prediction uncertainty may differ in different algorithms. The most popular way to measure the prediction uncertainty is the confidence level provided by the GP models (see Chap. 4), however, GP models are of high computation complexity. As CAL-SAPSO employs a committee of surrogate models, the prediction uncertainty can be measured in an alternative way, i.e., the discrepancies between the prediction of the surrogate models. To balance the exploration and exploitation, CAL-SAPSO adopts three different infill sampling criteria.

The first criterion is to select the most uncertain sample for exploration. It is based on the idea of query by committee (QBC) Seung et al. (1992) by defining the largest disagreement between the surrogate models as the measurement for uncertainty. Thus, $\mathbf{x}^u$ can be determined by optimizing the following criterion:

$$\mathbf{x}^u = \arg\max_{\mathbf{x}}(\max(\hat{f}_i(\mathbf{x}) - \hat{f}_j(\mathbf{x}))). \tag{6.19}$$

The second criterion is to select the most promising sample $\mathbf{x}^f$ predicted by the global surrogate model $\hat{f}^g_{ens}$, which can be written as:

$$\mathbf{x}^f = \arg\min_{\mathbf{x}} \hat{f}^g_{ens}(\mathbf{x}). \tag{6.20}$$

To further enhance the accuracy of the surrogate models in the promising regions, the third criterion is to select the most promising sample $\mathbf{x}^{ls}$ predicted by the local model $\hat{f}^l_{ens}$, which can be written as:

$$\mathbf{x}^{ls} = \arg\min_{\mathbf{x}} \hat{f}^l_{ens}(\mathbf{x}). \tag{6.21}$$

With the above three criteria, three types of samples $\mathbf{x}^u$, $\mathbf{x}^f$, and $\mathbf{x}^{ls}$ are selected and reevaluated by the real objective function to update the surrogate models.

6.4.3 Overall Framework

Different from most existing algorithms with individual-based model management that choose samples within their population according to the infill sampling criterion in each generation, CAL-SAPSO employs a canonical PSO algorithm Shi and Eberhart (1998) (see more details in Chap. 3) to separately optimize those three infill sampling criteria as described in Eqs. (6.19)–(6.21). In other words, the whole process of CAL-SAPSO is a loop of optimizing one criterion for infilling samples and updating the surrogate models with augmented training data.

Algorithm 5 Pseudo code of CAL-SAPSO.

Require: Offline data $\mathcal{D}$ with d samples.
1: **while** stopping criterion is not met **do**
2: **if** No new data has been sampled or the previous $\mathbf{x}^f$ is the best in $\mathcal{D}$ or the previous $\mathbf{x}^{ls}$ is not the best in $\mathcal{D}$ **then**
3: Re-build $\hat{f}^g_{ens}$ on a diverse subset of $\mathcal{D}$.
4: Search $\mathbf{x}^u$ by optimizing Equation (6.19 using PSO.
5: Evaluate $\mathbf{x}^u$ using the real objective function.
6: Add $\mathbf{x}^u$ to $\mathcal{D}$
7: Re-build $\hat{f}^g_{ens}$ on a diverse subset of $\mathcal{D}$.
8: Search $\mathbf{x}^f$ by optimizing Equation (6.20 using PSO.
9: Evaluate $\mathbf{x}^f$ using the real objective function.
10: Add $\mathbf{x}^f$ to $\mathcal{D}$.
11: **else**
12: Re-build $\hat{f}^l_{ens}$ on the best 10% samples in $\mathcal{D}$.
13: Search $\mathbf{x}^{ls}$ by optimizing Equation (6.21 using PSO.
14: Evaluate $\mathbf{x}^{ls}$ using the real objective function.
15: Add $\mathbf{x}^{ls}$ to $\mathcal{D}$.
16: **end if**
17: **end while**
Ensure: optimum in $\mathcal{D}$.

CAL-SAPSO controls the three different types of samples to balance the exploration and exploitation. As Algorithm 5 shows, a global or local surrogate model only is used until it cannot produce better solution (i.e. $\mathbf{x}^f$ or $\mathbf{x}^{ls}$ is not the best in $\mathcal{D}$), then CAL-SAPSO switches to the other model. Note that optimizing Eqs. (6.19)–(6.21) is performed separately using PSO in CAL-SAPSO.

6.4.4 Experimental Results on Benchmark Problems

To study the behavior of CAL-SAPSO, we compare CAL-SAPSO with three different data-driven EAs on five benchmark problems (Ellipsoid, Rosenbrock, Ackley, Griewank, and Rastrigin) with different number of decision variables ($n = 10, 20, 30$). Below is a brief introduction to the compared algorithms.

- GPEME Liu et al. (2014a): a GP-based data-driven EA that uses LCB to choose samples for re-evaluations. Note that the hyperparameter optimization for the GP model was done using GA in its original paper, which is very time-consuming. We only apply the Hooke-Jeeves method Kowalik and Osborne (1968) in the experiment.
- WTA1 Goel et al. (2007): an online data-driven EA that employs weighted surrogate ensemble of the PR, RBFN, and GP models. In WTA1, only the predicted optimum is chosen to be re-evaluated using the real objective function.
- GS-SOMA Lim et al. (2010): an online data-driven EA using a PR model and a ensemble model (PR, RBFN, and GP models).

All the compared algorithms start with $5n$ training data points and end with $11n$ training data points. The optimum obtained by the five algorithms over 20 independent runs are shown in Table 6.4, which are analyzed using the Friedman test and the p-values are adjusted according to the Hommel procedure Derrac et al. (2011b), where CAL-SAPSO is the control method.

In general, CAL-SAPSO outperforms other three compared algorithms on both uni- and multi-modal problems. This might be attributed to the folowing three reasons. Firstly, the surrogate ensemble is able to provide sufficiently robustness fitness predictions. Secondly, the global and local surrogate ensembles are able to better balance the global and local search. Thirdly, the QBC-based infill sampling criterion can economically improve the quality of surrogate models, which is based on the prediction of various surrogate models rather than the similarity of the data.

6.5 Hierarchical Surrogate-Assisted Multi-scenario Optimization

As explained in Chap. 1, the operating scenario of an engineer design may be subject to a large amount of uncertainty Beyer and Sendhoff (2007), and consequently, its performance in different scenarios may vary. Therefore, multi-scenario optimization is highly demanded for real-world engineer design to improve the robustness of the optimal solutions. In this section, we will introduce a hierarchical surrogate-assisted multi-scenario optimization algorithm (termed HSA-MSES) for RAE2822 airfoil design Wang et al. (2018b).

6.5.1 *Multi-scenario Airfoil Optimization*

Different from single-scenario airfoil optimization, multi-scenario airfoil optimization Wang et al. (2018b) considers a variety of scenarios (a range of lifts C_L and mach numbers M) rather than a fix lift C_L and a fix mach number M. Therefore, the formulation for multi-scenario optimization involves scenario variables, decision variables, and objective functions.

The decision variables in the multi-scenario airfoil optimization problem controls the airfoil geometry, which employs the following five Henne-Hicks functions Hicks and Henne (1978) to modify the baseline geometry.

$$f_1 = \frac{x^{0.5}(1-x)}{x^{15x}} \tag{6.22}$$

$$f_2 = \sin(\pi x^{0.25})^3 \tag{6.23}$$

Table 6.4 Optimal solutions, presented in terms of Avg±Std, obtained by CAL-SAPSO, GPEME, WTA1, GS-SOMA, and MAES-ExI, which are analyzed by the Friedman test

Problem	n	CAL-SAPSO	GPEME	WTA1	GS-SOMA
Ellipsoid	10	8.79e−01±8.51e−01	3.78e+01±1.53e+01	3.00e+00±2.02e−03	**1.77e−01±1.65e−01**
Ellipsoid	20	**1.58e+00±4.83e−01**	3.19e+02±9.03e+01	2.17e+01±8.51e+00	9.97e+00±3.41e+00
Ellipsoid	30	**4.02e+00±1.08e+00**	1.23e+03±2.24e+02	8.56e+01±1.17e+01	6.67e+01±1.11e+01
Rosenbrock	10	**1.77e+00±3.80e−01**	2.07e+01±7.44e+00	1.18e+01±2.13e-03	4.77e+00±1.14e+00
Rosenbrock	20	**1.89e+00±3.32e−01**	6.15e+01±2.19e+01	1.00e+01±3.90e−02	6.52e+00±1.13e+00
Rosenbrock	30	**1.76e+00±3.96e−01**	8.42e+01±2.79e+01	1.18e+01±8.19e−01	9.82e+00±1.10e+00
Ackley	10	2.01e+01±2.44e-01	**1.38e+01±2.50e+00**	1.90e+01±1.23e+00	1.84e+01±1.73e+00
Ackley	20	2.01e+01±0.00e+00	1.84e+01±9.19e−01	2.01e+01±0.00e+00	**1.83e+01±1.99e+00**
Ackley	30	1.62e+01±4.13e-01	1.95e+01±4.39e-01	**1.51e+01±6.98e−01**	1.61e+01±3.61e−01
Griewank	10	1.12e+00±1.21e−01	2.72e+01±1.13e+01	1.07e+00±2.04e−02	1.08e+00±1.78e−01
Griewank	20	**1.06e+00±3.66e−02**	1.37e+02±3.21e+01	2.00e+00±4.32e-01	1.17e+00±6.37e−02
Griewank	30	**9.95e−01±3.99e−02**	2.84e+02±5.25e+01	3.22e+00±3.25e−01	2.51e+00±4.06e−01
Rastrigin	10	8.88e+01±2.26e+01	**7.15e+01±1.27e+01**	9.58e+01±3.20e+00	1.05e+02±1.52e+01
Rastrigin	20	**7.51e+01±1.44e+01**	1.69e+02±2.86e+01	1.53e+02±3.12e+00	1.54e+02±4.23e+00
Rastrigin	30	**8.78e+01±1.65e+01**	2.86e+02±3.06e+01	2.54e+02±3.07e+01	2.81e+02±2.66e+01

The p-values are adjusted by the Hommel procedure (CAL-SAPSO is the control method, the significant level is 0.05)

$$f_3 = \sin(\pi x^{0.757})^3 \tag{6.24}$$

$$f_4 = \sin(\pi x^{1.357})^3 \tag{6.25}$$

$$f_5 = \frac{x^{0.5}(1-x)}{x^{10x}} \tag{6.26}$$

In the 2D geometry space, the upper and lower surfaces (y_u and y_l) of the geometry can be written as:

$$y_u = y_u^b + \sum_{i=1}^{5} a_i f_i, \tag{6.27}$$

$$y_l = y_l^b + \sum_{i=1}^{5} b_i f_i, \tag{6.28}$$

where a_i and b_i are ten decision variables to control the shape.

The full scenario space is defined by $C_L \in [0.3, 0.75] \times M \in [0.5, 0.76]$. To optimize the airfoil shape in such a large scenario space, the location of a drag divergence boundary B_{DD} Doherty (2017) is employed in the objective function. The boundary B_{DD} means the drag begins to rapidly increase with increasing M, which can be written as:

$$\frac{\partial C_{dw}}{\partial M} = 0.1, \tag{6.29}$$

where $C_{dw}(M, C_L)$ is the wave drag at scenario (M, C_L). Based on the design aim, we hope the airfoil operating at a high velocity requires a small thrust. Therefore, the objective function is defined to minimize the area between B_{DD} and the extreme scenario $(M_{\max}, C_{L\max})$.

The main challenge of this problem is the expensive detection of B_{DD}, which requires 35×15 CFD simulations in the whole scenario space as shown in Fig. 6.5. Therefore, the multi-scenario airfoil optimization problem is computationally extremely expensive.

6.5.2 *Hierarchical Surrogates for Multi-scenario Optimization*

As explained in the previous subsection, the multi-scenario airfoil optimization problem is an expensive 10-dimensional optimization problem. One full-scenario evaluation needs 35×15 CFD simulations, which makes most existing data-driven EAs unpractical.

The process of calculating the objective function is based on the 2-dimensional C_{dw} landscape using 35×15 CFD simulations, where the C_{dw} landscape can be

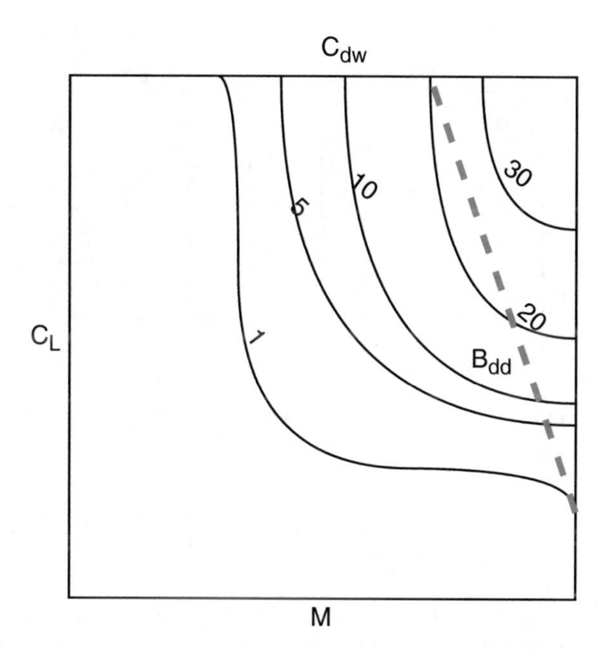

Fig. 6.5 Full-scenario evaluations of an airfoil

approximated by a surrogate model to reduce the computation cost. In other words, one surrogate model is built for the full-scenario evaluation of each candidate design during the optimization process. Thus, we transfer the hard task of building 10-dimensional surrogate model to multiple relatively easier tasks of building 2-dimensional surrogate models.

For each candidate solution, without using 35 × 15 CFD simulations, we assign 22 CFD simulations to approximate its C_{dw} landscape. To improve the accuracy of the approximated C_{dw} landscape near the B_{DD} boundary, we divide the 22 CFD simulations into 10 offline simulations (randomly sampled in the scenario space) $\mathcal{D}_{off}$ and 12 online simulations (controlled by the algorithm) $\mathcal{D}_{on}$.

As we can see in Fig. 6.6, the bottom right corner of the scenario space is of zero drag, and therefore we should not waste time to sample for those scenarios. Consequently, we propose a hierarchical surrogate model consisting a k-nearest neighbors (KNN) classifier based on Mahalanobis distances Altman (1992) and a kriging model with a Gaussian kernel, where the KNN model aims to find the boundary of the zero drag scenarios and the kriging model is meant to approximate the nonzero drag scenarios.

Firstly, a KNN classifier is trained using $\mathcal{D}_{off}$ to find the boundary of zero- and non-zero drag. Then, the classifier predicts zero-drag scenarios $\mathcal{D}_0$, which is added to $\mathcal{D}$ as the training data of the kriging model. Thus, no additional CFD simulations should be performed in the zero-drag area. To make the best use of the online samples, we aim to collect new data around B_{DD} and for improving the global accuracy of the surrogates. Consequently, the following modified LCB criterion is adopted.

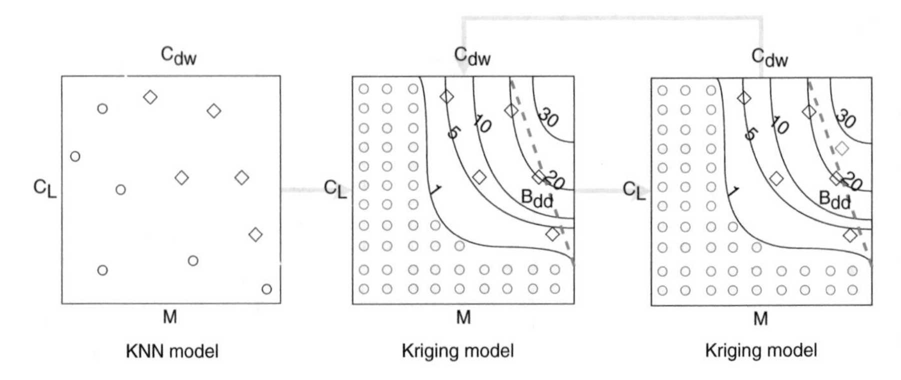

Fig. 6.6 Process of building the hierarchical surrogate model

$$\arg\min_{(M,C_L)} f_{lcb} = |\frac{\partial \hat{C}_{dw}(M, C_L)}{\partial M} - 0.1| - w\sigma_{dw}(M, C_L), \tag{6.30}$$

where $\hat{C}_{dw}(M, C_L)$ and $\sigma_{dw}(M, C_L)$ are the mean and standard deviation of the predicted C_{dw}, w in the following equation is to balance exploitation and exploration.

$$w = \frac{|\frac{\partial \hat{C}_{dw}(M,C_L)}{\partial M} - 0.1|}{\sigma_{dw}(M, C_L)}. \tag{6.31}$$

The online sampling for the kriging model repeats for 12 times. In each iteration, the kriging model is built based on $\mathcal{D} \cup \mathcal{D}_0$, and one new simulation is performed to augment $\mathcal{D}$ in the scenario with the minimal f_{lcb}. When 12 online simulations are done, the kriging model is constructed based on $\mathcal{D} \cup \mathcal{D}_0$, and B_{DD} is then detected based on the kriging model. The area size between B_{DD} and the extreme scenario $(M_{\max}, C_{L\max})$ is calculated as the fitness for the optimizer.

Even though the fitness of each candidate solution can be estimated by a surrogate model using 22 CFD simulations, the approximated fitness evaluation is still very expensive. Therefore, HSA-MSES employs the CMA-ES Hansen and Ostermeier (2001) as its optimizer, because it is based on a small population and well known for its fast convergence for continuous optimization.

As described in Chap. 3, CMA-ES generate λ offspring solutions in each generation g from an n-dimensional multivariate normal distribution $N(\mathbf{m}^g, \sigma^{\mathbf{g}^2}\mathbf{C}^{\mathbf{g}})$, where $\mathbf{m}^g$ is the mean, $\sigma^{\mathbf{g}}$ is the standard deviation, and $\mathbf{C}^{\mathbf{g}}$ is the covariance matrix between the decision variables. Then, the best μ individuals are selected as the parents of the next generation $(g + 1)$ and the distribution will be updated accordingly as follows:

$$\begin{aligned} \mathbf{m}^{g+1} &= \sum\nolimits_{i=1}^{u} w_i \mathbf{x}_i \\ w_i &= \frac{\ln(\mu + 0.5) - \ln i}{\sum_{j=1}^{u} \ln(\mu + 0.5) - \ln j}. \end{aligned} \tag{6.32}$$

The standard deviation and the covariance matrix are updated as described below:

$$\mathbf{C}^{g+1} = (1 - \frac{2}{n^2} - \frac{0.3\lambda}{n^2})\mathbf{C}^g + \frac{2}{n^2}\mathbf{p_c}^{g+1}(\mathbf{p_c}^{g+1})^T + \frac{0.3\lambda}{n^2}\sum_{i=1}^{u} w_i \frac{\mathbf{x}_i - \mathbf{m}^g}{\sigma^g}(\frac{\mathbf{x}_i - \mathbf{m}^g}{\sigma^g})^T \tag{6.33}$$

$$\sigma^{g+1} = \sigma^g \times \exp(\frac{4/n}{1+\sqrt{0.3\lambda/n}}(\frac{\left\|\mathbf{p}_\sigma{}^{g+1}\right\|}{E\left\|N(\mathbf{0},\mathbf{I})\right\|} - 1)) \tag{6.34}$$

where, $\mathbf{p_c}^{g+1}$ and $\mathbf{p}_\sigma{}^{g+1}$ are known as evolution paths and can be adapted according to the following equations:

$$\mathbf{p_c}^{g+1} = (1 - \frac{4}{n})\mathbf{p_c}^g + 1_{\{\|p_\sigma{}^g\|<1.5\sqrt{n}\}}\sqrt{1-(1-\frac{4}{n})^2}\sqrt{0.3\lambda}\frac{\mathbf{m}^{g+1} - \mathbf{m}^g}{\sigma^g}, \tag{6.35}$$

$$\mathbf{p}_\sigma{}^{g+1} = (1 - \frac{4}{n})\mathbf{p}_\sigma{}^g + \sqrt{1-(1-\frac{4}{n})^2}\sqrt{0.3\lambda}(\mathbf{C}^g)^{-0.5}\frac{\mathbf{m}^{g+1} - \mathbf{m}^g}{\sigma^g}, \tag{6.36}$$

where $1_{\{\|p_\sigma{}^g\|<1.5\sqrt{n}\}}$ is an indicator function that evaluates to 1 when $\|p_\sigma{}^g\| < 1.5\sqrt{n}$ or otherwise 0.

HSA-MSES is applied to multi-scenario RAE2822 airfoil design, where the parent and offspring populations have 6 and 11 geometries. After 50 generations, the exact full-scenario drag landscape of the optimal design obtained by HSA-MSES is shown in Fig. 6.7. Compared with the baseline design, B_{DD} has been pushed to the the extreme scenario ($M_{\max}$, $C_{L\max}$).

The optimization process of HSA-MSES uses 13,200 CFD simulations in total, which is comparable to 25 full-scenario evaluations. In fact, such computation cost only affords two generations of full-scenario CMA-ES, which is far from adequate to achieve satisfactory results (Fig. 6.7).

6.6 Adaptive Surrogate Selection

6.6.1 Basic Idea

Similar to the idea in hyper-heuristics, one may start using multiple surrogates at the same probability, and then adapt the probability of using each surrogate according

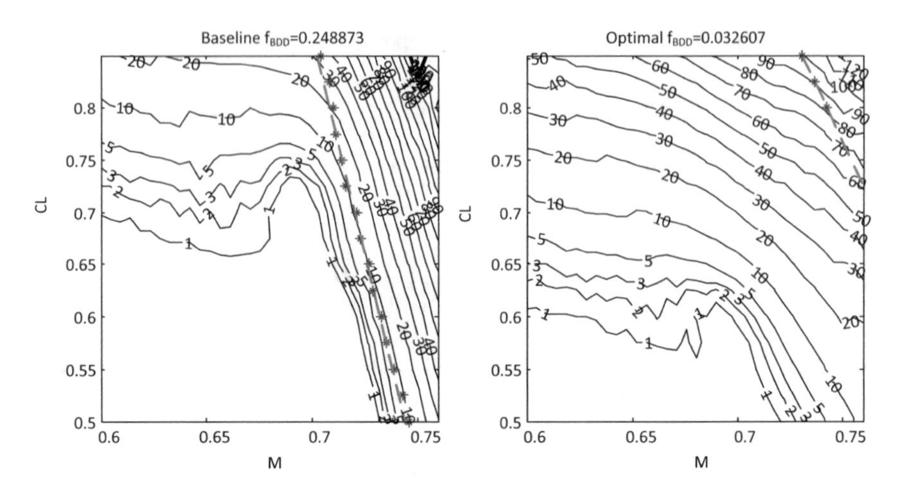

Fig. 6.7 Exact full-scenario drag landscape of the optimal design obtained by HSA-MSES and baseline design

to its performance or contribution to the fitness improvement in the search process. Let us consider a surrogate-assisted memetic algorithm as shown in Fig. 6.8 (Le et al. (2013)). In the first generations, a normal evolutionary algorithm is run for a few generation, in which the real fitness function is used for evaluations. One sufficient data is collected for reliable surrogate modelling, a number surrogates (K) are constructed. In the first round of surrogate-assisted search, one surrogate is randomly selected (at a probability of $1/K$) for each offspring individual and perform local search using the trust region based method. The solution found by the local search will be evaluated using the real fitness function, and the sampled data is added to the data base (Fig. 6.8).

6.6.2 Probabilistic Model for Surrogate Selection

The key component here is to determine the probability at which a surrogate is selected for assisting the trust-region based search. To this end, a performance measure, called evolvability measure of a surrogate M on individual solution $\mathbf{x}$ at generation for a minimization problem, denoted by p_M, is defined for the expected fitness improvement Le et al. (2013):

$$p_M(\mathbf{x}) = \mathrm{E}\left[f(\mathbf{x}) - f(\mathbf{y}^{\mathrm{opt}})|\mathbf{P}^t, \mathbf{x}\right] \tag{6.37}$$

$$= f(\mathbf{x}) - \int_y f(\phi_M(\mathbf{y})) \times P(\mathbf{y}|\mathbf{P}^t, \mathbf{x})d\mathbf{y}, \tag{6.38}$$

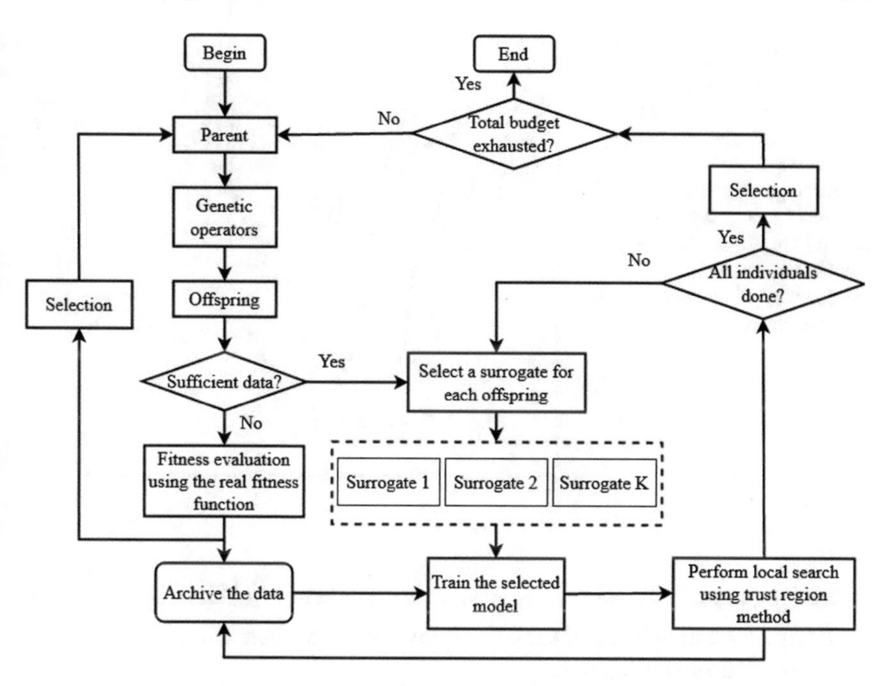

Fig. 6.8 A memetic algorithm in which a surrogate is adaptively selected for each offspring individual based on a probabilistic model and then trust-region based local search is carried out

where $P(\mathbf{y}|\mathbf{P}^t, \mathbf{x})$ is the conditional density function of the genetic operators applied on a parent individual $\mathbf{x}$ to get the offspring solution $\mathbf{y}$ at generation t, i.e., $\mathbf{y} \approx P(\mathbf{y}|\mathbf{P}^t, \mathbf{x})$, where $\mathbf{P}^t$ is the current reproduction pool consisting of solutions after environmental selection.

Let $\phi_M(\mathbf{y})$ be the optimal solution found for individual $\mathbf{y}$ by the local search assisted by surrogate M, which is evaluated using the real fitness function, then $D_M = \{(\mathbf{y}_i, \phi_M(\mathbf{y}_i))\}_{i=1}^K$ is a collection of all resultant solutions that represents the historical contribution of the surrogate M on the given problem. By means of a weighted sampling approach, we can get $\hat{p}_i(\mathbf{x})$ as an estimate of the expected fitness improvement, to define the probability of choosing a sample $(\mathbf{y}_i, \phi_M(\mathbf{y}_i))$. The approximated fitness improvement $p_i(\mathbf{x})$ represents the current probability of reaching offspring $\mathbf{y}_i$ in the database going from solution $\mathbf{x}$ through genetic operator $P(\mathbf{y}|\mathbf{P}^t, \mathbf{x})$, as illustrated in Fig. 6.9. Considering $\{(\mathbf{y}_i, \phi_M(\mathbf{y}_i))\}_{i=1}^K$ as distinct samples from the current distribution $P(\mathbf{y}|\mathbf{P}^t, \mathbf{x})$, the following holds for the estimated improvement for all surrogates $\hat{p}_i(\mathbf{x})$ associated with sample $(\mathbf{y}_i, \phi_M(\mathbf{y}_i))$:

$$\sum_{i=1}^{K} \hat{p}_i(\mathbf{x}) = 1, \tag{6.39}$$

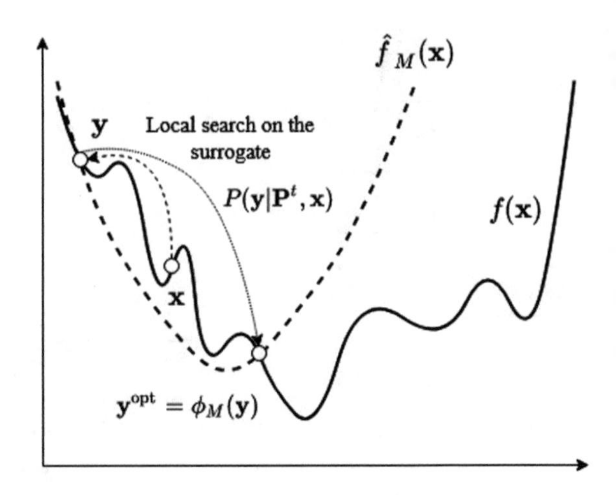

Fig. 6.9 Expected improvement via local search on a surrogate, where $\mathbf{x}$ is a parent individual in the current population, $\mathbf{y}$ is the offspring generated by genetic search, and $\mathbf{y}^{\text{opt}}$ is the optimal solution resulting from local search on the surrogate based on the trust region method

where $\hat{p}_i(\mathbf{x})$ is proportional to $\int_{\mathbf{V}(\mathbf{y}_i)} P(\mathbf{y}_i|\mathbf{P}^t, \mathbf{x})dy$, where $\mathbf{V}(\mathbf{y}_i)$ is an arbitrarily small bin ϵ around solution $\mathbf{y}_i$ and the integral is taken over the interval $[\mathbf{y}_i^{(k)} - \epsilon, \mathbf{y}_i^{(k)} + \epsilon]$ on each dimension k (Fig. 6.9).

The conditional density function $P(\mathbf{y}|\mathbf{P}_t, \mathbf{x})$ is modeled based on the properties of the genetic operators that reflect the current search dynamics. Since calculation of $\int_{\mathbf{V}(\mathbf{y}_i)} P(\mathbf{y}|\mathbf{P}^t, \mathbf{x})dy$ is computationally intensive, weight $\hat{p}_i(\mathbf{x})$ is estimated in the following way:

$$\hat{p}_i(\mathbf{x}) = \frac{P(\mathbf{y}_i|\mathbf{P}^t, \mathbf{x})}{\sum_{j=1}^{K} P(\mathbf{y}_j|\mathbf{P}^t, \mathbf{x})}. \tag{6.40}$$

Using the archived samples in $\phi_M = \{(\mathbf{y}_i, \phi_M(\mathbf{y}_i))\}_{i=1}^{K}$ and weights in Equation (6.40), the expected improvement can be estimated as follows:

$$p_M(\mathbf{x}) = f(\mathbf{x}) - \sum_{i=1}^{K} f(\phi_M(\mathbf{y}_i)) \times \hat{p}_i(\mathbf{x}). \tag{6.41}$$

Based on the estimated fitness improvement, a probability of selecting a particular surrogate can be adapted to maximize the search efficiency. Several experiments performed on benchmark problems as well as an aerodynamic optimization of a car validate that adaptive selection of multiple surrogates performs better than a single surrogate. For example, on a 30-D Ackley function Le et al. (2013), three surrogates, one radial-basis-function (RBF) model, one Gaussian process (GP), and one polynomial (PR) model are adopted as surrogates for assisting local search. In the beginning, each surrogate is selected at a probability of 1/3. As the evolution proceeds, the probability of using the RBF becomes much larger (about 0.8) than that of using the GP or PR models, as shown in Fig. 6.10. Note however this adaptation

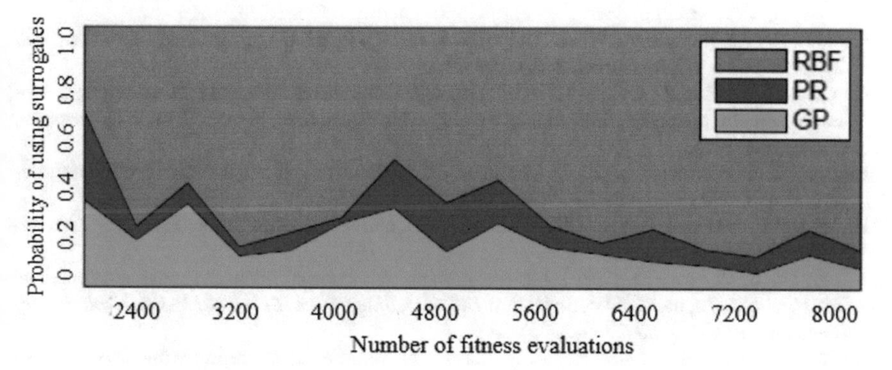

Fig. 6.10 Adaptation of the probability of using the three surrogates

process is problem specific. For example on the 30-D Rosenbrock function, the final probability of selecting the GP or RBF becomes approximately 0.4, while the probability of using the PR model drops to less than 0.2 (Fig. 6.10).

6.7 Summary

In many cases, it is helpful to use multiple surrogates in assisting the evolutionary search or local search in a memetic algorithm. Different surrogates may be used for different purposes, ranging from plain ensemble model for accuracy improvement and prediction reliability detection, to local and global models, to models for that approximates different components in a complex optimization problem. It should be noted that surrogates may be served as other purposes. For example, surrogates can be used to approximate the constraint functions at various controlled levels of complexity Jin et al. (2010) so that the feasible regions can be adapted to facilitate the solution of highly constrained optimization problems.

References

Altman, N. S. (1992). An introduction to kernel and nearest-neighbor nonparametric regression. *The American Statistician*, *46*(3), 175–185.

Beyer, H.-G., & Sendhoff, B. (2007). Robust optimization-a comprehensive survey. *Computer Methods in Applied Mechanics and Engineering*, *196*(33), 3190–3218.

Derrac, J., García, S., Molina, D., & Herrera, F. (2011b). A practical tutorial on the use of non-parametric statistical tests as a methodology for comparing evolutionary and swarm intelligence algorithms. *Swarm and Evolutionary Computation*, *1*(1), 3–18.

Doherty, J. J. (2017). Transonic airfoil study using sonic plateau, optimization and off-design performance maps. In *35th AIAA Applied Aerodynamics Conference* (p. 3056).

Goel, T., Haftka, R. T., Shyy, W., & Queipo, N. V. (2007). Ensemble of surrogates. *Structural and Multidisciplinary Optimization*, *33*(3), 199–216.

Guo, D., Jin, Y., Ding, J., & Chai, T. (2018). Heterogeneous ensemble-based infill criterion for evolutionary multiobjective optimization of expensive problems. *IEEE Transactions on Cybernetics*, *49*(3), 1012–1025.

Hansen, N., & Ostermeier, A. (2001). Completely derandomized self-adaptation in evolution strategies. *Evolutionary Computation*, *9*(2), 159–195.

Hicks, R. M., & Henne, P. A. (1978). Wing design by numerical optimization. *Journal of Aircraft*, *15*(7), 407–412.

Jin, Y., Oh, S., & Jeon, M. (2010). Incremental approximation of nonlinear constraint functions for evolutionary constrained optimization. In *Proceedings of the IEEE Congress on Evolutionary Computation (CEC)* (pp. 2966–2973). IEEE.

Jin, Y. & Sendhoff, B. (2004b). Reducing fitness evaluations using clustering techniques and neural network ensembles. In *Genetic and Evolutionary Computation Conference* (pp. 688–699). Springer.

Jin, Y. (2005). A comprehensive survey of fitness approximation in evolutionary computation. *Soft Computing*, *9*(1), 3–12.

Kowalik, J. S., & Osborne, M. R. (1968). *Methods for unconstrained optimization problems*. North-Holland: Springer.

Le, M. N., Ong, Y. S., Menzel, S., Jin, Y., & Sendhoff, B. (2013). Evolution by adapting surrogates. *Evolutionary Computation*, *21*(2), 313–340.

Lim, D., Jin, Y., Ong, Y.-S., & Sendhoff, B. (2010). Generalizing surrogate-assisted evolutionary computation. *IEEE Transactions on Evolutionary Computation*, *14*(3), 329–355.

Liu, B., Zhang, Q., & Gielen, G. G. (2014a). A Gaussian process surrogate model assisted evolutionary algorithm for medium scale expensive optimization problems. *IEEE Transactions on Evolutionary Computation*, *18*(2), 180–192.

Olea, R. A. (2000). Geostatistics for engineers and earth scientists. *Technometrics*, *42*(4), 444–445.

Seung, H. S., Opper, M., & Sompolinsky, H. (1992). Query by committee. In *Proceedings of the fifth annual workshop on Computational Learning Theory* (pp. 287–294). ACM.

Shi, Y., & Eberhart, R. (1998). A modified particle swarm optimizer. *Proceedings of the IEEE Congress on Evolutionary Computation (CEC)* 69–73.

Suganthan, P. N., Hansen, N., Liang, J. J., Deb, K., Chen, Y.-P., Auger, A., et al. (2005). Problem definitions and evaluation criteria for the CEC 2005 special session on real-parameter optimization. *KanGAL Report*, *2005005*, 2005.

Sun, C., Jin, Y., Zeng, J., & Yu, Y. (2015). A two-layer surrogate-assisted particle swarm optimization algorithm. *Soft Computing*, *19*(6), 1461–1475.

Wang, H., Doherty, J., & Jin, Y. (2018b). Hierarchical surrogate-assisted evolutionary multi-scenario airfoil shape optimization. In *2018 IEEE Congress on Evolutionary Computation (CEC)* (pp. 1–8). IEEE.

Wang, H., Jin, Y., & Doherty, J. (2017a). Committee-based active learning for surrogate-assisted particle swarm optimization of expensive problems. *IEEE Transactions on Cybernetics*, *47*(9), 2664–2677.

Chapter 7
Surrogate-Assisted Multi-objective Evolutionary Optimization

7.1 Evolutionary Multi-objective Optimization

7.1.1 Hypothesis and Methodologies

As discussed in Sect. 1.3, many real-world problems have more than one objective, which cannot simultaneously achieve their optimums due to an inherent conflicting nature between the objectives. These optimization problems are known as multi-objective optimization problems (MOPs) and they usually have a solution set, known as Pareto optimal solution set.

Traditional optimization algorithms such as gradient-based methods are not able to solve MOPs directly. Instead, MOPs need at first be converted into one or multiple single-objective optimization problems (SOPs) (Miettinen, 1999). Consider the MOP defined in Eq. (1.15) and assume there is no equality and inequality constraints, the following approaches can be used to convert an MOP into a single-objective optimization problem.

- *Weighted aggregation methods.* The simplest approach is to linearly aggregate the multiple objective functions into one scalar function using a set of predefined weights:

$$F(\mathbf{x}) = \sum_{i=1}^{m} w_i f_i(\mathbf{x}), \tag{7.1}$$

  where $0 \leq w_i \leq 1$ is the weight for the i-th objective function and satisfies:

$$\sum_{i=1}^{m} w_i = 1. \tag{7.2}$$

  The above approach suffers from several practical limitation. First, one needs to run multiple runs to achieve multiple solutions. Second, Pareto optimal solutions

Y. Jin et al., *Data-Driven Evolutionary Optimization*,
Studies in Computational Intelligence 975,
https://doi.org/10.1007/978-3-030-74640-7_7

on the concave part of the Pareto front of an MOP cannot be obtained. That is to say, for a concave Pareto front, only two extreme solutions can be approximated regardless how the weights are tuned. Third, even if the weights are evenly divided between 0 and 1, the resulting Pareto optimal solutions may distribute unevenly. These are known as non-uniform Pareto fronts.

It has been revealed that the weighted aggregation can achieve all Pareto optimal solutions only if $F(\mathbf{x})$ is a strict monotonically increasing function of each objective $f_i(\mathbf{x})$. This can be achieved by changing the curvature of the aggregated objective function using e.g., the following form:

$$F(\mathbf{x}) = \sum_{i=1}^{m} w_i \left(f_i(\mathbf{x})\right)^{s_i}, \tag{7.3}$$

where $s_i > 1$ is a positive integer.

A more general approach to converting an MOP to an SOP is the so-called weighted metrics method (Miettinen, 1999):

$$F(\mathbf{x}) = \left(\sum_{i=1}^{m} w_i [f_i(\mathbf{x}) - z_i^\star]^p\right)^{\frac{1}{p}}, \tag{7.4}$$

where $p \in (0, \infty)$, $\mathbf{z}^\star$ is the so-called ideal point, where each objective achieves its minimum. If $p = 1$ and $\mathbf{z}^\star = [z_1^\star, z_2^\star, \cdots, z_m^\star]$ is set to the origin, then (7.4) reduces to the linear weighted aggregation function. If $p = \infty$, it becomes the Chebyshev scalarizing function:

$$F(\mathbf{x}) = \max_i [w_i |f_i - z_i^\star|]. \tag{7.5}$$

- *ϵ-constraint method.* This method converts all objectives except for one objective into constraints:

$$\min f_j(\mathbf{x}), \tag{7.6}$$

$$\text{s.t. } f_i(\mathbf{x}) \leq \epsilon_i, i \neq j, \tag{7.7}$$

where ϵ_i is a constraint for the i-th objective function, and the j-th function is to me minimized.

No matter which method is used, traditional optimization methods need to pre-defined either a weight vector or constraints to convert an MOP to an SOP to obtain one Pareto optimal solutions. If multiple Pareto optimal solutions are desirable, one needs to solve multiple SOPs by defining different sets of weights or constraints.

Meta-heuristics, such as evolutionary algorithms are population based search methods, making it possible to achieve a set of Pareto optimal solutions in one optimization run. Indeed, evolutionary multi-objective optimization is probably the most successful area evolutionary optimization since it became popular in the 1990s and a huge number of multi-objective evolutionary algorithms (MOEAs) have been proposed (Fonseca & Fleming, 1995, Deb, 2001, Coello et al., 2002). Although evolutionary multi-objective algorithms can be divided into *a priori*, *a posteriori* and interactive approaches, the majority of MOEAs are *a posteriori* methods that aim to achieve a representative subset of the Pareto optimal solutions, and the size of the obtained solution set equals the size of population, or the size of an external archive. Consequently, all techniques designed in these algorithms aim to accelerate the convergence speed, and improve the diversity, i.e., an even distribution over the whole Pareto front. The widely used performance indicators are described in Sect. 1.5.2.

7.1.2 Decomposition Approaches

The early ideas behind the decomposition approaches to multi-objective optimization (Zhang & Li 2007, Giagkiozis & Fleming, 2015, Cheng et al., 2016) to evolutionary multi-objective optimization are similar to mathematical optimization. In other words, these algorithms convert an MOP into a number of SOPs and then simultaneously solve multiple SOPs using an evolutionary algorithm, typically each individual using a different set of weights, or the weights change over the generations. The weights in the scalarized MOP in Eq. (7.1) can be determined as follows (Murata et al., 1996, 2001):

$$w_i = \frac{\text{random}_i}{\text{random}_1 + \text{random}_2 + \cdots + \text{random}_m}, \tag{7.8}$$

where random_i generates a non-negative random real number. A set of m random weights are then generated for calculating the fitness of any offspring individuals. That is, each individual uses a set of randomly generated weights. Similarly, a set of random weights are generated for a bi-objective (μ, λ)-ES in (Jin et al., 2001a, b):

$$w_1^i = \frac{\text{random}(\lambda)}{\lambda}, \tag{7.9}$$

$$w_2^i = 1 - w_1^i, \tag{7.10}$$

where $i = 1, 2, \cdots, \lambda$, $\text{random}(\lambda)$ generates a uniformly distributed random number between 0 and λ.

A slightly different idea is proposed in (Jin et al., 2001a, b), where in the current generation t, all individuals in the population use the same set of weight, which, however, changes over the generations:

$$w_1(t) = |\sin(2\pi t/F)|, \tag{7.11}$$

$$w_2(t) = 1 - w_1(t), \tag{7.12}$$

where F is the frequency of weight change. This approach is called dynamic weighted aggregation.

A set of evenly distributed weights have also been suggested (Murata et al., 2001):

$$w_1^i = \frac{i-1}{N-1}, i = 1, 2, \cdots, N, \tag{7.13}$$

$$w_2^i = 1.0 - w_1^i. \tag{7.14}$$

where N is the population size. Thus, N sets of weights that are evenly distributed between 0 and 1 are generated and each individual will use one of these evenly distributed weights for fitness evaluation. Illustrative examples for bi-objective and three-objective optimization are shown in Fig. 7.1. This is called the cellular multi-objective genetic algorithm, C-MOGA for short. In addition to the fixed, evenly distributed weight vectors, a neighborhood is defined for choosing parents for generating offspring. For example in Fig. 7.1b, for generating offspring for the individual residing in cell A, two parent individuals in its neighboring cells (those five shaded cells) will be randomly chosen for crossover to generate offspring.

A decomposition based algorithm that became most popular is the multi-objective evolutionary algorithm based on decomposition (MOEA/D) (Zhang & Li, 2007). MOEA/D also uses a set of evenly distributed weights, a neighborhood for reproduction, and a local search strategy. The performance is further improved when the linear weighted sum based scalarizing function is replaced by the Chebyshev scalarizing function, and in particular the penalty boundary intersection (PBI) function:

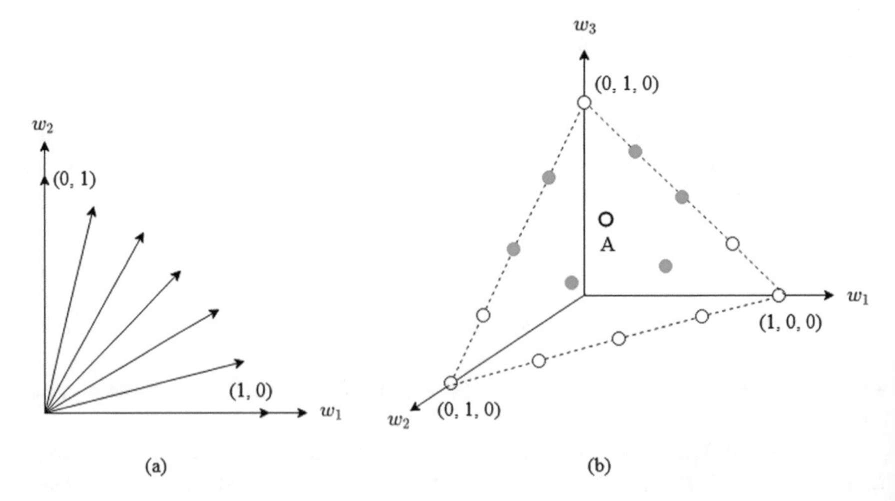

Fig. 7.1 Decomposition based method using evenly distributed weights. **a** Bi-objective optimization; **b** three-objective optimization

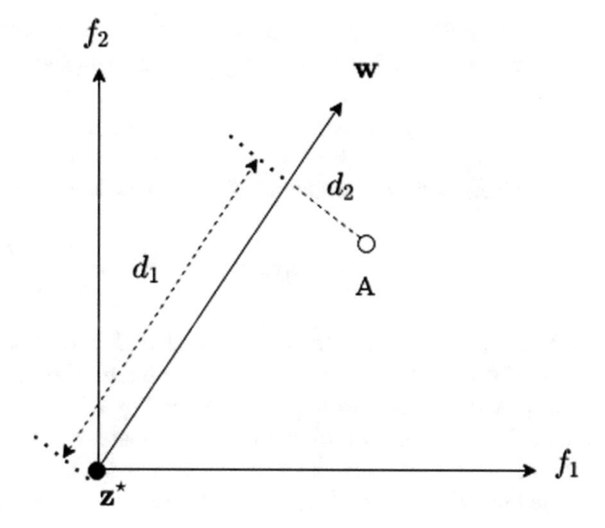

Fig. 7.2 Definition of penalty boundary intersection based scalarizing function

$$F(\mathbf{x}) = d_1 + \theta\, d_2, \tag{7.15}$$

$$d_1 = |\mathbf{f} \frac{\mathbf{w}}{\| \mathbf{w} \|}|, \tag{7.16}$$

$$d_2 = \| \mathbf{f} - d_1 \frac{\mathbf{w}}{\| \mathbf{w} \|} \| . \tag{7.17}$$

where $\mathbf{z}^\star$ is the ideal point, d_1 is the perpendicular Euclidean distance from solution A to the weight vector $\mathbf{w}$, d_2 is the distance from the projected point to the ideal point, and θ is a hyperparameter. Clearly, d_1 reflects the convergence performance of the solution, while d_2 reflects the diversity performance of the solution (Fig. 7.2).

A large number of variants have been proposed for the decomposition based approach, many of which focus on the modification of the generation of the weights (Trivedi et al., 2017), or the definition of the scalarizing function (Chugh, 2019), in order to achieve a set of evenly distributed and converged Pareto set for MOPs with different types of Pareto fronts. More discussions about these two aspects will be provided in Sect. 8.1.

7.1.3 Dominance Based Approaches

For two- or three-objective optimization, dominance based or Pareto based approaches are quite natural since they use the dominance relationship to rank different solutions and then select the better ones. Early ideas include the assignment of a fitness rank according to the number of solutions that dominate a solution (Fonseca & Fleming, 1995, Deb, 2001). For the i-th solution ($i = 1, 2, \ldots, P$) in a population, where P is the population size, assume the number of solutions that can dominate solution i is n_i, then the rank assigned to the i-th solution will be

$$r_i = 1 + n_i. \tag{7.18}$$

The following scalar fitness value can then be assigned to the i-th individual:

$$f_i = f_{\min} + (f_{\max} - f_{\min}) \frac{(r_i - r_{\min})}{(r_{\max} - r_{\min})}, \tag{7.19}$$

where $r_{\min}$ and $r_{\max}$ are the minimum and maximum ranks assigned in the population, respectively, and $f_{\min}$ and $f_{\max}$ are the minimum and maximum fitness values, respectively. An illustrative example is given in Fig. 7.3, where there are five solutions. Of these five solutions, no solutions can dominate solutions 1 and 2, i.e., $n_1 = n_2 = 0$, and solution 3 is dominated by solutions 1 and 2, i.e., $n_3 = 2$, and solution 4 is dominated by solutions 2, i.e., $n_4 = 1$ and solution 5 is dominated by solutions 1, 2, 3, and 3, i.e., n_5 =4. Thus, the assigned to the five solutions are: $r_1 = 1, r_2 = 1, r_3 = 3$, $r_4 = 2$, and $r_5 = 5$. If the minimum and maximum fitness values are set to 0 and 10, respectively, then the fitness value of the five individuals are:

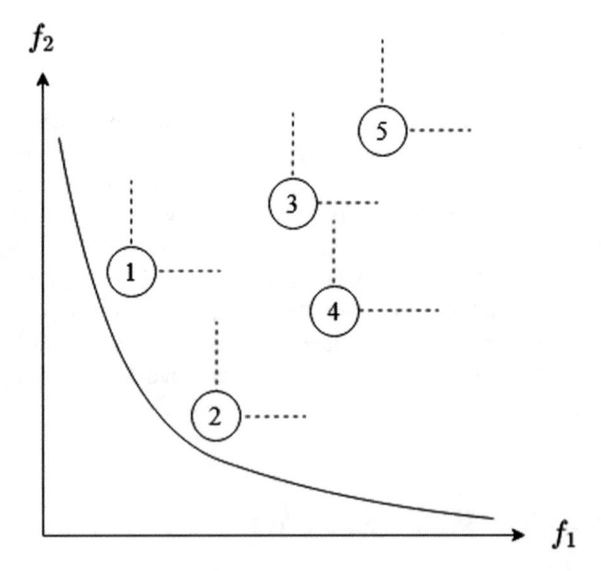

Fig. 7.3 An illustrative example of a population consisting of five solutions. The dominance relationship can be visually judged based on their relative position and a niche share (neighborhood of size d) is defined for calculating the niche count

$$f_1 = 0.0, \tag{7.20}$$
$$f_2 = 0.0, \tag{7.21}$$
$$f_3 = 5.0, \tag{7.22}$$
$$f_4 = 2.5, \tag{7.23}$$
$$f_5 = 10.0. \tag{7.24}$$

In this way, a bi-objective objective optimization is converted into a single-objective optimization problem and any selection methods for single-objective optimization can be used for selection.

However, recall that the goal of multi-objective optimization is to achieve a subset of representative solutions of the true Pareto front. Thus, the diversity (even distribution) of the solutions should also play a role in selection. Thus, the fitness assigned according to the above method must be scaled according to the crowdedness of the solutions. The crowdedness can be measured, e.g., by a niche count, denoted by nc_i. To calculate the niche count, define a niche size d and then calculate its normalized distance to all individuals having the same rank:

$$d_{ij} = \sqrt{\sum_{k=1}^{m} \left(\frac{f_k^{(i)} - f_k^{(j)}}{f_k^{\max} - f_k^{\min}} \right)^2}, \tag{7.25}$$

where m is the number of objectives, typically m equals 2 or 3, $f_k^{(i)}$ and $f_k^{(j)}$ are the k-objective of solution i and j, respectively, and $f_k^{\min}$ and $f_k^{\max}$ are the minimum and maximum of k-th objective of all non-dominated solutions. Then the fitness sharing function is defined by:

$$\text{sh}(d_{ij}) = \begin{cases} 1 - d_{ij}/d, & \text{if } d_{ij} < d, \\ 0, & \text{otherwise.} \end{cases} \tag{7.26}$$

where d_{ij} is the distance between solutions i and j in the objective space. Then, the niche count can be calculated as follows:

$$\text{nc}_i = \sum_{j=1}^{s_i} \text{sh}(d_{ij}), \tag{7.27}$$

where s_i is the number of solutions in the niche of the i-th solution. Note that nc_i is a number larger than 1.

A more popular way for facilitating selection in multi-objective optimization is known as crowded non-dominated sorting (Deb et al., 2002). Non-dominated sorting sorts the individuals in a population into a number of non-dominated fronts. The most straightforward approach to non-dominated sorting can be as follows:

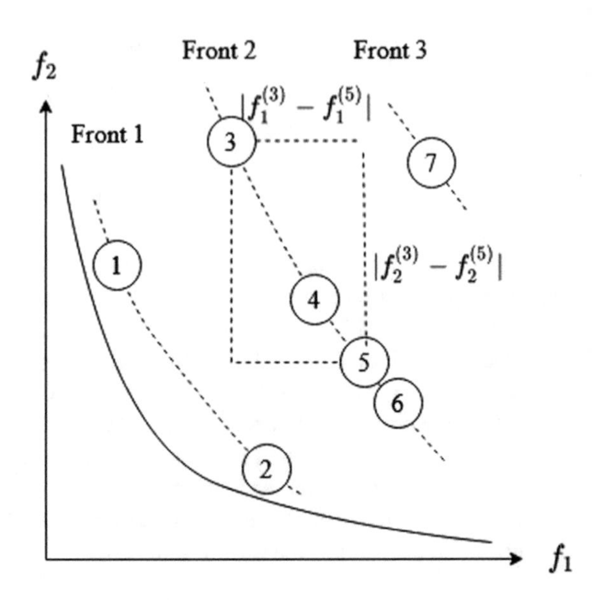

Fig. 7.4 Seven solutions for a bi-objective optimization problems are sorted into three non-dominated fronts. The crowding distance is calculated according to a solution's two neighbors

- Find out the non-dominated solutions and assign a front number 1.
- Remove all non-dominated solutions, and find again the non-dominated solutions in the rest of the solutions, which belong to non-dominated front 2. Assign a front number 2.
- Repeat the above procedure until all solutions in the population are assigned to a non-dominated front.

Figure 7.4 shows one example, in which seven solutions are sorted into three non-dominated fronts. The above non-dominated sorting method is conceptually simple, but computationally intensive, which has a time complexity of $\mathbf{O}(mN^3)$, where m is the number of objectives, and N is the population size. Therefore, many computationally more efficient sorting method has been suggested.

- *Fast non-dominated sorting* (Deb et al., 2002). The fast non-dominated sorting sort the in the following steps:

  1. For each solution p, record the number of solutions that dominate p, which is denoted as n_p, and collect a list of solutions dominated by p, which is denoted as $\mathbf{S}_p$.
  2. For all solutions p whose $n_p = 0$, assign rank 1 to them, and they form the front 1 in the set F_1. Set front counter $i = 1$.
  3. For each solution p in F_i
     For each solution in $\mathbf{S}_p$, $n_p = n_p - 1$.
  4. For the solutions with $n_p = 0$, assign rank $i + 1$, and $i = i + 1$;
  5. Repeat Steps 3–4 until $n_p = 0$ for all solutions, i.e., all solutions are assigned to a non-dominated front.

The computational complexity of the fast non-dominated sorting is $\mathbf{O}(mN^2)$.

- *Efficient non-dominated sorting* (Zhang et al., 2015). The fast non-dominated sorting sorts the population according to one of the objective at first and rank them from the best to worst. The first solution is then assigned to front 1. Then, the second solution is compared with the first, and if not dominated, the second solution is assigned to the same front; otherwise to the second front. This process repeats until all solutions are assigned to a front. The computational trick here is, when a solution is to be assigned, always compare it with the last solution (denoted the n_k-th) on the k-th front, $k = 1, 2, \cdots, K$, n_k is the number of solutions currently on the k-th front, K is the current number of fronts. If not dominated, then compared it with the $n_k - 1$-th solution; if none of the solutions on the k-th front can dominate it, then assign it to the k-th so that becomes the $n_k + 1$-th solution on the k-th front. If the solution is dominated by any of the solutions on the k-th front, compare it to the last solution of the $k + 1$-th front. If the solution is dominated by the last solution on the last front (K-th), then assign it to the $K + 1$-th front. The above process repeats until all solutions are assigned to a front.
Depending on the sorting method for sorting the solutions according to one function before the non-dominated sorting, the time complexity of the efficient non-dominated sorting method can be form $\mathbf{O}(mN\sqrt{N})$ to $\mathbf{O}(mN^2)$. This, the worst case of the efficient non-dominated sorting equals that of the fast non-dominated sorting.

An overview of the popular non-dominated sorting methods can be found in (Tian et al., 2017b).

One the individuals in the population are sorted, a crowding distance for controlling the diversity of the solutions is calculated layer-wise. To distinguish solutions on the same front, the crowding distance of a solution is measured by the distance of its two neighboring solutions on the same front in the objective space:

$$d_i = \sum_{k=1}^{m} \frac{|f_k^{(i-1)} - f_k^{(i+1)}|}{f_k^{\max} - f_k^{\min}}, \tag{7.28}$$

where d_i is the crowding distance of the i-th solution on the front, $f_k^{(i-1)}$ and $f_k^{(i+1)}$ are the two neighboring solutions of the i-th solution on the same front, $f_k^{\min}$ and $f_k^{\max}$ are the minimum and maximum of the k-th objective of all solutions on the same front.

In selection, the larger the crowding distance, the more preferred the solution is. Since calculation of the crowding distance involves to neighboring solutions, for those extreme solutions of each front that have only one neighbor, a sufficiently large crowding distance is assigned to them, so that they are always prioritized in selection,

For example in Fig. 7.4, seven solutions are sorted into three non-dominated fronts. For the two solutions on the first front, a large value is assigned to be their crowding distance. For solution 4 on the second front, the sum of the normalized distance between all objectives of solutions 3 and 5, here $|f_1^{(3)} - f_1^{(5)}| + |f_2^{(3)} - f_2^{(5)}|$, is used as the crowding distance of solution 4.

The elitist non-dominated sorting genetic algorithm, called NSGA-II (Deb et al., 2002), is the most popular dominance based evolutionary algorithm for multi-objective optimization. NSGA-II starts with generating a random parent population P_0 of size N. At generation t, an offspring population O_t is generated using crossover and mutation from the parent population P_t. Here, if a binary GA is used, crossover and flip mutation can be used; the SBX and polynominal mutation can be used for real-coded GA, instead. The offspring population is then combined with parent population to form a combined population R_t. Afterwards, the combined population is sorted into a number of non-dominated fronts based on non-dominated sorting, and the crowding distance is calculated for the individuals on the same front so that they are sorted in an descending order according to the crowding distance. During the environmental selection, the better half of the sorted combined population R_t are selected as the parent population of generation $(t + 1)$. This process repeats until the convergence criterion is met. An illustrative example of environmental selection based on non-dominated sorting and crowding distance sorting is given in Fig. 7.5, and a diagram of the overall NSGA-II algorithm is shown in Fig. 7.6.

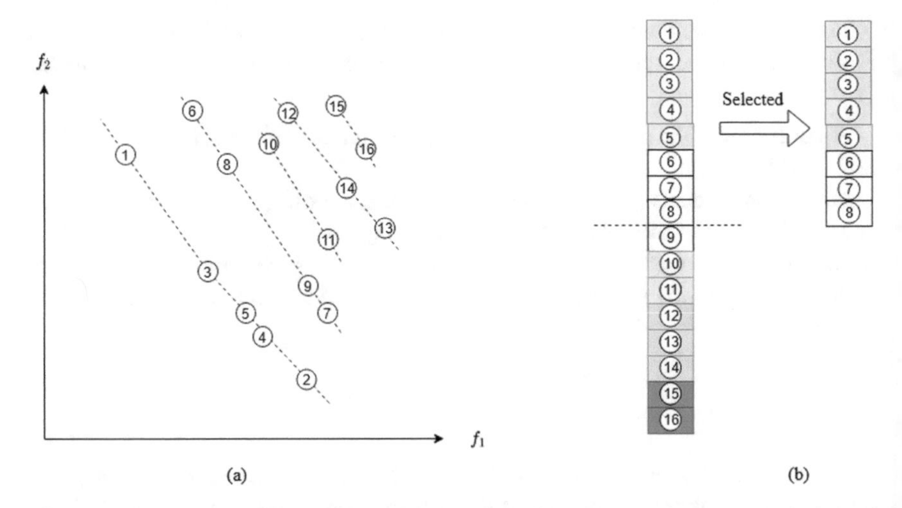

Fig. 7.5 An illustrative example of environmental selection based on non-dominated sorting and crowding distance sorting. There are 16 solutions in the combined population (**a**), which are sorted into five non-dominated fronts. Then a crowding distance is calculated for solutions on the same front so that they can be sorted according in a descending order, as shown in (**b**). Note that the order of the two boundary solutions can be exchanges, since they both have an assigned very large distance. Finally, the top eight solutions are selected as the parents of the next generation

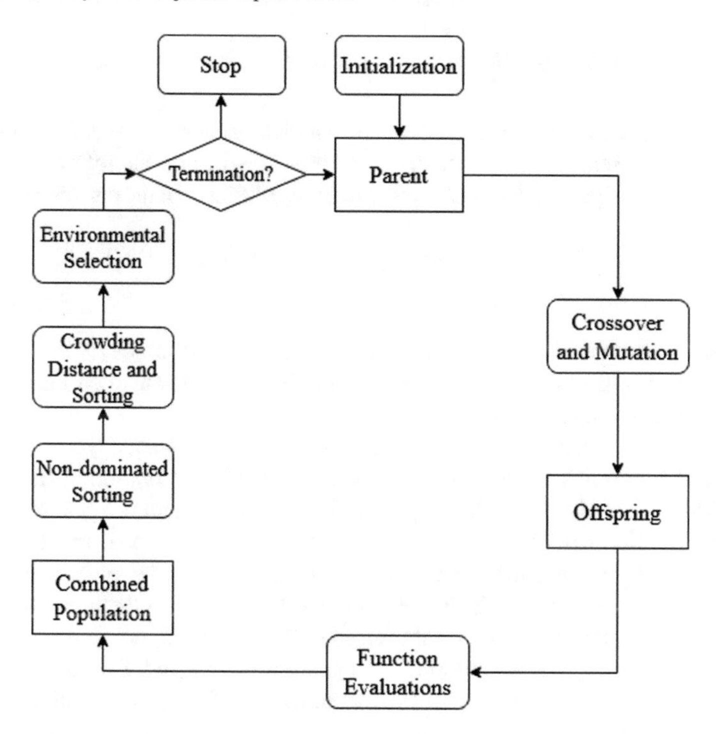

Fig. 7.6 A diagram of the NSGA-II

7.1.4 Performance Indicator Based Approaches

Most performance indicators in multi-objective optimization calculate a scalar value to measure the accuracy and diversity of a non-dominated solution set. This makes it possible to use a performance indicator to compare the quality of a solution and use this as the fitness of a solution. The question that needs to be answered is, how to measure the quality of a single solution using a performance indicator, while it is meant for measuring the quality of a solution set.

A general indicator-based evolutionary algorithm (IBEA) uses a binary performance indicator to compare solutions pairwise and then assigns a fitness to each individual as done in Sect. 7.1.3 for environmental selection (Zitzler & Künzli, 2004). The main difference between IBEA and the dominance based method is that no additional measure for diversity needs to be considered, since diversity has already been taken into account in the binary performance indicator. The binary performance indicator can be seen as a generalized dominance relationship, if the binary performance indicator is dominance preserving. By dominance preserving, it is meant that for any solutions $\mathbf{x}_1$, $\mathbf{x}_2$, and $\mathbf{x}_3$, if $\mathbf{x}_1 \prec \mathbf{x}_2$, then $I(\mathbf{x}_1, \mathbf{x}_2) < I(\mathbf{x}_2, \mathbf{x}_1)$, and if $\mathbf{x}_1 \prec \mathbf{x}_2$, then $I(\mathbf{x}_3, \mathbf{x}_1) \geq I(\mathbf{x}_3, \mathbf{x}_2)$ for all $\mathbf{x}_1, \mathbf{x}_2, \mathbf{x}_3$. To this end, the following binary additive ϵ-indicator indicator is dominance preserving:

$$I_{\epsilon^+}(A, B) = \min_{\epsilon} \{\forall \mathbf{x}_2 \in B \, \exists \mathbf{x}_1 \in A : f_i(\mathbf{x}_1) - \epsilon \leq f_i(\mathbf{x}_2) \text{ for } i \in \{1, \cdots, n\}\}. \tag{7.29}$$

Actually, several other binary performance indicator can also be used, as presented in (Hansen & Jaszkiewicz, 1998). Given the binary indicator, the following method can be used to assign a scalar fitness value to each solution $\mathbf{x}$ in the current population P:

$$F(\mathbf{x}) = \sum_{\mathbf{x}' \in P \backslash \mathbf{x}} -e^{-I(\mathbf{x}', \mathbf{x})}, \tag{7.30}$$

where $\mathbf{x}'$ is any solution in the current population P except for $\mathbf{x}$.

Then, the multi-objective minimization problem can be converted into a single-objective maximization problem.

Another early performance based algorithm is the steady-state $(\mu + 1)$-EMOA, also called $\mathcal{S}$ metric selection EMOA (SMS-EMOA) was proposed in (Emmerich, 2005). This is a steady-state multi-objective algorithm, in that in each generation, only the worst solution will be discarded and replaced with a newly generated offspring. This is done by sorting the current population into a number of non-dominated fronts using non-dominated sorting, and then, one solution on the last front should be removed. To determine which solution should be removed, one can calculate the change of the hypervolume if the particular solution is removed. Denote the solution set on the last front (the worst one) as I, then, the contribution of solution $\mathbf{x}$ can be calculated as follows:

$$F(\mathbf{x}) = \mathcal{S}(I) - \mathcal{S}(I \backslash \mathbf{x}), \tag{7.31}$$

where $\mathcal{S}(I)$ and $\mathcal{S}(I \backslash \mathbf{x})$ are the hypervolume of solution set R and the set after removing solution $\mathbf{x}$. Thus, the solution having the minimum $F(\mathbf{x})$ can be removed. Figure 7.7 shows an example of calculating an solution's contribution to the hypervolume of a solution set. In this example, the non-dominated front consists of four solutions, i.e., $I = \{x_1, x_2, x_3, x_4\}$, and R is the reference point. Then, reduction of the hypervolume by removing one of the four solutions can be calculated by computing the area of the shaded region on the top right of each solution.

A comprehensive survey of indicator based multi-objective evolutionary algorithms is given in (Falcon-Cardona & Coello Coello, 2020).

7.2 Gaussian Process Assisted Randomized Weighted Aggregation

7.2.1 Challenges for Surrogate-Assisted Multi-objective Optimization

Although there have been many evolutionary multi-objective optimization algorithms which can be employed for surrogate-assisted multi-objective optimization, it is still

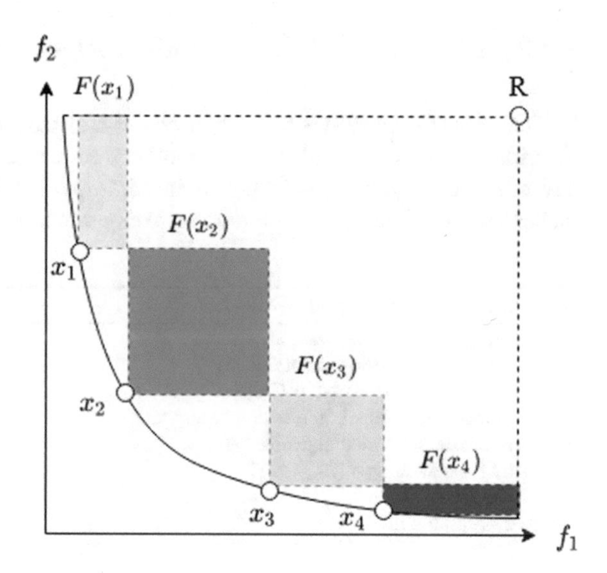

Fig. 7.7 An illustrative example of each solutions contribution to the hypervolume of a solution set

very challenging to design effective surrogate-assisted multi-objective evolutionary algorithm. The main challenges come from the multiple objective functions in MOPs, as summarized below:

- **Approximation errors**: It is relatively straightforward to build one surrogate model for each objective function to replace the real function evaluations, however, the accumulative approximation error of multiple surrogate models may seriously mislead the multi-objective solver. To alleviate this negative effect, the original MOP can be converted into an SOP based on an aggregation function or a performance indicator, as described in the previous Sections.
- **Diversity maintenance**: Different from single-objective optimization, multi-objective optimization requires to search a solution set with good convergence as well as diversity. Therefore, many existing surrogate-assisted evolutionary algorithms for SOPs cannot be directly extended to MOPs, where the diversity maintenance should be more carefully considered.

In this section, we introduce a popular surrogate-assisted multi-objective evolutionary algorithm (termed ParEGO), where an MOP is converted into a number of SOPs and a well-known surrogate-assisted single-objective optimization algorithm, i.e., the efficient global optimization (EGO) (Jones et al., 1998), is employed. Note that EGO also adopts infill criteria similar to those acquisition functions in Bayesian optimization.

7.2.2 Efficient Global Optimization Algorithm

EGO (Jones et al., 1998) is a widely used GP-assisted single-objective optimization algorithm for expensive black-box problems. As mentioned in Sect. 4.2, the main advantage of GP models is its uncertainty information [(i.e. ε in Eq. (4.24)]. EGO makes use of ε to balance exploration and exploitation for the search.

Algorithm 1 Pseudo code of EGO.

1: Sampling initial training data $\mathcal{D}$.
2: **while** stopping criterion is not met **do**
3: Build a GP model based on $\mathcal{D}$.
4: Optimize the acquisition function to find $\mathbf{x}$.
5: Evaluate $\mathbf{x}$ for its real objective value y.
6: Add the new sample $(\mathbf{x}, y)$ to $\mathcal{D}$.
7: **end while**
Output: Optimum in $\mathcal{D}$.

In fact, EGO is a classic online data-driven optimization algorithm, which includes the steps of offline data initialization, surrogate model building, and online data sampling, similar to Bayesian optimization.

As shown in Algorithm 1, EGO starts with collecting an initial training data set $\mathcal{D}$, which is usually generated by the Latin hypercube sampling (LHS) method (Husslage et al., 2011). After a GP model is built based on the training data $\mathcal{D}$, an optimization algorithm is applied to solve an acquisition function, as described in Sect. 5.4 (such as EI, PI, and LCB) to determine the new online sample $\mathbf{x}$ for a real function evaluation. The evaluated sample is added to $\mathcal{D}$ and the GP model is updated. Again, the acquisition function is optimized for identify a new sample, and new sample is evaluated using the real objective function to augment $\mathcal{D}$ until the computation budget is exhausted. Finally, EGO outputs the optimum in $\mathcal{D}$ as its result.

7.2.3 Extension to Multi-objective Optimization

ParEGO (Knowles, 2006) can be seen as a variant of EGO for solving MOPs. The main idea of ParEGO is to repeated convert an MOP into a set of SOPs and then employs EGO to solve the SOPs. A flowchart of ParEGO is shown in Fig. 7.8, which contains a loop of assigning problems, building a GP model, optimizing the acquisition function, and sample a new solution.

After the offline sampling for $\mathcal{D}$ using LHS, ParEGO assigns one converted SOP to the EGO solver, where the following augmented Chebyshev function is employed for the transformation:

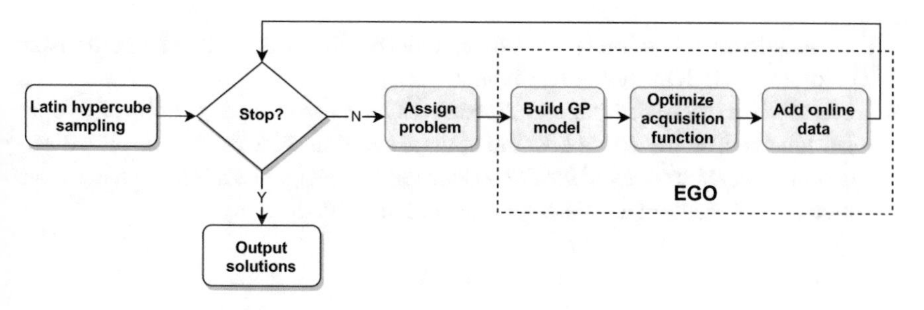

Fig. 7.8 A flowchart of ParEGO

$$f_\lambda(\mathbf{x}) = \max_{j=1}^{m}(\lambda_j f_j(\mathbf{x})) + \rho \sum_{j=1}^{m} \lambda_j f_j(\mathbf{x}), \tag{7.32}$$

where λ is an m-dimensional weight vector ($\sum_{j=1}^{m} \lambda_j = 1$), and ρ is a small value (0.05). To achieve a set of diverse solutions, the MOP is converted into SOPs using a number of uniformly distributed weight vectors at the very beginning of ParEGO. In each iteration, a weight vector λ is randomly chosen from the pre-defined weight set for Eq. (7.32).

Then, ParEGO adopts a modified EGO to optimize the assigned SOP. To build a GP model to approximate the assigned SOP, the fitness value of the samples in the training data $\mathcal{D}$ is calculated according to Eq. (7.32). ParEGO uses EI as the acquisition function for the assigned SOPs and an EA optimizer to optimize EI to find a new sample for a real function evaluation. It is worth noting that the EA optimizer restarts for a different SOP during the whole process of ParEGO. In other words, for each assigned SOP, the EA optimizer starts with a random population and returns a chosen sample to enrich $\mathcal{D}$.

When all the computation budget is exhausted, ParEGO outputs the non-dominated solutions in $\mathcal{D}$ as the final result.

ParEGO can be seen as a straightforward implementation of an evolutionary multi-objective Bayesian optimization algorithm for solving expensive MOPs. The source code[1] is available in PlatEMO (Tian et al., 2017a). However, such a simple methodology suffers from several weaknesses as stated below.

- **Random decomposition**: ParEGO does not use any sophisticated evolutionary multi-objective optimization techniques as previously described. Instead, it uses a random decomposition method to convert an MOP into SOPs. Therefore, ParEGO may fail to obtain some optimal solutions.
- **Low efficiency**: In fact, ParEGO sequentially solves a set of SOPs and the SOPs are treated as independent. The GP models for these SOPs are built separately, and

[1]https://github.com/BIMK/PlatEMO/tree/master/PlatEMO/Algorithms/Multi-objective %20optimization/ParEGO.

their acquisition function is solved separately without any knowledge transfer. Therefore, ParEGO is not very efficient.

- **Lack of diversity maintenance**: Since ParEGO is based on single-objective optimization, its diversity control for the optimal solution set is weak. The only diversity maintenance strategy in ParEGO is the random weight vector in each iteration, which cannot result in a satisfactory performance on diversity.

7.3 Gaussian Process Assisted Decomposition-Based Multi-objective Optimization

Since ParEGO optimizes a random set of SOPs aggregated by the Chebyshev function in each iteration, ParEGO cannot guarantee the diversity of the obtained solutions. To improve the diversity, MOEA/D-EGO (Zhang et al., 2010) employs MOEA/D (Zhang & Li, 2007) as its baseline optimizer, which decomposes the original MOP into a series of sub-problems by the Chebyshev function. Due to the parallelism of MOEA/D, MOEA/D-EGO has better diversity than ParEGO.

7.3.1 MOEA/D

MOEA/D (Zhang & Li, 2007) is a classical decomposition-based MOEA. As shown in Fig. 7.9, MOEA/D contains three main steps: decomposition, search, and collaboration.

At the very beginning, MOEA/D generates a set of N evenly distributed weight vectors λ. With those weight vectors λ, the original MOP can be decomposed into a number of SOPs using the Chebyshev scalarizing function:

$$g^{te}(\mathbf{x}, \lambda) = \max_{j=1}^{m}(\lambda_j(\hat{f}_j(\mathbf{x}) - z_j)), \tag{7.33}$$

where z_j is the minimum of the j-th objective.

Then, MOEA/D initializes its population with N individuals and assigns each individual one SOP $g^{te}(\mathbf{x}, \lambda)$. Also, it defines a neighborhood for each weight vector, i.e., similar weight vectors are neighbors, which share similar SOPs.

In each generation, MOEA/D employs existing evolutionary search operators, such as crossover and mutation to generate offspring in the population. The best solution found so far for each SOP is saved in the population.

Since the SOPs in one neighborhood are similar, a collaboration in solving the neighboring SOPs can accelerate the search. The collaboration in MOEA/D has two steps: mating selection and replacement. In mating selection for crossover, each individual borrows solutions from its neighbors for generating new solutions. In

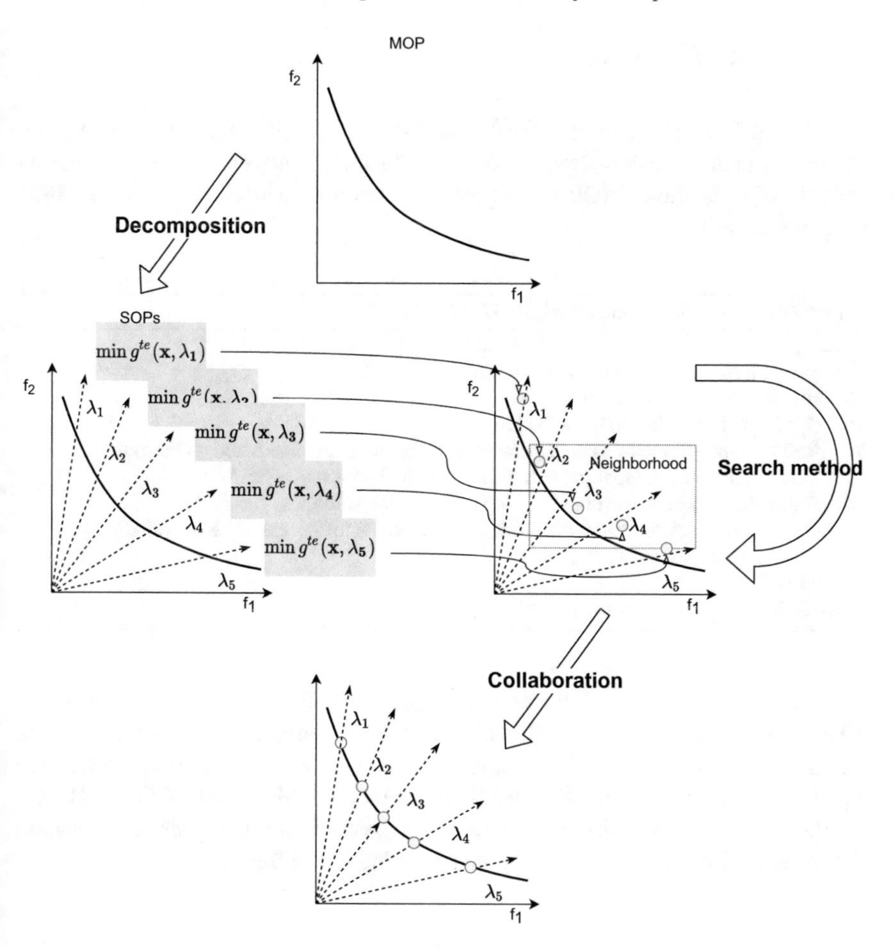

Fig. 7.9 A flowchart of MOEA/D

replacement step, each individual sends new solutions to its neighbors to check if the new solution can perform well on the SOPs in its neighbors.

MOEA/D is a combination of traditional mathematical optimization method and evolutionary search, where both the convergence and diversity are considered. Its collaboration mechanism makes good use of the population in EAs and the similarity of decomposed SOPs. Such an implicit parallelism can effectively speed up the multi-objective optimization search process.

7.3.2 Main Framework

MOEA/D-EGO (Zhang et al., 2010) is an MOEA/D variant for solving expensive MOPs. As an online data-driven EA, MOEA/D-EGO employs a GP model to approximated each objective, MOEA/D as the optimization method, and EI as the infill sampling criterion.

Algorithm 2 Pseudo code of MOEA/D-EGO.

1: Sampling initial training data $\mathcal{D}$ as the initial population.
2: Generate a set of uniform weights λ.
3: **while** stopping criterion is not met **do**
4: Apply the fuzzy clustering method to $\mathcal{D}$.
5: Build a local GP model to approximate each objective function based on the clustered $\mathcal{D}$.
6: Maximizing infill sampling criterion (EI) for all SOPs using MOEA/D.
7: Apply the k-means algorithm to group the offspring solutions.
8: Choose one solution $\mathbf{x}$ in each cluster, and evaluate $\mathbf{x}$ for its real objective value y.
9: Add new samples $(\mathbf{x}, y)$ to $\mathcal{D}$.
10: **end while**
Output: Non-dominated solutions in $\mathcal{D}$.

As shown in Algorithm 2, the process of MOEA/D-EGO follows the main step of MOEA/D. In the initialization step, MOEA/D-EGO samples the offline data as the training data $\mathcal{D}$ and the initial population, it also generates a set of evenly distributed weights λ as done in MOEA/D. Then, in the main loop, MOEA/D-EGO builds local GP models for all the objective functions and select infilling samples for updating GP models. The details are given in the following two sub-sections.

7.3.3 Local Surrogate Models

MOEA/D-EGO employs GP models to approximate each objective function in parallel. Since the time complexity of training one GP model is $O(d^3)$, MOEA/D-EGO builds local models if the number of training data d is too large in order to enhance the local accuracy and reduce the computation time. To achieve this, two parameters L_1 and L_2 are defined:

- L_1 is the threshold of the number of training data points of building multiple local models. If d is larger than L_1, local models are built, otherwise, only one global model is built for each objective function.
- L_2 is the threshold of the number of training data points of adding one local model.

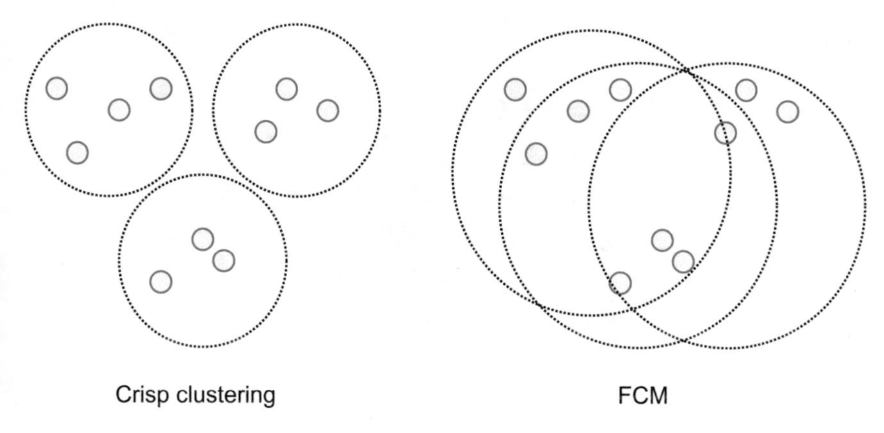

Fig. 7.10 Differences between crisp clustering and FCM

Therefore, the number of local models N_C for each objective function can be calculated as:

$$N_C = 1 + \left\lceil \frac{d - L_1}{L_2} \right\rceil \tag{7.34}$$

The training data $\mathcal{D}$ into N_C clusters by a fuzzy c-mean clustering method (FCM) (Bezdek, 2013). Different from other clustering method like crisp clustering, FCM may assign one data point to multiple clusters, as shown in Fig. 7.10. Thus, m GP models ($\hat{f}_1(\mathbf{x})$,..., $\hat{f}_m(\mathbf{x})$) are built in each cluster for m objective functions. MOEA/D-EGO trains mN_C GP models in total in each generation.

7.3.4 Surrogate Management

Unlike ParEGO, MOEA/D-EGO does not use GP models to approximate the aggregated SOPs, which make the existing EI unable to be directly adopted for sampling. With those GP models, a sub-problem $\hat{g^{te}}(\mathbf{x}, \lambda)$ can be written as:

$$\hat{g^{te}}(\mathbf{x}, \lambda) = \max_{j=1}^{m}(\lambda_j(\hat{f}_j(\mathbf{x}) - z_j)), \tag{7.35}$$

which requires a composition of m GP models. In the case of $m = 2$, the mean $\hat{g^{te}}(\mathbf{x}, \lambda)$ is

$$E(g^{te}(\mathbf{x}, \lambda)) = \mu_1\Phi(\alpha) + \mu_2\Phi(-\alpha) + \tau\phi(\alpha), \tag{7.36}$$

where

$$\begin{aligned}
\mu_i &= \lambda_i((\widehat{f_i}(\mathbf{x}) - z_i) \\
\tau &= \sqrt{[\lambda_1 \widehat{s_1}(\mathbf{x})]^2 + [\lambda_2 \widehat{s_2}(\mathbf{x})]^2} \\
\alpha &= (\mu_1 - \mu_2)/\tau \\
\phi(t) &= (2\pi)^{-0.5} e^{(-t^2/2)} \\
\Phi(t) &= \int_{-\infty}^{t} \phi(\theta) d\theta,
\end{aligned} \tag{7.37}$$

and variance is

$$\widehat{s^{te}}(\mathbf{x})^2 = E(g^{te}(\mathbf{x}, \lambda)^2) - E(g^{te}(\mathbf{x}, \lambda))^2, \tag{7.38}$$

where

$$\begin{aligned}
E(g^{te}(\mathbf{x}, \lambda)^2) &= (\mu_1^2 + \sigma_1^2)\Phi(\alpha) \\
&\quad + (\mu_2^2 + \sigma_2^2)\Phi(-\alpha) + (\mu_1 + \mu_2)\phi(\alpha) \\
\sigma_i^2 &= [\lambda_i \widehat{s_i}(\mathbf{x})]^2
\end{aligned} \tag{7.39}$$

. In the case of $m = 3$, the approximated Chebyshev function can be written as below to infer its mean and variance function.

$$\hat{g^{te}}(\mathbf{x}, \lambda) = \max\{\max\{\lambda_1(\widehat{f_1}(\mathbf{x}) - z_1), \lambda_2(\widehat{f_2}(\mathbf{x}) - z_2)\}, \lambda_3(\widehat{f_3}(\mathbf{x}) - z_3)\}, \tag{7.40}$$

Thus, the EI criterion of $\hat{g^{te}}(\mathbf{x}, \lambda)$ can be written as the following equation.

$$\begin{aligned}
EI(\mathbf{x}, \lambda) &= [g^{te}_{\min}(\mathbf{x}, \lambda) - \widehat{g^{te}}(\mathbf{x}, \lambda)]\Phi\left(\frac{g^{te}_{\min}(\mathbf{x}, \lambda) - \widehat{g^{te}}(\mathbf{x}, \lambda)}{\widehat{s^{te}}(\mathbf{x})}\right) \\
&\quad + \widehat{s^{te}}(\mathbf{x})\phi\left(\frac{g^{te}_{\min}(\mathbf{x}, \lambda) - \widehat{g^{te}}(\mathbf{x}, \lambda)}{\widehat{s^{te}}(\mathbf{x})}\right)
\end{aligned} \tag{7.41}$$

MOEA/D-EGO uses the EI criterion of each $\hat{g^{te}}(\mathbf{x}, \lambda)$ in Eq. (7.41) to choose new samples, i.e., maximizing $EI(\mathbf{x}, \lambda)$ is the SOP for each individual.

After the selection operation in each generation, MOEA/D-EGO filters out different solutions from the training data $\mathcal{D}$ in the population. Then, the k-means algorithm is applied to divide the those solutions into K_E clusters, one solution $\mathbf{x}$ in each cluster is selected to calculate its real objective values and added to the training data $\mathcal{D}$. When the computational budget exhausted, MOEA/D-EGO outputs the non-dominated solutions in $\mathcal{D}$.

7.3.5 *Discussions*

MOEA/D-EGO outperforms ParEGO in terms of both convergence and diversity, which indicates that a multi-objective optimizer is more efficient than a single-objective optimizer in data-driven MOEAs. However, due to the max function, it becomes complex to combine more than two GP models to the Chebyshev function. In other words, when the number of objective functions is large, a GP-based Chebyshev function is hard to be built. We therefore recommend to use other aggregation functions in MOEA/D-EGO. The source code of MOEA/D-EGO [2] has been included in PlatEMO (Tian et al., 2017a).

7.4 High-Dimensional Multi-objective Bayesian Optimization

7.4.1 *Main Challenges*

Bayesian optimization is a powerful method for black-box expensive optimization and the acquisition functions, also known as infill criteria, have widely been used for assisting expensive multi-objective optimization. However, Bayesian optimization usually relies on Gaussian processes (GPs) but GP-assisted evolutionary optimization has been applied only to low-dimensional problems (up to ten decision variables) (Chugh et al., 2018), mainly due to the fact that the computational cost of constructing the GP is $O(N^3)$, where N is the number of training data. By contrast, other surrogate models can handle a much higher dimension up to 100. Thus, it is of great importance to seek a computationally scalable replacement of the Gaussian process while still being able to use the infill criteria (acquisition functions). Another challenge is to extend the single-objective Bayesian optimization to multi-objective optimization. To be able to use the acquisition function for model management, surrogates that are able to replace Gaussian processes should be computationally scalable to the increase in the amount of training data (and therefore to the increase in the number of decision variables) and to estimate the uncertainty of the predicted fitness values.

One idea is to use a machine learning ensemble, which consists of multiple base learners (Guo et al., 2018). If the base learners are both accurate and diverse, it is then feasible to estimate the uncertainty of a prediction based on the variance of the outputs of the base learners. Thus, the central point is to construct a high-quality ensemble that can ensure both accurate prediction (based on the average of the outputs of the base learners) and a reliable estimate of the uncertainty.

---

[2]https://github.com/BIMK/PlatEMO/tree/master/PlatEMO/Algorithms/Multi-objective%20optimization/MOEA-D-EGO.

7.4.2 Heterogeneous Ensemble Construction

To construct a diverse and accurate ensemble, we present here a heterogeneous ensemble that uses different features as well as different model types. To construct different feature spaces, we use both feature selection and feature extraction techniques, which are two popular methods for dimension reduction. While feature selection aims to select the most compact and relevant subset of the given features, feature extraction transforms the existing features into a lower dimensional space so that the new features are nonredundant and informative.

- Feature selection. By removing the irrelevant and redundant features, feature selection makes it possible to build a compact model, thereby enhancing the generalization capacity and the efficiency of the model. Feature selection is essentially a combinatorial optimization problem, and it becomes challenging when the original feature space is large. According to the way in which a feature selection strategy is combined with a machine learning model, feature selection methods can be divided into wrapper, embedded and filter approaches. Although the resulting learning performance of filter methods is typically not as good as that of the other two methods, they are computationally more efficient and more robust. In (Gu et al., 2018) proposes to use the competitive swarm optimization algorithm (Cheng & Jin, 2014), which is an efficient variant of particle swarm optimizer for large-scale optimization, for high-dimensional feature selection. If the original feature space is not very large, e.g., up to 100, one can simply use the standard particle swarm optimizer to achieve faster convergence.
- Feature extraction. As discussed in Sect. 4.1.2, there are linear and nonlinear feature extraction methods. Depending on the correlation relationships between the decision variables, either a linear or nonlinear feature extraction method can be used. Here, the most popular principle component analysis method can be adopted for computational simplicity.

Once feature selection and feature extraction have been completed, three groups of base learners can be constructed: One based on the original feature space, the second on the selected features, and the third based on the extracted features. For each group of base learners, again different models can be used, e.g., an SVM and an RBFN. As mentioned in Sect. 4.2.8, different learning algorithm can also result in slightly different models. For instance, we can the gradient based method or the least square method, respectively, to train the RBFN. Figure 7.11 provides an example of heterogeneous ensemble using three groups of models with different inputs and different models. Consequently, diversity can be guaranteed.

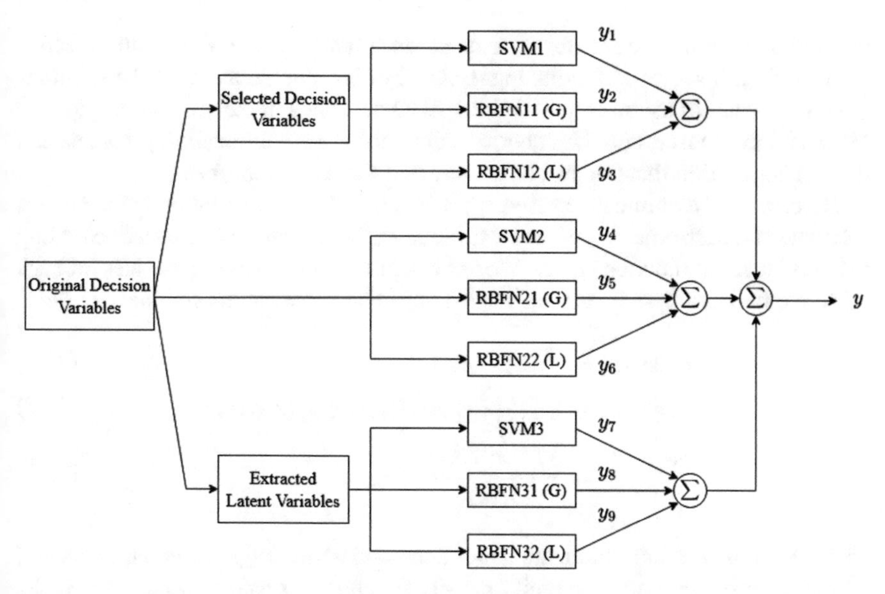

Fig. 7.11 Construction of a heterogeneous ensemble consisting of nine base learners, three with the selected subset as the input, three with the extracted features and three original features

In this example, the mean (predicted fitness) and variance of the surrogate can be calculated as follows:

$$\hat{f}(\mathbf{x}) = \frac{1}{9}\sum_{i=1}^{9} y_i, \tag{7.42}$$

$$\hat{s}^2(\mathbf{x}) = \frac{1}{9-1}\sum_{i=1}^{9}(y_i - \hat{f}(\mathbf{x}))^2. \tag{7.43}$$

where $\hat{f}(\mathbf{x})$ is the predicted fitness of solution $\mathbf{x}$.

Many other approaches to diverse ensembles can also be used, as described in Sect. 4.2.8.

7.4.3 Pareto Approach to Multi-objective Bayesian Optimization

As discussed in Sect. 5.4.3 and illustrated in Fig. 5.9, Bayesian evolutionary optimization uses an acquisition function (also known as infill criterion) to determine which individuals in a population should be sampled, i.e., evaluated using the expensive objective functions. For single-objective optimization, we can simply pick out the individual with the maximum acquisition function value. For multi-objective optimization, however, different approaches can be adopted to select the solutions to be sampled. For example in MOEA/D-EGO described in the previous section, an MOP is decomposed into a number of single-objective optimization problems and then an acquisition function can be constructed for each subproblem.

Here, we use a dominance based approach, i.e., the acquisition function value is calculated for each objective of the solutions, then these solutions are sorted according to the acquisition function values. We can calculate either LCB or EI values for each objective function based on the estimated objective value and its confidence level:

$$\mathrm{LCB}_j(\mathbf{x}) = \hat{f}_j(\mathbf{x}) - \beta \hat{s}_j(\mathbf{x}), \tag{7.44}$$

$$\mathrm{EI}_j(\mathbf{x}) = (y_j^{\min} - \hat{f}_j(\mathbf{x}))\Phi(z_j) + \hat{s}_j(\mathbf{x})\phi(z_j), \tag{7.45}$$

$$z_j = \frac{y_j^{\min} - \hat{f}_j(\mathbf{x})}{\hat{s}_j(\mathbf{x})}. \tag{7.46}$$

where β is positive for minimization problems, and where $\Phi(\cdot)$ and $\phi(\cdot)$ are standard normal distribution and probability density function, respectively, and $y_j^{\min}$ is the minimum of the j-th objective values of the existing sampled data, i.e., the best value of the j-th objective that has achieved so far, $j = 1, \cdots, m$, m is the number of objectives, typically either $m = 2$ or $m = 3$ for multi-objective optimization.

It is possible to sample all non-dominated solutions, which may consume too much of the limited computational budget. To reduce the number of solutions to be sampled, we can cluster the non-dominated solutions in the objective or decision space using a simple clustering algorithm, for instance the K-means clustering algorithm to reduce the number of solutions to be sampled. Typically, we can set the number of clusters to five for two- or three-objective optimization.

7.4.4 *Overall Framework*

The overall framework using a heterogeneous ensemble for assisting multi-objective optimization, called HeE-MOEA, is summarized below. It distinguishes itself from the other surrogate-assisted MOEAs mainly in that HeE-MOEA uses a heterogeneous ensemble instead of the GP for fitness approximation and the variances of the outputs of the ensemble members are used to estimate the confidence intervals. Initially, $11d - 1$ training data points are generated using the Latin hypercube sampling, where d is the dimension of the decision space.

A filter method based on the binary PSO (Cervante et al., 2012) is used to select input features for the generation of different inputs. The selection function is defined by minimizing the redundancy of the selected features and maximizing their relevance to the objective values as follows:

$$\min R = \alpha R_1 - (1 - \alpha) R_2, \tag{7.47}$$

$$R_1 = \frac{1}{d_1} \sum_{i=1}^{d_1} \mathrm{dCor}(x_i, \mathbb{C}_{X_s} x_i), \tag{7.48}$$

$$R_2 = \mathrm{dCor}(X_s, \mathbf{y}), \tag{7.49}$$

where $d_1 < d$ is the number of features in selected subset X_s, x_i is the i-th feature of X_s, and $\mathbb{C}_{X_s} x_i$ denotes all elements in X_s excluding x_i. R_1 calculates the mean of distance correlations between one feature in X_s and others, describing the redundancy of the selected features. R_2 indicates the dependency of the selected features on the function values using the distance correlation of X_s to Y. The distance correlation is calculated according to the distance covariance. Let X_s and Y be K samples of random variables $(\mathbf{x}_s, \mathbf{y})$, and then two $K \times K$ distance matrices A and B can be calculated as follows:

$$a_{i,j} = \| \mathbf{x}_s^i - \mathbf{x}_s^j \|^2, \tag{7.50}$$

$$b_{i,j} = \| \mathbf{y}^i - \mathbf{y}^j \|^2, \tag{7.51}$$

$$A_{i,j} = a_{i,j} - \bar{a}_{i\cdot} - \bar{a}_{\cdot j} - \bar{a}_{\cdot\cdot}, \tag{7.52}$$

$$B_{i,j} = b_{i,j} - \bar{b}_{i\cdot} - \bar{b}_{\cdot j} - \bar{b}_{\cdot\cdot}, \tag{7.53}$$

where $i, j = 1, , \cdots, K$, $\bar{a}_{i\cdot}$ and $\bar{a}_{\cdot j}$ are the mean of i-th row and j-th column, respectively, and $\bar{a}_{\cdots}$ is the grand mean of matrix a. The distance covariance can then be calculated as the average of the dot product of A and B:

$$\mathrm{dCov}^2(X_s, Y) = \frac{1}{K^2} \sum_{i=1}^{K} \sum_{j=1}^{K} A_{i,j} B_{i,j}. \tag{7.54}$$

Consequently, the distance correlation can be calculated by

$$\mathrm{dCor}(X_s, Y) = \frac{\mathrm{dCov}(X_s, Y)}{\sqrt{\mathrm{dCov}(X_s, X_s)\, \mathrm{dCov}(Y, Y)}}. \tag{7.55}$$

For feature extraction, the PCA can be used. As shown in Fig. 7.11, an ensemble consists of nine base learners can then be built. In the framework of evolutionary multi-objective Bayesian optimization, an MOEA, for instance, the NSGA-II (Deb et al., 2002) introduced in Sect. 7.1.3 can be employed to maximize the LCB or EI of each objective. The LCB or EI value of a given solution can be calculated using the equation defined in Equations (7.44) or (7.45) as the objective values of these solutions. Note that if LCB is used, the goal of the optimization is to minimize the LCB with respect to the different objectives, or to maximize the EI. To exploit the new samples as much as possible, new solutions close to the existing samples will be first deleted. For example, two solutions are considered to be close of their Euclidean distance in decision space is smaller than 10^5. This threshold, of course, depends on the search range of the decision variable. Then, the k-means clustering is used to choose k_m solutions from the non-dominated solutions, typically $k_m = 5$. After that, the objective values of these new k_m solutions will be calculated using the real expensive objectives (in practice, time-consuming numerical simulations, or physical experiments, or other data collection methods). These new data will then used to update the ensemble surrogates.

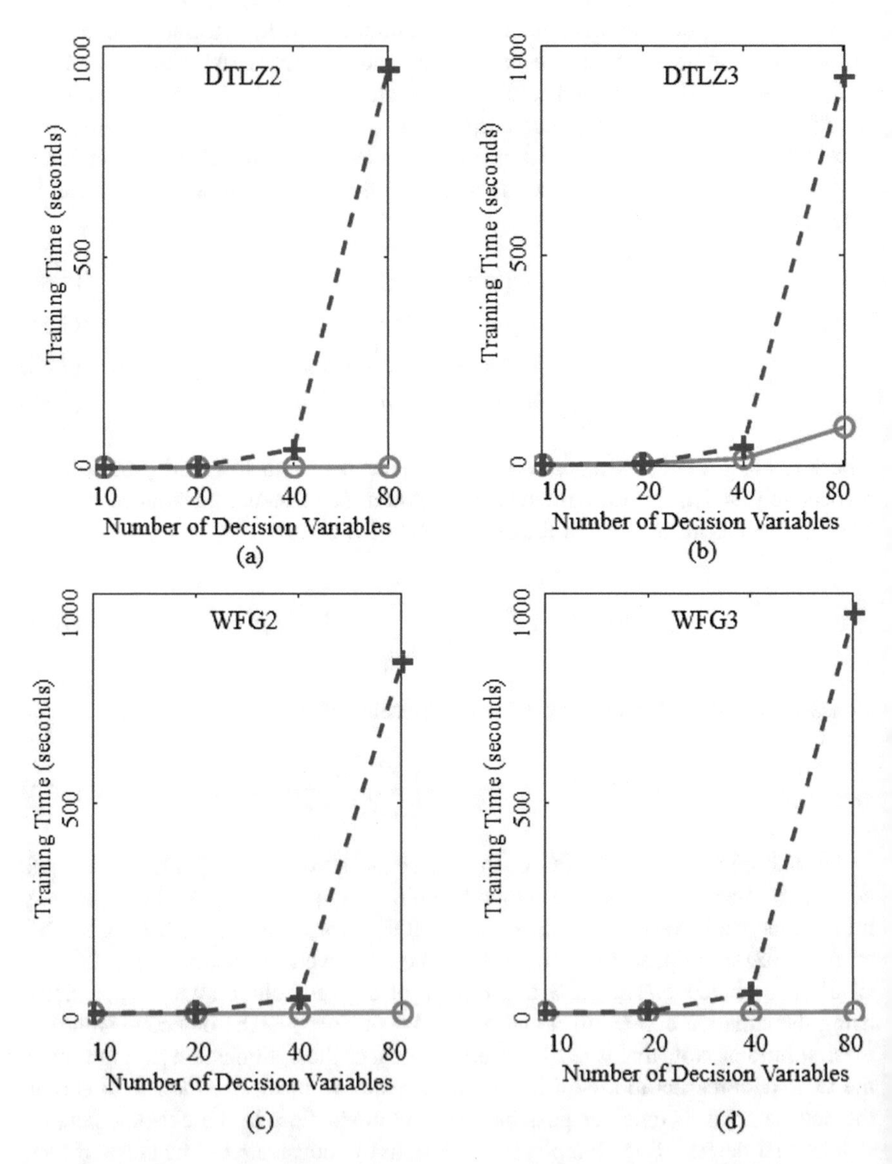

Fig. 7.12 Runtime for training the heterogeneous ensemble (denoted by circles) and the Gaussian process (denoted by pluses) as the dimension of decision variables increases from 20 to 80. **a** DTLZ2; **b** DTLZ3; **c** WFG2; and **d** WFG3

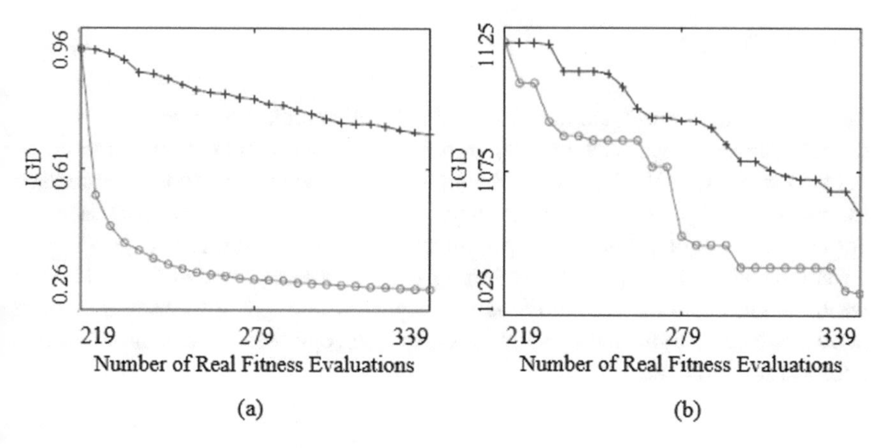

Fig. 7.13 The change of the mean IGD value of the HeE-MOEA (denoted by circles) and the GP-MOEA (denoted by pluses) over the number of real function evaluations when the EI is used as the acquisition function on **a** 20D DTLZ2 and **b** 20 D DTLZ3

Note that in HeE-MOEA, feature selection is performed once only before the evolutionary optimization since it is quite time-consuming, whereas feature extraction by PCA is performed in each iteration as PCA is computationally very cheap. In addition, not all collected samples are used to update the ensemble. If the number of the whole training data exceeds an upper bound, only a subset of the data will be picked out for updating the ensemble: the first half of the training data are directly chosen according to non-dominated sorting and crowding distance, and the other half are randomly sampled from the database without replacement.

Figure 7.12 presents some comparative results of the runtime for training the heterogeneous ensemble and the Gaussian process, respectively, for DTLZ2, DTLZ3, WFG2, and WFG3, four test problems used in evolutionary multi-objective optimization. We can see that the runtime for training the Gaussian process becomes much larger than for training ensemble, in particular when the number of decision variables is 80.

While it is able to reduce the computation time to reduce the Gaussian process with the heterogeneous ensemble, the optimization performance is also competitive on most of the benchmark problems widely used in the multi-objective optimization community. Figure 7.13 shows the change of the mean of the IGD value of the HeE-MOEA and the Gaussian process assisted MOEA (GP-MOEA) averaged over 20 runs on 20D DTLZ2 and 20D DTLZ3. More comparative results can be found in (Guo et al., 2018).

7.5 Summary

The basic ideas of surrogate-assisted single-objective optimization can be extended to multi-objective optimization, although it is not always straightforward to do so in many cases, in particular when the infill criteria or acquisition functions are used for model management. This can mainly be attributed to the fact that Bayesian optimization was originally developed for single-objective optimization, and several ideas have been developed for converting a multi-objective optimization problem into a single objective one. In the next Chapter, we are going to discuss an even more challenging topic that uses surrogates to assist optimization of many-objective optimization problems.

References

Bezdek, J. C. (2013). *Pattern recognition with fuzzy objective function algorithms*. Springer Science & Business Media.

Cervante, L., Xue, B., Zhang, M., & Shang, L. (2012). Binary particle swarm optimisation for feature selection: A filter based approach. In *Congress on Evolutionary Computation*.

Cheng, R., & Jin, Y. (2014). A competitive swarm optimizer for large scale optimization. *IEEE Transactions on Cybernetics*, *45*(2), 191–204.

Cheng, R., Jin, Y., Olhofer, M., & Sendhoff, B. (2016). A reference vector guided evolutionary algorithm for many objective optimization. *IEEE Transactions on Evolutionary Computation*, *20*(5), 773–791.

Chugh, T. (2019). *Scalarizing functions in bayesian multiobjective optimization*. arXiv:1904.05760.

Chugh, T., Jin, Y., Miettinen, K., Hakanen, J., & Sindhya, K. (2018). A surrogate-assisted reference vector guided evolutionary algorithm for computationally expensive many-objective optimization. *IEEE Transactions on Evolutionary Computation*, *22*, 129–142.

Coello, C. A. C., Lamont, G. B., & Veldhuizen, D. A. V. (2002). *Evolutionary algorithms for solving multi-objective problems*. Springer.

Deb, K. (2001). *Multi-objective optimization using evolutionary algorithms*. Wiley.

Deb, K., Pratap, A., Agarwal, S., & Meyarivan, T. (2002). A fast and elitist multiobjective genetic algorithm: NSGA-II. *IEEE Transactions on Evolutionary Computation*, *6*(2), 182–197.

Emmerich, M. (2005). *Single-and multi-objective evolutionary design optimization assisted by Gaussian random field metamodels*. Ph.D. dissertation. University of Dortmund.

Falcon-Cardona, J. G., & Coello Coello, C. A. (2020). Indicator-based multi-objective evolutionary algorithms: A comprehensive survey. *ACM Computing Surveys*, *53*(2), 29.

Fonseca, C. M., & Fleming, P. J. (1995). An overview of evolutionary algorithms in multiobjective optimization. *Evolutionary computation*, *1*(3), 1–16.

Giagkiozis, I., & Fleming, P. J. (2015). Methods for multi-objective optimization: An analysis. *Information Sciences*, *293*, 338–350.

Gu, S., Cheng, R., & Jin, Y. (2018). Feature selection for high dimensional classification using a competitive swarm optimizer. *Soft Computing*, *22*(3), 811–822.

Guo, D., Jin, Y., Ding, J., & Chai, T. (2018). Heterogeneous ensemble-based infill criterion for evolutionary multiobjective optimization of expensive problems. *IEEE Transactions on Cybernetics*, *49*(3), 1012–1025.

Hansen, M. P., & Jaszkiewicz, A. (1998). Evaluating the quality of approximations of the non-dominated set. Technical Report M-REP-1998-7, Institute of Mathematical Modeling,Technical University of Denmark.

Husslage, B. G., Rennen, G., Van Dam, E. R., & Den Hertog, D. (2011). Space-filling latin hypercube designs for computer experiments. *Optimization and Engineering*, *12*(4), 611–630.

Jin, Y., Okabe, T., & Sendhoff, B. (2001a). Adapting weighted aggregation for multiobjective evolution strategies. *Proceedings of the first international conference on evolutionary multi-criterion optimization* (pp. 96–110).

Jin, Y., Okabe, T., & Sendhoff, B. (2001b). Dynamic weighted aggregation for evolutionary multi-objective optimization: Why does it work and how? *Genetic and evolutionary computation conference* (pp. 1042–1049).

Jones, D. R., Schonlau, M., & Welch, W. J. (1998). Efficient global optimization of expensive black-box functions. *Journal of Global Optimization*, *13*(4), 455–492.

Knowles, J. (2006). ParEGO: A hybrid algorithm with on-line landscape approximation for expensive multiobjective optimization problems. *IEEE Transactions on Evolutionary Computation*, *10*(1), 50–66.

Miettinen, K. (1999). *Nonlinear multiobjective optimization*. Springer.

Murata, T., Ishibuchi, H., & Gen, M. (2001). Specification of genetic search directions in cellular multi-objective genetic algorithms. *Proceedings of the first international conference on evolutionary multi-criterion optimization* (pp. 82–95).

Murata, T., Ishibuchi, H., & Tanaka, H. (1996). Multi-objective genetic algorithm and its applications to flowshop scheduling. *Computers & Industrial Engineering*, *30*(4), 957–968.

Tian, Y., Cheng, R., Zhang, X., & Jin, Y. (2017a). PlatEMO: A MATLAB platform for evolutionary multi-objective optimization. *IEEE Computational Intelligence Magazine*, *12*(4), 73–87.

Tian, Y., Wang, H., Zhang, X., & Jin, Y. (2017b). Effectiveness and efficiency of non-dominated sorting for evolutionary multi-and many-objective optimization. *Complex & Intelligent Systems*, *3*(4), 247–263.

Trivedi, A., Srinivasan, D., Sanyal, K., & Ghosh, A. (2017). A survey of multiobjective evolutionary algorithms based on decomposition. *IEEE Transactions on Evolutionary Computation*, *21*(3), 440–462.

Zhang, Q., & Li, H. (2007). MOEA/D: A multiobjective evolutionary algorithm based on decomposition. *IEEE Transactions on Evolutionary Computation*, *11*(6), 712–731.

Zhang, Q., Liu, W., Tsang, E., & Virginas, B. (2010). Expensive multiobjective optimization by MOEA/D with Gaussian process model. *IEEE Transactions on Evolutionary Computation*, *14*(3), 456–474.

Zhang, X., Tian, Y., Cheng, R., & Jin, Y. (2015). An efficient approach to non-dominated sorting for evolutionary multi-objective optimization. *IEEE Transactions on Evolutionary Computation*, *19*(6), 761–776.

Zitzler, E. and Künzli, S. (2004). Indicator-based selection in multiobjective search. In *Proceedings of the parallel problem solving from nature-PPSN* (pp. 832–842). Springer.

Chapter 8
Surrogate-Assisted Many-Objective Evolutionary Optimization

8.1 New Challenges in Many-Objective Optimization

8.1.1 Introduction

The early multi-objective evolutionary algorithms (MOEAs) are developed mainly for problems having two or three objectives. Many of these algorithms will suffer from different issues when the number of objectives increases. In the evolutionary optimization field, MOPs having four or more objectives are known as many-objective optimization (Purshouse & Fleming, 2003, Ishibuchi et al., 2008, Li et al., 2015a). For the three main categories of evolutionary algorithms, the following challenges arise in dealing with many-objective optimization problems.

- For Pareto or non-dominance based MOEAs, selection pressure will dramatically decrease as the number of objectives increase (Fleming et al., 2005, Purshouse & Fleming, 2007). This can be attributed to the fact that for a limited population size, the number of dominated solutions will become smaller, making it increasingly harder to distinguish better solutions from worse ones based on dominance comparisons. As a result, convergence will be seriously decreased.
- For indicator based MOEAs, in particular those based on the hypervolume, computational complexity will become intractable as the number of objectives increases, although convergence will not be much affected. In addition, the diversity of the obtained solutions will become harder to assess. One remedy is to use approximate methods for calculating the hypervolume to reduce the computation time, and the other approach is to use computationally scalable performance indicators, such as the inverse generational distance or its variants.
- Decomposition based approaches can mostly still be able to converge to the Pareto front, however, the search efficiency may drastically deteriorate for problems

Y. Jin et al., *Data-Driven Evolutionary Optimization*,
Studies in Computational Intelligence 975,
https://doi.org/10.1007/978-3-030-74640-7_8

whose Pareto front does not cover the whole objective space. This is mainly because most decomposition based MOEAs pre-define a set of evenly distributed reference vectors or weights, or reference points. Thus, a large portion of these reference vectors may become idle.

Several remedies for alleviating the issue of Pareto dominance based approaches have been proposed. In addition to reducing the number of objectives by removing redundant and unimportant objectives, several ideas have proposed to help increase the selection pressure. These include:

- *Reduction of objectives*. This is a straightforward approach that removes redundant and / or unimportant objectives. An objective may be redundant if it is strongly correlated with others. One can also use weighted aggregation or scalarizing approaches to reduce the number of objectives.
- *Modification of Pareto dominance*. One main difficulty for dominance based approaches to many-objective optimization is that most solutions in a population are non-dominated, making it hard for the population to converge to the Pareto front. To resolve this issue, several ideas have been proposed to modify the Pareto dominance to make the dominance stronger, being about to dominate more solutions. For example, ϵ-dominance and α-dominance presented in Sect. 1.3.2 are two strengthened dominance relationships. Several other variants, L-dominance (Zou et al., 2008), θ-dominance (Yuan et al., 2016), L-dominance and strengthened dominance relation (Tian et al., 2019) have also proposed.
- *Introduction of additional criteria*. To increase the selection pressure, additional selection criteria can be introduced on top of the Pareto dominance. For example, the distance to the hyperplane constructed based on the extreme solutions is employed (Zhang et al., 2015b), while in (Li et al., 2015b) a decomposition based selection criterion is combined with the dominance comparison.

8.1.2 Diversity Versus Preferences

In two- or three-objective optimization, the basic hypothesis is that a population (or archive) is used to approximate a set of Pareto optimal solutions, which can represent the theoretical Pareto front. Thus, the diversity of the obtained solution set, or even distribution of the solutions on the whole Pareto front becomes essential. Even distribution or diversity is actually based on an implicit assumption that the number of solutions in the set is sufficient to represent the Pareto front, since a very small number of solutions, even of they are evenly distributed on the whole front, cannot represent the theoretical Pareto front. Thus, a diverse set of solutions is considered to be equivalent to a representative set of solutions only if there is a sufficient number of solutions. This implicit assumption holds for two- or three-objective optimization, in which the Pareto front is either a one-dimensional curve or a two-dimensional surface, which can be sufficiently represented by a population of, typically 100–200 solutions. Unfortunately, this implicit assumption becomes less true as the number of

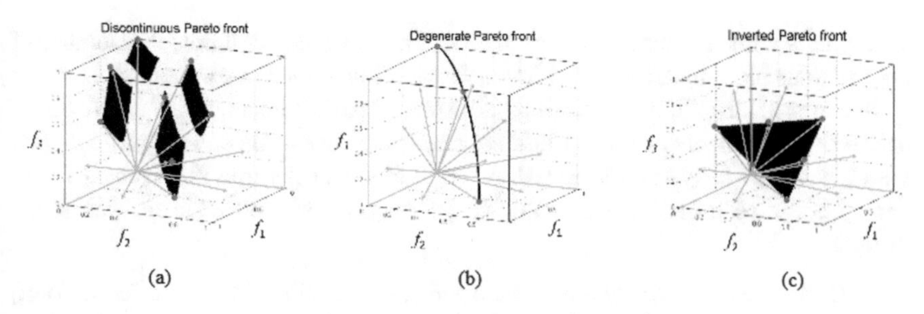

Fig. 8.1 Examples of irregular Pareto fronts. **a** Discrete; **b** degenerate; **c** inverted

objectives increases, and it is not likely to use a limited population to approximate a representative subset of many-objective optimization problems, as discussed in (Yu et al., 2019).

Apart from the limited number of solutions, which makes it hardly possible to represent a high-dimensional Pareto front, another factor that makes diversity to be less useful for quality assessment for many-objective optimization is the irregular distribution of the Pareto optimal solutions. Most decomposition based approaches define a set of evenly distributed reference vectors or reference solutions as the goals, implicitly assuming that the Pareto front covers the whole objective space. However, the Pareto front of a wide class of multi- and many-objective optimization problems is irregular, that is, the Pareto is either degenerate, or discrete, or inverted. Usually, the dimension of the Pareto front is $m - 1$, where m is the number of objectives. If the dimension is smaller than $m - 1$, the Pareto front is called degenerate. Example of discrete, degenerate and inverted Pareto fronts are given in Fig. 8.1.

Due to the above two reasons, it has been argued in (Yu et al., 2019) that diversity is no longer an adequate or appropriate performance indicator for many-objective optimization problems. Instead, user preferences are more suited to guide the search process in addition to convergence in solving many-objective optimization problems. Preference modeling methods described in Sect. 1.3.3 can be employed for identifying preferred solutions.

8.1.3 Search for Knee Solutions

While it is highly desirable to incorporate user preferences in many-objective optimization, preferences are not always available to the user, in particular when there is little knowledge about the problem is available. In this case, it is quite natural to search for knee points in many-objective optimization to target for desired Pareto optimal solutions. It should be pointed out that identifying knee points in a high-dimensional objective space is non-trivial, mainly because defining knee points becomes more challenging, and the number of knee points may be exponentially increase. In addition, one needs different performance indicators to evaluate the quality of obtained knee points with respect to the closeness to the true knee solutions, the number of knee regions identified, and the solution distribution in the knee regions.

In (Yu et al., 2020), the following performance indicators are suggested for evaluation of the performance of evolutionary multi- or many-objective algorithms that search for knee points and knee regions, provided that the true knee points K are known and a set of reference solutions that evenly cover all knee regions R are defined.

- KGD. KGD evaluates the convergence performance (closeness to the Pareto front) of the obtained solutions to the reference points in the knee regions, which is defined as follows:

$$KGD = \frac{1}{|S|}\sum_{i=1}^{|S|} d(x_i, R), \tag{8.1}$$

  where S is the obtained solution set, x_i is a solution in S, $d(x_i, R)$ means the Euclidean distance between the solution x_i in S to its closest reference point in R. A smaller KGD indicates that the solutions have converged to the knee regions.
- KIGD. KIGD evaluates the evenness of the distribution of the achieved solutions in the knee regions:

$$KIGD = \frac{1}{|R|}\sum_{i=1}^{|R|} d(x_i^r, S), \tag{8.2}$$

  where $d(x_i^r, S)$ is the Euclidean distance between a reference point x_i^r in R and the solution closest to this reference point in R. A smaller KIGD means an even coverage of the knee region.
- KD. KD assesses if all knee points can be found by an algorithm:

$$KD = \frac{1}{|K|}\sum_{i=1}^{|K|} d(x_i^k, S), \tag{8.3}$$

  where $d(x_i^k, S)$ is the Euclidean distance between a true knee point x_i^k in K to its closest solution in S for evaluating the completeness in identifying all knee points.

The above performance indicators assume that the location and number of knee points of a given MOP is exactly known. To make this possible, a suite of test problems dedicated to benchmarking evolutionary algorithms for identifying knee points is presented in (Yu et al., 2020). The designed test problems can specify the number, the location and the size of the knee regions on the Pareto front.

Evolutionary algorithms for identifying knee solutions for two- or three-objective optimization algorithms have been proposed. In (Branke et al., 2004), the angle of a Pareto optimal solution between the two lines connection its two neighbors is used as an additional measure for prioritizing knee solutions. An algorithm that uses the weighted sum of the objectives to find the knee points is suggested in (Rachmawati & Srinivasan, 2009), which, based on a rough approximation to the Pareto front, optimizes a linear weighted sums of the original objective functions by carefully designating the weights. A recent algorithm that integrates a localized α-dominance and and a localized knee-oriented dominance relationship in presented in (Yu et al., 2020). By localized α-dominance, it is meant that the objective space is at first divided into a number of subspaces using a set of reference vectors and then the α-dominance is applied for environmental selection within each space. Given two solutions $\mathbf{x}_1$ and $\mathbf{x}_2$, $\mathbf{x}_1$ is said to *knee-oriented-dominate* solution $\mathbf{x}_2$, if the following conditions hold:

$$\mu(\mathbf{x}_1, \mathbf{x}_2) < 0, \tag{8.4}$$

s.t.

$$\mu(\mathbf{x}_1, \mathbf{x}_2) = \angle(\mathbf{x}^*\mathbf{x}_1, \mathbf{x}_1\mathbf{x}_2) - \tau(\max_{i=1,\cdots,m} \{\delta_i(\mathbf{x}_1)\} + \min_{i=1,\cdots,m} \{\delta_i(\mathbf{x}_1)\}, \tag{8.5}$$

$$\delta_i(\mathbf{x}_1) = \arctan \frac{\sqrt{\sum_{j=1,j\neq i}^{m} (f_j(\mathbf{x}_1) - f_j(\mathbf{x}^*))^2}}{|f_i(\mathbf{x}_1) - \max f_i(E) - \epsilon|}, \tag{8.6}$$

where $\mathbf{x}^*$ is the ideal point, m is the number of objectives, and $\angle(\mathbf{x}^*\mathbf{x}_1, \mathbf{x}_1\mathbf{x}_2)$ is the acute angle between the line connecting $\mathbf{x}^*$ and $\mathbf{x}_1$ and that connecting $\mathbf{x}_1$ and $\mathbf{x}_2$, and the ideal point is defined by $f_j(\mathbf{x}^*) = \min f_j(E) - \epsilon$, where $E = \{E^i | i = 1, \cdots, m\}$ are the extreme points, and ϵ is a small positive constant to avoid division by zero. Typically, $\tau \in [0.5, 1]$. Similarly, a localized dominance-oriented dominance means that this dominance relationship is applied to a subspace.

8.1.4 *Solving Problems with Irregular Pareto Fronts*

As previously discussed, the Pareto front of MOPs with irregular fronts cannot be efficiently solved using the standard decomposition based methods that adopt a set of evenly distributed reference vectors or reference solutions. Methodologies that tackle MOPs with irregular Pareto fronts can largely be divided into three categories, namely adaptation of reference vectors, adaptation of reference points, or clustering methods (Hua et al., 2021). For example, a hierarchical clustering algorithm is applied on top of the non-dominated sorting to group the solutions on the last non-dominated into a number of clusters, based on which environmental selection is performed (Hua et al., 2019). Multiple sets of reference vectors are introduced in (Hua et al., 2020) to locate Pareto optimal solutions of a degenerate Pareto front. An improved growing neural gas network based adaptive reference vector based algorithm is reported in (Liu et al., 2020) for solving many-objective optimization problems with irregular Pareto fronts.

8.2 Evolutionary Many-Objective Optimization Algorithms

8.2.1 Reference Vector Guided Many-Objective Optimization

As discussed in Sect. 8.1, it becomes more challenging to solve problems with more than three objective, as called many-objective optimization. Remedying measures include the reduction of the number of objectives, modification of the dominance relationship, introducing a second selection criterion, among many others.

Reference vector guided many-objective optimization, called RVEA, is a popular MOEA for many-objective optimization (Cheng et al., 2016). One main difference between RVEA and other scalarizing function based decomposition algorithms is that it uses a set of reference vectors in the objective space to specify the preferred solutions, making it more straightforward to reflect the user's preferences, and easier to adapt the preferences if needed. This is because the mapping from the weight space to the objective space may be non-uniform, i.e., evenly distributed weights may result in unevenly distributed Pareto optimal solutions in the objective space. In addition, RVEA uses the angle between vectors instead of the Euclidean distance to measure the diversity, which is more scalable to the increase in the number of objectives. Finally, RVEA introduces adaptation mechanisms for adapting the reference vectors and tuning the selection pressure between convergence and diversity.

8.2.1.1 Reference Vector Generation

All reference vectors are unit vectors inside the first quadrant with the origin being the initial point. Theoretically, a unit vector can be generated by dividing an arbitrary vector by its norm. Similar to most general purpose decomposition based algorithms, RVEA also assumes that the Pareto front cover the overall objective space and therefore, a set of evenly distributed vectors will be generated in the objective space. To generate evenly distributed reference vectors, RVEA adopts the approach introduced in (Cheng et al., 2015). At first, RVEA generates a set of evenly distributed points on a unit hyperplane using the canonical simplex-lattice design method (Cornell, 2011):

$$\begin{cases} \mathbf{w_i} = (w_i^1, w_i^2, ..., w_i^m), \\ w_i^j \in \{\frac{0}{H}, \frac{1}{H}, ..., \frac{H}{H}\}, \sum_{j=1}^{m} w_i^j = 1, \end{cases} \tag{8.7}$$

where $i = 1, ..., N$, N is the number of reference vectors to be generated, m is the number of objectives, and H is a positive integer for the simplex-lattice design. Then, the corresponding unit reference vectors $\mathbf{v_i}$ can be obtained as follows:

$$\mathbf{v_i} = \frac{\mathbf{w_i}}{\|\mathbf{w_i}\|}, \tag{8.8}$$

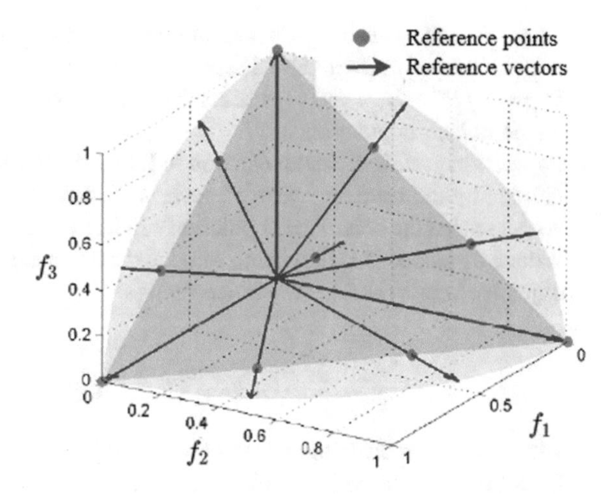

Fig. 8.2 An example of a set of reference points and the corresponding set of reference vectors in a three-objective space

which maps the reference points from a hyperplane to a hypersphere. Figure 8.2 shows one example. Based on the simplex-lattice design, given H and m, $N = \binom{H+m-1}{m-1}$ evenly distributed reference vectors in total can be generated.

As mentioned above, RVEA uses the angle to measure the relationship between different solutions. To this end, RVEA measures the spatial relationship between two reference vectors by means of the cosine value of the acute angel between these two vectors. For any two reference vectors $\mathbf{v}_i$ and $\mathbf{v}_j$, where $i, j \in N$, we have:

$$\cos\theta_{ij} = \frac{\mathbf{v}_i \cdot \mathbf{v}_j}{\|\mathbf{v}_i\|\|\mathbf{v}_j\|}, \tag{8.9}$$

where $\|\cdot\|$ calculates the length of the vector. Since reference vectors are normalized, the length of the reference vectors equals to 1.

8.2.1.2 Angle-Penalized Distance Based Environmental Selection

In each generation, any genetic operators, depending on the coding used, can be employed to generate the same number of offspring individuals from the parents. Then the offspring population is combined with the parent population. Before selection, the following steps are carried out.

- Objective translation. Here, each objective value is translated based on the minimum value of each objective in the current combined population:

$$\mathbf{f}_i' = \mathbf{f}_i - \mathbf{f}^{\min}, \tag{8.10}$$

where i is the i-th individual in the combined population, $\mathbf{f}_i$, $\mathbf{f}'_i$ are the objective vectors of individual i before and after the translation, respectively, and $\mathbf{f}^{\min} = (f_1^{\min}, \ldots, f_m^{\min})$ represents the minimal objective values calculated from the combined population.

- Population partition. Since selection in RVEA is performed within each subspace divided by the reference vectors, each individual will be assigned to a reference vector so that the combined population will be divided into N subpopulations. This is done by calculating the cosine value of the acute angel between the solution (the objective vector) and the reference vectors:

$$\cos\theta_{i,j} = \frac{\mathbf{f}'_i \cdot \mathbf{v}_j}{\|\mathbf{f}'_i\|}, \tag{8.11}$$

where $\theta_{i,j}$ represents the angle between objective vector of the i-th individual $\mathbf{f}'_i$ and the j-th reference vector $\mathbf{v}_j$, where $i = 1, \ldots, |R|$, $|R|$ is the size of the combined population, $j = 1, \ldots, N$. Thus, individual i will be assigned to the reference vector having the minimum $\theta_{i,j}$.

- Angle penalized distance (APD). For the i-th individual in the subpopulation associated with the j-th reference vector $\mathbf{v}_j$, the angle penalized distance is calculated as follows:

$$d_{i,j}(t) = P(\theta_{i,j}(t)) \, \|\mathbf{f}'_i\|, \tag{8.12}$$

where $P(\theta_{i,j}(t))$ is a penalty function related to $\theta_{i,j}(t)$:

$$P(\theta_{i,j}(t)) = 1 + m \left(\frac{t}{t_{\max}}\right)^{\alpha} \theta_{i,j}(t), \tag{8.13}$$

where m is the number of objectives, t_{max} is the predefined maximum number of generations t is the current generation, and α is a user defined parameter controlling the rate of change of $P(\theta_{i,j}(t))$. The smaller $P(\theta_{i,j}(t))$ is, the closer $\mathbf{f}'_i(t)$ is to reference vector $\mathbf{v}_{t,j}$, and vice versa. As evolution proceeds, the weight of diversity (specified by $\theta_{i,j}(t)$) will increase. Consequently, more effort will be paid to diversity in the later stage of the selection.

8.2.1.3 Reference Vector Adaptation

The even distribution of the reference vectors is based on the assumption that the Pareto front covers the whole objective space, and that an even distribution of the reference vectors will lead to an even distribution of the obtained solution. Neither of the assumptions is necessarily true. The former case belongs to MOPs with irregular Pareto fronts, and many MOEAs dedicated to such problems have been designed (Hua et al., 2021). The second problem, as illustrated in Fig. 8.3, can be avoided if all objectives can be normalized into the same range, which, however, does not work for RVEA.

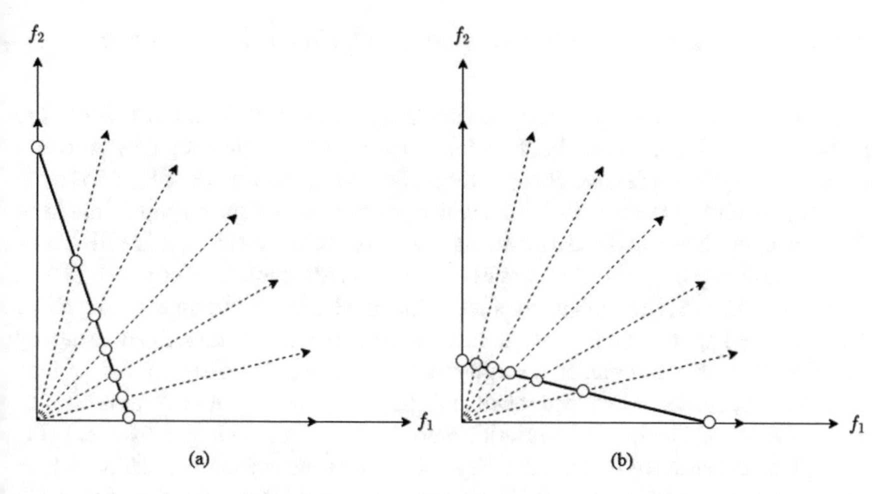

Fig. 8.3 Two examples in which an evenly distributed reference vector does not lead to an evenly distributed Pareto optimal solution set

To address this issue, RVEA adapts the reference vectors according to the ranges of the objective values as follows:

$$\mathbf{v}_i(t+1) = \frac{\langle \mathbf{v}_i(0)|\mathbf{f}^{\max}(t+1) - \mathbf{f}^{\min}(t+1)\rangle}{\|\langle \mathbf{v}_i(0)|\mathbf{f}^{\max}(t+1) - \mathbf{z}^{\min}(t+1)\rangle\|}, \tag{8.14}$$

where $i = 1, 2, ..., N$, $\mathbf{v}_i(t+1)$ denotes the i-th adapted reference vector for generation $t+1$, $\mathbf{v}_i(0)$ is the i-th originally generated evenly distributed reference vector, $\mathbf{f}^{\max}(t+1)$ and $\mathbf{f}^{\min}(t+1)$ represent the maximum and minimum values of each objective function in the current generation, respectively. The operator $\langle \mathbf{a}|\mathbf{b}\rangle$, known as the Hadamard product that element wisely multiplies two vectors (or matrices) of the same size. In other words, it generates a vector where each dimension in **a** and **b** are multiplied, respectively, i.e.:

$$\langle \mathbf{a}|\mathbf{b}\rangle = (a_1b_1, a_2b_2, ..., a_nb_n), \tag{8.15}$$

where n denotes the length of **a** and **b**. With the reference vector adaptation method described above, the proposed RVEA will be able to obtain evenly distributed solutions, even if the objective functions are not normalized to the same range.

It should be pointed out that the reference vector adaptation mechanism should not be employed frequently during the evolution to ensure a fast convergence. It is recommend that the reference vectors can be adapted in every 20 generations (Cheng et al., 2016).

8.2.2 A Knee-Driven Many-Objective Optimization Algorithm

The knee-driven many-objective optimization algorithm (KnEA) aims to address the problem of dominance based MOEAs for many-objective optimization by introducing a second criterion (Zhang et al., 2015c). The overall framework of KnEA is quite similar to NSGA-II (Deb et al., 2002), which generates an offspring population from the parent population using crossover and mutation. Note, however, that KnEA uses a modified binary mate selection strategy that selects two parents for crossover. Then, the parent and offspring populations are combined before environmental selection. The combined population is sorted according to the non-domination relationship, by e.g., the efficient non-dominated sorting method (Zhang et al., 2015b) as described Sect. 7.1.3. Instead of using the crowding distance on top of the non-dominated sorting in (Deb et al., 2002), KnEA prioritizes solutions that have a larger distance to the hyperplane constructed by the boundary solutions of the current population, which can be seen as the knee points of the current front. Note that these knee points are not necessarily the true knees on the Pareto front. In the following, we details the two distinct components in KnEA, the weighted distance based binary tournament mate selection and knee point based environmental selection.

8.2.2.1 Weighted Distance Based Binary Tournament Mate Selection

At first, two individuals are randomly chosen from the parent population. If one solution dominates the other, then the former solution is selected. If the two solutions are non-dominated with each other, then KnEA will check whether they are both knee points. The details of determining knee points will be provided later on. If only one of them is a knee point, then this solution will be chosen. If both of them are knee points or neither of them is a knee point, then a weighted distance will be calculated for comparing the two solutions, and the one having the larger weighted distance will be selected. If both solutions have the same weighted distance, one of them will be selected randomly.

The weighted distance of a solution $\mathbf{x}$ in a population based on the k-nearest neighbors is defined as follows:

$$DW(\mathbf{x}) = \sum_{i=1}^{k} w_{\mathbf{x}_i} \operatorname{dis}(\mathbf{x}, \mathbf{x}_i), \tag{8.16}$$

$$w_{\mathbf{x}_i} = \frac{r_{\mathbf{x}_i}}{\sum_{i=1}^{k} r_{\mathbf{x}_i}} \tag{8.17}$$

$$r_{\mathbf{x}_i} = \frac{1}{|\operatorname{dis}(\mathbf{x}, \mathbf{x}_i) - \frac{1}{k}\sum_{i=1}^{k} \operatorname{dis}(\mathbf{x}, \mathbf{x}_i)|} \tag{8.18}$$

where $\mathbf{x}_i$ represents the i-th nearest neighbor of $\mathbf{x}$ in the population, $w_{\mathbf{x}_i}$ represents the weight of $\mathbf{x}_i$, $\text{dis}(\mathbf{x}, \mathbf{x}_i)$ represents the Euclidean distance between solution $\mathbf{x}$ and $\mathbf{x}_i$, and $r_{\mathbf{x}_i}$ represents the rank of distance $\text{dis}(\mathbf{x}, \mathbf{x}_i)$ among all the distances $\text{dis}(\mathbf{x}, \mathbf{x}_j)$, $1 \leq j \leq k$.

8.2.2.2 Knee Point Based Environmental Selection

Like in NSGA-II, the solutions on the first fronts will be selected until at a point when the number of solutions on the corresponding non-dominated front (known as the critical front) is larger than the number of solutions that are needed. Instead of selecting solutions having a larger crowding distance, KnEA aims to select solutions based on a *knee distance* in the following steps.

- *Construction of the hyperplane using the boundary solutions*. For example in Fig. 8.4, A and B are two boundary solutions of a bi-objective optimization problem.
- *Calculation the distance to the hyperplane*. For a bi-objective minimization problem, the hyperplane L can be defined by $af_1 + bf_2 + c = 0$, where the parameters can be determined by the two extreme solutions. Then the distance from a solution $(f_1(\mathbf{x}), f_2(\mathbf{x}))$ to L can be calculated as follows:

$$d(\mathbf{x}, L) = \frac{|af_1(\mathbf{x}) + bf_2(\mathbf{x}) + c|}{\sqrt{a^2 + b^2}}. \tag{8.19}$$

where $f_1(\mathbf{x})$ and $f_2(\mathbf{x})$ are the first and second objective values of solution $\mathbf{x}$.

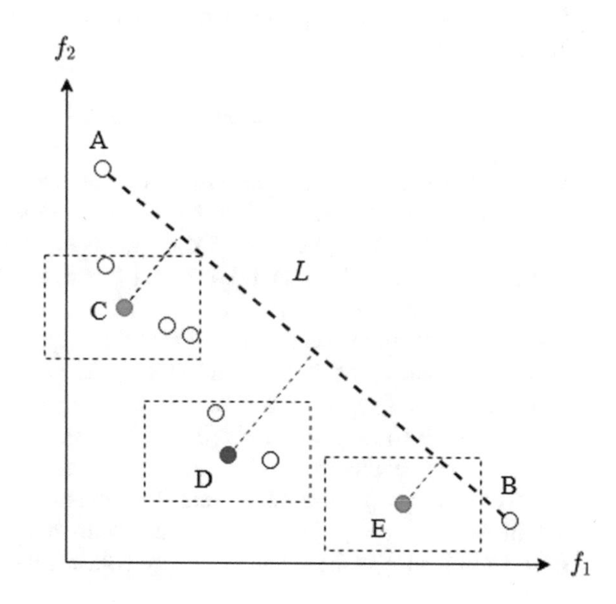

Fig. 8.4 An illustrative example consisting of ten solutions, in which A and B are boundary solutions, with which the hyperplane can be constructed, and C, D and E are the knee point of each given neighborhood

For minimization problems, only solutions in the convex knee regions are of interest. Therefore, the distance measure in Eq. (8.19) can be modified as follows to identify knee points:

$$d(\mathbf{x}, L) = \begin{cases} \frac{|af_1(\mathbf{x})+bf_2(\mathbf{x})+c|}{\sqrt{a^2+b^2}}, & \text{if } af_1(\mathbf{x}) + bf_2(\mathbf{x}) + c < 0 \\ -\frac{|af_1(\mathbf{x})+bf_2(\mathbf{x})+c|}{\sqrt{a^2+b^2}}, & \text{otherwise} \end{cases} \quad (8.20)$$

The above distance measure for identifying knee points can be extended to optimization problems with more than two objectives, where the extreme line will become a hyperplane.

- *Identification of the knee points.* As indicated in Fig. 8.4, the knee solution is the one having the maximum distance to the hyperplane within a local neighborhood. In other words, different neighborhood sizes will result in different knee points. For instance, if a neighborhood covers all ten solutions in Fig. 8.4, then only solution D will be identified as the knee point. In addition, the neighborhood size has also great influence on the diversity of the selected solutions.
Assume the combined population at generation t contains K non-dominated fronts, each of which has a set of non-dominated solutions denoted by F_i, $1 \leq i \leq K$. The neighborhood of a solution is defined by a hypercube of size $R_1(t) \times \ldots \times R_j(t) \times \ldots \times R_m(t)$, where $1 \leq j \leq m$, m is the number of objectives. The, the size of the neighborhood with respect to objective j, $R_j(t)$, is obtained by:

$$R_j(t) = (f_j^{\max}(t) - f_j^{\min}(t)) \cdot r(t) \quad (8.21)$$

where $f_j^{\max}(t)$ and $f_j^{\min}(t)$ denote the maximum and minimum of the j-th objective at the t-th generation in set F_i, and $r(t)$ is the ratio of the size of the neighborhood to the span of the j-th objective in non-dominated front F_i at generation t, which is adapted by:

$$r(t) = r(t-1) * e^{-\frac{1-\gamma(t-1)/T}{m}}, \quad (8.22)$$

where $r(t-1)$ is the ratio of the size of the neighborhood to the span of the j-th objective of the solutions in F_i at the $(t-1)$-th generation, m is the number of objectives, $\gamma(t-1)$ is the ratio of knee points to the number of non-dominated solutions in the i-th front, and $0 < T < 1$ is a threshold controlling the ratio of knee points in the solution set F_i.
Equation (8.22) aims to ensure that $r(t)$ will be significantly reduced when $\gamma(t-1)$ is much smaller than the given threshold T. Meanwhile, $r(t)$ will decrease more slowly as $\gamma(t-1)$ becomes larger. $r(t)$ will keep constant when $\gamma(t-1)$ equals T. Note that $\gamma(0)$ is set to 0 and $r(0)$ is initialized to 1.
Typically, the neighborhood size decreases quickly in the early search stage of the evolutionary optimization so that the number of identified knee points will significantly increase. The ratio of knee points to all non-dominated solutions $(\gamma(t))$ will reduce as the evolution proceeds, while the size of the neighborhoods

will gradually decrease. When $\gamma(t)$ approaches T, the size of the neighborhoods will converge and keep constant.
Once the neighborhood is determined, a knee solution can be identified. In addition, all solutions within the same neighborhood can be sorted according to a descending order of the distance in Eq. (8.21). Note that before determining the knee points, the combined population has already been sorted into N_F non-dominated fronts: $F_i, 1 \leq i \leq N_F$.

- *Environmental selection.* KnEA starts to select the non-dominated solutions in F_1, which contains $|F_1|$. If $|F_1| < N$, KnEA then selects $(N - |F_1|)$ solutions in the second non-dominated front F_2. This process continues until the so-called the critical front F_c, in which the number of solutions $|F_c|$ is larger than $(N - \sum_{i=1}^{c-1} |F_i|)$. In this case, the knee points of F_c will be selected at first. Let the number of knee points in F_c be NP_c. In case NP_c is larger than $(N - \sum_{i=1}^{c-1} |F_i|)$, then those knee points having a larger distance to the hyperplane are selected to fill the population; Otherwise, NP_c knee points are selected together with $(N - \sum_{i=1}^{c-1} |F_i| - NP_c)$ other solutions in F_c that have a larger distance to the hyperplane of F_c.

8.2.3 A Two-Archive Algorithm for Many-Objective Optimization

As a kind of special MOPs, many-objective optimization problems have more than three objective functions, which poses hardness to many-objective evolutionary algorithms on both convergence and diversity.

- **Convergence**: As mentioned, the Pareto dominance fails on many-objective optimization problems, because the percentage of non-dominated solutions in a random population is almost 100%, which makes Pareto-based MOEAs cannot converge. Existing MOEAs solely based on the Pareto dominance cannot satisfactorily solve many-objective optimization problems. Therefore, non-Pareto-based MOEAs like aggregation- and indicator-based algorithms can be a solution for many-objective optimization, where the original problem is transferred into a number of SOPs to improve the convergence.
- **Diversity**: The diversity in a high-dimensional objective space is hard to be maintained using a limited number of solutions, where their similarity cannot be measured.

In the previous section, RVEA is an example of aggregation-based MOEAs for many-objective optimization. We will introduce an alternative many-objective evolutionary algorithms based on indicators and Pareto dominance.

Two_Arch (Praditwong & Yao, 2006) is the first MOEA for many-objective optimization problems using two archives for convergence and diversity separately (termed CA and DA). However, it is still a Pareto-based MOEA and cannot perform well on many-objective optimization problems. In 2015, an improved Two_Arch (named Two_Arch2) (Wang et al., 2015) follows the two-archive structure but

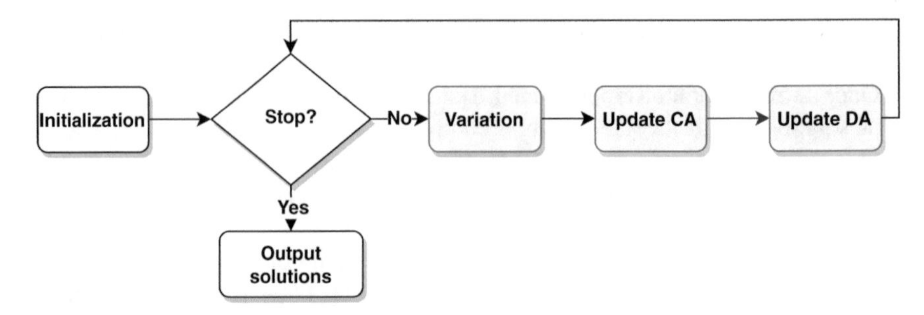

Fig. 8.5 Flow-chart of Two_Arch2

changes the selection methodologies, whose main flow chart is shown in Fig. 8.5. Two_Arch2 has the genetic EA process: initialization, variation, and selection. Due to the dual-archive structure, the variation and selection operation in Two_Arch2 needs to consider the interaction between and inside both CA and DA.

In Two_Arch2, CA employs an $I_{\varepsilon+}$-based selection to accelerate the convergence, which was used in IBEA (Zitzler & Künzli, 2004). The definition of $I_{\varepsilon+}$ can be written as Eq. (8.23), which means the minimum distance that $\mathbf{x}_1$ needs in order to dominate$\mathbf{x}_2$ in the objective space. Then, their fitness is assigned as Eq. (8.24) to select good converging solutions.

$$I_{\varepsilon+}(\mathbf{x}_1, \mathbf{x}_2) = min_{\varepsilon}(f_i(\mathbf{x}_1) - \varepsilon \leq f_i(\mathbf{x}_2), 1 \leq i \leq m) \tag{8.23}$$

$$Fitness(\mathbf{x}_1) = \sum_{\mathbf{x}_2 \in P \setminus \{\mathbf{x}_1\}} -e^{-I_{\varepsilon+}(\mathbf{x}_2, \mathbf{x}_1)/0.05} \tag{8.24}$$

In the step of updating CA, all the generated offspring are added to CA. Then, the extra solutions is deleted from CA according to Eq. (8.24), where the solution with the smallest $I_{\varepsilon+}$ loss is removed from CA, and the $I_{\varepsilon+}$ values of the remaining solutions are updated as shown in Algorithm 1. Such a loop stops when a pre-defined number of solutions are obtained in CA.

Algorithm 1 Pseudo-code of updating CA.

1: **Parameters**: n_{CA}-the fixed size of CA.
2: **while** $|CA| > n_{CA}$ **do**
3: Find the individual $\mathbf{x}^*$ with the minimal $F(\mathbf{x}^*)$.
4: Delete $\mathbf{x}^*$ from CA.
5: Update the remaining individual by $F(\mathbf{x}) = F(\mathbf{x}) + e^{-I_{\varepsilon+}(\mathbf{x}^*, \mathbf{x})/0.05}$.
6: **end while**

DA uses the Pareto dominance to select non-dominated solutions, where only the non-dominated solutions can be kept. The overflowed solutions are selected based on the similarity, i.e. diverse non-dominated solutions are selected. The selection in

DA is shown in Algorithm 2. When DA is full, boundary solutions (solutions with maximal or minimal objective values) are selected in DA. Then, the most different solution from the current DA is selected until DA is full.

Algorithm 2 Pseudo-code of updating DA.

```
1: Parameters: n_DA-the fixed size of DA.
2: Set DA empty.
3: Find solutions with maximal or minimal objective values in the set, and move them to DA.
4: while DA is not full do
5:    for each member i in the set do
6:       Similarity[i] = min(distance(i, j)), j ∈ DA
7:    end for
8:    I = argmax(Similarity), move solution I to DA.
9: end while
```

L_2-norm similarity in a high-dimensional space cannot qualitatively measure the distance in (Aggarwal et al., 2001, Morgan & Gallagher, 2013), whereas the fractional distances (L_p-norm, $p < 1$) works. Therefore, in Two_Arch2, the similarity is measured by $L_{1/m}$-norm-based distance, where m is the number of objective functions.

To further simultaneously improve the convergence and diversity of Two_Arch2, the interaction between CA and DA plays a key role. Therefore, the mutation operation is applied to CA only, which aims to speed up the convergence; and the crossover operation is applied between CA and DA, which aims to exchange the advantages of both archives.

With the clear optimization targets in CA and DA, Two_Arch2 shows effective on many-objective optimization problems without pre-setting any reference vectors. The main advantages of Two_Arch2 are two-fold.

- **A hybrid MOEA**: Two_Arch2 assigns different dominance relations to the selection of CA and DA, where CA is based on the $I_{\varepsilon+}$ indicator and DA is based the Pareto dominance. Such a setting inherits the advantages of indicator- and Pareto-based MOEAs (fast convergence and good diversity).
- **$L_{1/m}$-norm-based diversity maintenance**: An $L_{1/m}$-norm-based diversity maintenance is used in DA of Two_Arch2, which outperforms existing L_2-norm distance-based diversity maintenance methods, because the Euclidean distances lose their performance in a high-dimensional space.

8.2.4 Corner Sort for Many-Objective Optimization

Non-dominated sorting, a process of assigning solutions to different ranks $Front_j$, is an essential part in most MOEAs (Tian et al., 2017) for their selection operation. The whole process is based on the Pareto dominance for solution comparisons.

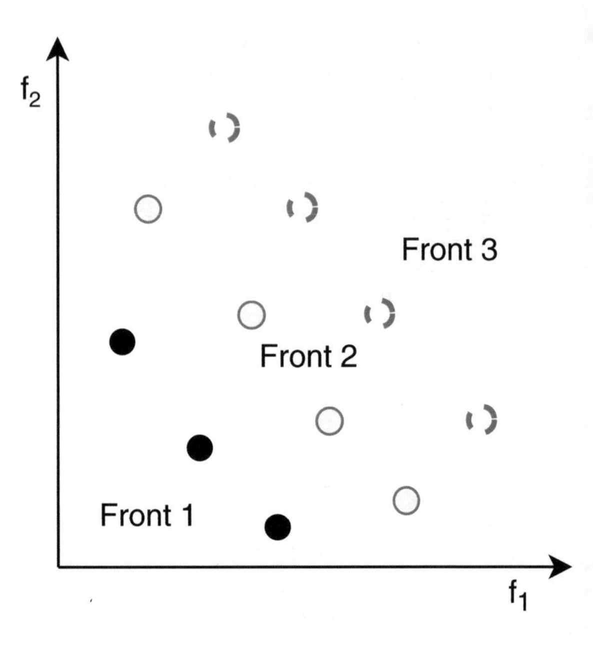

Fig. 8.6 An example population after non-dominated sorting with three fronts

To compare two solutions $\mathbf{a} = (a_1, ..., a_i, ..., a_m)$ and $\mathbf{b} = (b_1, ..., b_i, ..., b_m)$ with m objectives of a minimization MOP, **a** is said to dominate **b** ($\mathbf{a} \prec \mathbf{b}$), if $a_i \leq b_i$ $(1 \leq i \leq m)$ and $\mathbf{a} \neq \mathbf{b}$. Thus, N solutions in a population P are comparable by non-dominated sorting.

The aim of non-dominated sorting (Deb et al., 2002) is to obtain a division as $P = \{Front_1, ..., Front_j, ..., Front_n\}$ with n disjoint subsets, ranks, or fronts as shown in Fig. 8.6. The ranks of those fronts show the selection priority in MOEAs, where the subset with a smaller rank means better convergence. Therefore, for two solutions $\mathbf{x}_1$ and $\mathbf{x}_2$, the following conditions hold:

$$\forall \mathbf{x}_1, \mathbf{x}_2 \in Front_j (1 \leq j \leq L), \mathbf{x}_1 \not\prec \mathbf{x}_2, \tag{8.25}$$

$$\forall \mathbf{x}_1 \in Front_1, \forall \mathbf{x}_2 \in P, \mathbf{x}_2 \not\prec \mathbf{x}_1, \tag{8.26}$$

$$\forall \mathbf{x}_1 \in Front_j, \exists \mathbf{x}_2 \in Front_{j-1} (2 \leq j \leq L), \mathbf{x}_2 \prec \mathbf{x}_1. \tag{8.27}$$

In each rank $Front_j$, solutions are non-dominated to each other. The solutions with the first rank $Front_1$ are non-dominated solutions. The solution in $Front_j$ must be dominated by one solution in $Front_{j-1}$.

Naturally, non-dominated sorting is iterative process to find the non-dominated solutions of the remaining unranked solution, which is shown in Algorithm 3. the time complexity of a non-dominated sorting is $O(mN^3)$, where m is the number of objectives and N is the population size.

Algorithm 3 Pseudo code of natural non-dominated sorting.

Input: P-population, N-population size
1: $Rank[1:N] = 0$.
2: $i = 1$.
3: **for** all unranked solutions $P[j]$ **do**
4:   **for** all unranked solutions $P[k]$ **do**
5:     **if** $P[k] \prec P[j]$ **then**
6:       Break
7:     **end if**
8:   **end for**
9:   **if** $P[j]$ is non-dominated **then**
10:     $Rank[j] = i$
11:   **end if**
12: **end for**
13: $i = i + 1$.
Output: $Rank$.

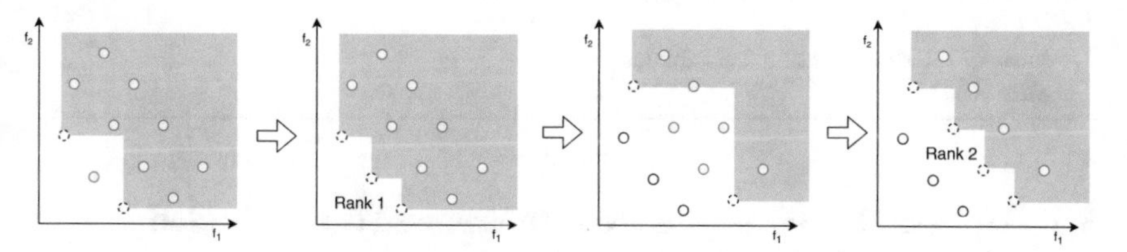

Fig. 8.7 An example of the process of corner sort

To reduce such a high time complexity, fast non-dominated sorting (Deb et al., 2002) sacrifices the space complexity. It travels all the solution pairs to record their dominance relation. Then, the time complexity is reduced to $O(mN^2)$.

However, as the number of objectives increases, the complexity of non-dominated sorting raises, because a large number of objective comparisons are needed to determine their dominance relation. To save objective comparisons in non-dominated sorting on many-objective optimization problems, corner sort (Wang & Yao, 2014) uses corner solutions which are the solutions with the best objective values for a fast non-dominated solution detection, because corner solutions must be non-dominated.

The main process of corner sort can be shown as the example in Fig. 8.7. The corner solutions can be quickly found by only comparing one objective, they are included in the current rank $Front_j$. Then, the comparisons with their dominating solutions (i.e. the solutions in the gray area) can be saved. As Algorithm 4 shows, corner sort has two loops: the outer loop is to increase the rank and the inner loop is to find corner solutions and ignore the comparisons with dominating solutions. When all the solutions are ranked, corner sort outputs the rank results.

Although the time complexity of corner sort is still $O(mN^2)$, it saves a number of objective comparisons on many-objective optimization problems. The comparative experiments in (Wang & Yao, 2014) shows its efficiency.

Algorithm 4 Pseudo code of corner sort.

Input: P-population, N-population size, $Rank$-rank result, m-number of objectives.
1: $Rank[1:N] = 0$.
2: $i = 1$.
3: **repeat**
4: Unmark all the unranked solutions.
5: $j = 1$
6: **repeat**
7: Find solution $P[q]$ of the best objective f among the unmarked ones
8: mark q, $Rank[q] = i$
9: $j = (j + 1)\%m + 1$
10: **for** $k = 1 : N$ **do**
11: **if** $P[k]$ is unmarked and $P[q] \prec P[k]$ **then**
12: Mark $P[k]$
13: **end if**
14: **end for**
15: **until** all the solutions in P are marked
16: $i = i + 1$.
17: **until** all the solutions in P are ranked.
Output: $Rank$.

8.3 Gaussian Process Assisted Reference Vector Guided Many-Objective Optimization

Not many surrogate-assisted evolutionary algorithms dedicated to many-objective optimization have been reported so far. Recently, an Gaussian process assisted RVEA (which is introduced in Sect. 8.2.1) has become popular for its attractive performance (Chugh et al., 2018). The Gaussian process assisted RVEA is called K-KRVA since Gaussian process is also known as kriging model. K-RVEA can be seen as a Bayesian evolutionary optimization framework, in which RVEA is run for a number of generations ($t_{\max}$) based on the surrogates. Before optimization starts, a set of initial samples (N_I) are collected using the Latin hypercube sampling strategy. According to (Jones et al., 1998; Knowles, 2006), N_I can be set to $11n - 1$, where n is the number of decision variables (search dimension). Each objective function is modeled by a Gaussian process and therefore, m Gaussian process models are built. Once one round of RVEA based optimization has completed, K-RVEA uses the predicted fitness (the angle penalized distance in RVEA) and the estimated uncertainty of the predicted objective values for selecting solutions obtained in the last generation of the current round of optimization to be sampled, i.e., to be evaluated using the real expensive

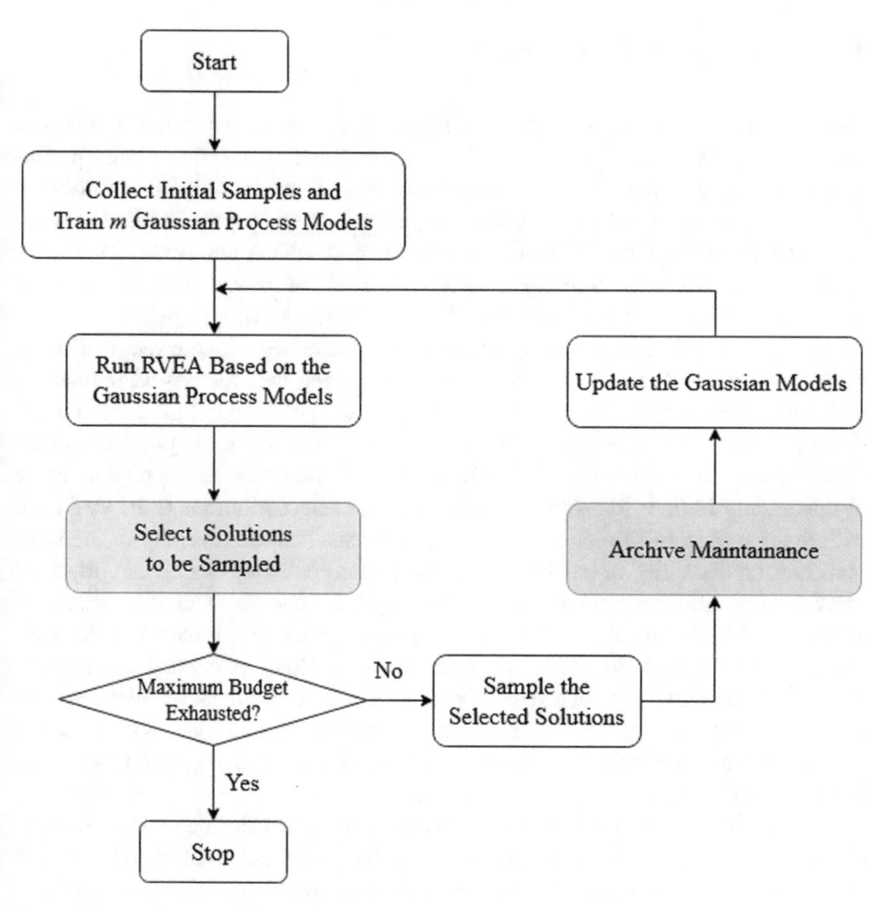

Fig. 8.8 A diagram of the K-RVEA algorithm

objective functions. After the selected solutions are sampled, they are added to the training database, which is to be maintained to keep the size of the training data to be constant (to be the same as the initial data size N_I. Finally, all Gaussian process models are updated, and a new round of optimization based on the updated surrogates and the RVEA is started. Therefore, the main new components in K-RVEA are a model management strategy according to the balance between diversity and convergence, and a strategy for maintaining the archive that stores the training data.

A diagram of the overall K-RVEA algorithm is provided in Fig. 8.8.

8.3.1 Surrogate Management

The main mechanism for surrogate management determine whether the solutions with high predicted quality (based on the APD representing the convergence performance), or those with a high degree of uncertainty, should be sampled according to the diversity of the population. To this end, a set of fixed, evenly distributed reference vectors (denoted by $\mathbf{v}^f$) is introduced. Then, K-RVEA compares the number of empty reference vectors in the fixed reference set before and after this round of surrogate update, which are denoted by $k_{\mathbf{v}^f}(t_u - 1)$ and $k_{\mathbf{v}^f}(t_u)$, respectively. A reference vector is called empty when no solution is associated with the vector. Thus, if $k_{\mathbf{v}^f}(t_u) - k_{\mathbf{v}^f}(t_u - 1) > \delta$, where δ is a positive integer, then there is a big increase in the number of empty vectors in $\mathbf{v}^f$, meaning that the diversity of the solutions has significantly decreased compared with that of the solutions in the previous round. Consequently, the priority should be given to the solutions that have a higher degree of uncertainty in their fitness prediction when we select solutions to be evaluated with the real objective function. Otherwise, it means that the diversity of the solutions has not decreased or has increased, and therefore, the algorithm can prioritize solutions that help convergence when selecting solutions for evaluation using the expensive objective functions. Note that the convergence performance of the solutions, namely the angle penalized distance (APD) should be calculated according to the RVEA's adaptive reference vector ($\mathbf{v}^a$), since RVEA adapts its reference vectors at a certain frequency to achieve a set of evenly distributed solutions, as discussed in Sect. 8.2.1. Note that in the first round, APD is used by default in selecting solutions to be sampled.

A remaining question is how many solutions should be selected in each round of updating the surrogates. Since the available total budget, usually measured in terms of the number of expensive fitness evaluations, denoted by $FE_{\max}$, is limited, K-RVEA groups the solutions into a number of groups based on $\mathbf{v}^a$. The number of groups, denoted by u, is specified by the user. Like in HeE-MOEA presented in Sect. 7.4.2, we can empirically set $u = 5$ by default. Then, the active adaptive vectors $\mathbf{v}^a$ are clusters into u clusters using k-means clustering. Finally, the solution closest to each active reference vectors grouped in the same cluster are compared with respect to their APD values or their degrees of uncertainty, depending on whether convergence or diversity is used as the criterion for selecting individuals. As a result, u solutions are chosen for real fitness evaluations. Note that if the number of active adaptive reference vectors is smaller than u, then each active reference vector will be treated as one cluster.

8.3.2 Archive Maintenance

Maintenance of the archive storing the training data, or selection of data for updating the Gaussian processes, serves two purposes. First, the training data size should be controlled due to the cubic time complexity of Gaussian processes to the number of training data. Second, the GP models should be updated in such a way that the updated surrogates can more effectively assist the K-RVEA's search on the Gaussian process based surrogates.

At first the u recently evaluated solutions are added into the archive for storing the training data. Duplicate solutions, if any, are removed. Here, a threshold is needed for defining duplicate solutions in the decision space. If the number of training data is still larger than N_I, some data that are evaluated in the previous rounds of updates will be removed. To achieve this, at first, the u recently evaluated solutions are assigned to the adaptive reference vectors. Then, the rest solutions in the archive that are sampled in the previous rounds will be assigned to the *inactive* adaptive reference vectors, denotes as V^a_{inactive}. Then, the reference vectors in V^a_{inactive} that become active can be identified, which will be further grouped into $(N_I - u)$ clusters using e.g., the k-means clustering algorithm. Finally, one solution from each of these $(N_I - u)$ clusters will be randomly selected to be stored in the archive. An illustrative example of solution assignment and reference vector clustering for archive maintenance is provided in Fig. 8.9.

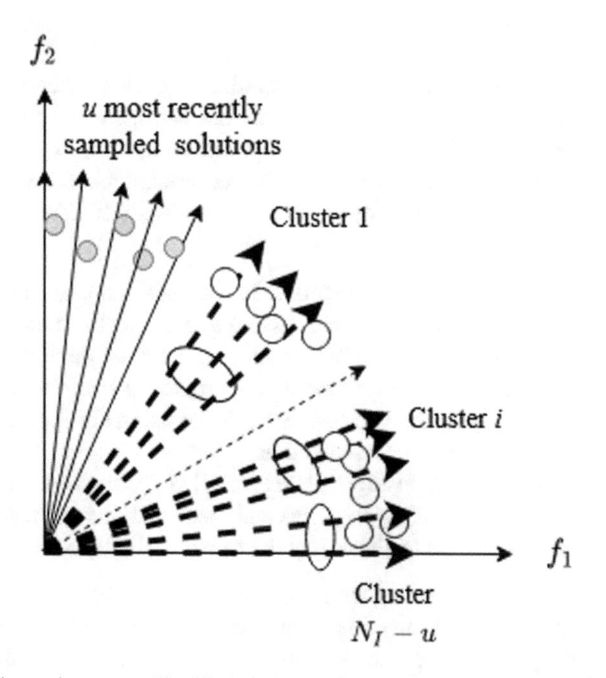

Fig. 8.9 Archive maintenance. The filled circles are the u solutions sampled in this round. Therefore, the five reference vectors on the left indicated by solid lines are active. The rest solutions in the archive (indicated by the circles) are then assigned to the inactive reference vectors, and the inactive reference vectors indicated by the thick dashed line become active while only one vector indicated by the thin dashed line remains inactive. The inactive vector that become active (indicated by the thick dashed line are then clustered into $N_I - u$ clusters and one solution will be randomly selected to be stored in the archive

8.4 Classification Surrogate Assisted Many-Objective Optimization

8.4.1 Main Framework

Most surrogates directly predict the objective values or the fitness values in both single- and multi-objective evolutionary optimization. However, such strategies are not scalable to the increase in the number of objectives, in particular when construction of a single surrogate is computationally intensive, such as the Gaussian process. For multi- or many-objective optimization, it is highly desirable if a surrogate is able to directly predict the dominance relationship or the ranking of the solutions to be compared (Loshchilov et al., 2010; Zhang et al., 2015a), or the ranking of the solutions (Lu & Tang, 2012, Bhattacharjee & Ray, 2010). However, prediction of the non-dominance relationship is extremely challenging, since the hyperplane separating dominated and non-dominated solutions changes over the decision space. A classification based surrogate-assisted evolutionary algorithm (CSEA) for expensive many-objective optimization is presented in (Pan et al., 2018). Instead of predicting the dominance relationship between the candidate solutions, CSEA relies on a surrogate to learn the dominance relationship between the candidate solutions and selected reference solutions, rather than between the candidate solutions themselves, because the former is more predictable. As a result, only one surrogate is required regardless the number of objectives. Like in Bayesian optimization, a degree of uncertainty is estimated using a validation data set, which is then used in model management.

CSEA distinguishes itself with other surrogate-assisted MOEAs in mainly in the following aspects:

- CSEA consists of two optimization loops. In the main loop, parent solutions for the next generation are selected according to the real objective functions, while in the second loop, promising solutions are selected based on the surrogate.
- A radial projection based selection method is employed both for selecting reference points from the training data (those evaluated using the expensive objective functions), and for environmental selection.
- A surrogate based on a multi-layer perceptron (MLP) is adopted for predicting the dominance relationship between the candidate solutions and a set of selected reference solutions that are evaluated by the true objective functions.

The pseudo code of the overall algorithm is listed in Algorithm 5.

Before the main loop starts, an initial population P consisting of $11n - 1$ solutions are generated using Latin hypercube sampling, where n is the search dimension. An MLP with H hidden neurons is initialized using randomly generated weights, and the sigmoid function is adopted as the activation function is the sigmoid function. Like in other surrogate-assisted MOEAs, $11n - 1$ functional evaluations are made using the real objective functions and these solutions are stored in archive Arc as the training data for the MLP.

Algorithm 5 The Framework of CSEA.

Input:

N (population size), K (number of reference solutions), H (number of hidden neurons), $t_{\max}$ (maximum number of FEs), $g_{\max}$ (maximum number of surrogate predictions before being updated).

Output:

P (final population).

1: $P \leftarrow$ Initialize the population with $11n - 1$ solutions using Latin hypercube sampling method
2: $t \leftarrow 11n - 1$ /* n is the number of decision variables */
3: $MLP \leftarrow$ Initialize the MLP with H hidden neurons /*m is the number of objectives*/
4: $Arc \leftarrow P$
5: **while** $t \leq t_{\max}$ **do**
6: $P_R \leftarrow$ UpdateReference(P, K)
7: $C \leftarrow$ Classify(P_R, Arc)
8: $rr \leftarrow$ The rate of category II solutions in C
9: $tr \leftarrow \min\{rr, 1 - rr\}$
10: $[D_{\text{train}}, D_{\text{validation}}] \leftarrow$ DataPartition(Arc, C)
11: $MLP \leftarrow$ Train$(MLP, D_{\text{train}}, T)$
12: $[p_1, p_2] \leftarrow$ Validation$(MLP, D_{\text{validation}})$
13: $Q \leftarrow$ SurrogateAssistedSelection$(P, P_R, p_1, p_2, g_{\max}, tr)$
14: $Arc \leftarrow Arc \cup Q$
15: $P \leftarrow$ EnvironmentalSelection$(P \cup Q, N)$
16: $t \leftarrow t + |Q|$
17: **end while**

The main loops of CSEA is very similar to a standard MOEA, except that its offspring will be further evolved in the second loop and only a subset of preselected individuals which are predicted to be promising will be combined with the parent population, from which parents for the next generation will be selected using the radial projection based selection to be discussed in detail in Sect. 8.4.2.

The second loop consists of three key steps of CSEA as follows:

- *Reference Set Selection* A set of reference solutions P_R is selected from the existing samples that are evaluated using the expensive objective functions. Here, the radial projection based selection method is adopted. This reference set is used to construct a classifier for prediction.
- *Surrogate Management* To manage the surrogate, solutions in the archive are grouped into two categories. 75% of these solutions are saved in the training data set and the rest as the validation data. The validation set is for cross-validation to estimate the reliability of the surrogate, which is a classifier in CSEA.
- *Surrogate-Assisted Selection* Crossover and mutation are then applied on the parent solutions to create offspring. Parents for the next generation are selected based on the surrogate as well as the reliability of classification results in the second loop, which is similar to the level of uncertainty in Bayesian optimization.

In the following, we detail the radial projection based selection, a method for prediction of dominance relationship based on a reference set, reference set selection, and surrogate management.

8.4.2 Radial Projection Based Selection

The radial projection based selection proposed in (He et al., 2017) is adopted in CSEA both for reference set selection and environmental selection. The radial projection based selection maps a set of normalized m-dimensional objective vectors onto a 2-dimensional radial space, each occupying a grid. Then, the diversity of the population can be promoted by selecting solutions from different grids, while an adaptive penalty based approach proposed in RVEA is proposed to select a better converged solution from the grid with multiple solutions.

Given a solution $\mathbf{x}$ in an m-objective space, $F(\mathbf{x}) = (f_1(\mathbf{x}), \ldots, f_m(\mathbf{x}))$, where $f_i(\mathbf{x}) \in [0, 1]$, that is, all objective values are normalized to [0, 1]. Then, two weight vectors $\mathbf{w}_1$ and $\mathbf{w}_2$ can be defined as follows:

$$\mathbf{w}_1 = (\cos\theta_1, \ldots, \cos\theta_m)^T \quad (8.28)$$

$$\mathbf{w}_2 = (\sin\theta_1, \ldots, \sin\theta_m)^T \quad (8.29)$$

where θ_i is the angle corresponding to dimension i and $\theta_i = 2\pi(i-1)/m$. Then the coordinate of solution A in the radial space, denoted by $(y_1(\mathbf{x}), y_2(\mathbf{x}))$, where $y_1(\mathbf{x}) = F(\mathbf{x})\mathbf{w}_1(F(\mathbf{x})\,\mathbf{1})^{-1}$, and $y_2(\mathbf{x}) = F(\mathbf{x})\mathbf{w}_2(F(\mathbf{x})\,\mathbf{1})^{-1}$, where $\mathbf{1}$ is an $m \times 1$ vector with all elements being 1.

It has been shown theoretically that the Euclidean distance between two solutions in a high-dimensional space can be properly reflected by that of the projected solutions in the radial space. Hence, the radial projection is able to maintain the diversity of the solutions in the m-dimensional objective space when they are mapped to the 2-dimensional radial space.

Subsequently, K solutions will be selected one by one using a criterion account for both convergence and diversity. In the radial projection based selection, function Con is used to indicate the convergence quality of a solution based on its Euclidean distance to the ideal point, and function Div is used to assess the diversity property of a solution according to the number of selected solutions in the same grid. The details of the radial projection based selection are presented in Algorithm 6, and its component that determines the grids in radial projection is given in Algorithm 7.

Note that K in Algorithm 6 is replaced with N, which is the population size.

8.4.3 Reference Set Based Dominance Relationship Prediction

In dominance based MOEAs, it is theoretically not necessary to know the exact objective function values for selection, so long as the dominance relationship between the solutions can be figured out. However, predicting the non-dominance relationship is hard, if not impossible, since dominance relationship is a relative measure, which changes over the decision space. Therefore, it is essential to define a learnable classification boundary that can be used for predicting the dominance relationship of

Algorithm 6 RadialSelection(P, K)

Output:
P_R (reference population).
1: $[Y, G] \leftarrow$ RadialGrid(P, K)
2: $C \leftarrow 0$
3: $Con(P) \leftarrow ||\frac{P-\min P}{\max P-\min P}||$
4: $P_R \leftarrow \arg_{P_R \in P} \min Con(P)$
5: $Div(P_R) \leftarrow 1$
6: **while** $|P_R| < K$ **do**
7: $Q \leftarrow \arg\min Div(P)$
8: $Fit(Q, P_R) \leftarrow 0.1 \cdot m \cdot Con(Q) - \min ||Y_Q - Y_{P_R}||$
9: $P_q \leftarrow \arg_{q \in K} \min Fit(Q, P_R)$
10: $P_R \leftarrow P_R \cup \{P_q\}$
11: $P \leftarrow P \backslash \{P_q\}$
12: $Div(q) \leftarrow Div(q) + 1$
13: **end while**

Algorithm 7 RadialGrid(P, K)

Output:
Y (coordinate in radial space), G (rectangle labels).
1: $P_N \leftarrow \frac{P-\min P}{\max P-\min P}$
2: $\theta_i \leftarrow 2\pi(i-1)/m$ /*$i = 1, 2, \ldots, m$*/
3: $\mathbf{w}_1 \leftarrow (\cos(\theta_1), \ldots, \cos(\theta_m))$,
$\mathbf{w}_2 \leftarrow (\sin(\theta_1), \ldots, \sin(\theta_m))$
4: $Z \leftarrow (P_N \mathbf{1})^{-1}$
5: $Y \leftarrow (P_N \mathbf{w}_1 Z, P_N \mathbf{w}_2 Z)$
6: $n \leftarrow \lfloor \sqrt{K} \rfloor$
7: $B_l \leftarrow \min Y, B_u \leftarrow \max Y$
8: $G \leftarrow \lfloor n(Y - B_l)/(B_u - B_l) \rfloor$
/*Calculate the label of each projected point*/

unseen solutions. To this end, CSEA defines a set of reference solutions to construct the Pareto dominance boundary, which divides solutions into two classes. For a bi-objective minimization problem, Fig. 8.10 provides an example for illustrating the proposed idea, where four reference solutions (shown in the objective space) are used to classify ten candidate solutions into two groups to indicate there dominance relationship. Solutions on the right side of the non-dominance boundary are denoted as category I, and the ones on the left side are denoted as category II.

The next question is which of the given solutions that have been evaluated using the real objective functions should be selected as the reference set for defining the classification boundary. The reference solutions play a key role in the performance of CSEA, since as we can see from Fig. 8.10, the boundary specified by the reference solutions determines which solutions are classified as category I and which are category II, which will influence not only the convergence but also the diversity of the selected solutions.

Usually, a classification boundary specified by a small number of reference solutions will more likely to classify solutions into category I, resulting in better conver-

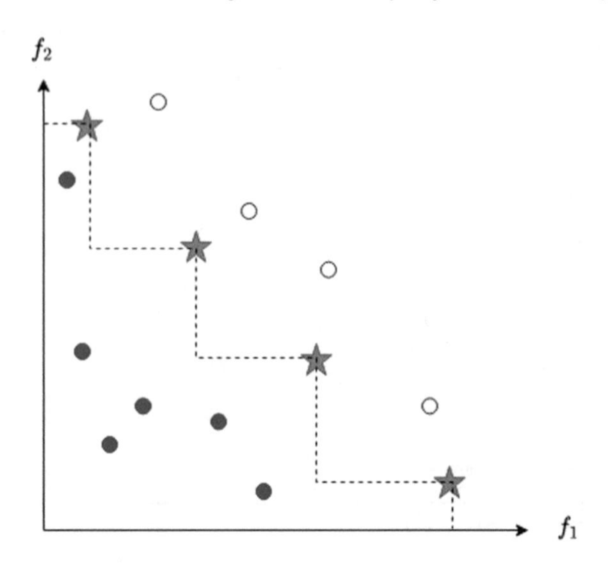

Fig. 8.10 Illustration of non-dominance boundary for classification for a bi-objective minimization problem. The stars indicate the defined reference solutions that are evaluated by the real objective functions. Based on the reference set, a dominance boundary (indicated by the dashed line) can then be determined, which can then classify the new candidate solutions (denoted by circles) into category I (unfilled circles) and category II (filled circles)

gence performance but a poor diversity. By contrast, a boundary resulted from a large reference set may group more solutions into category II, leading to poor convergence but a good degree of diversity. Thus, determining the number of reference solutions provides strike a balance between convergence and diversity of the selected category II solutions.

The right number of reference solutions K may be related to the relationship between the number of objectives. For problems with three to four objectives, CSEA is meant to emphasize on diversity maintenance, and therefore a relative large K can be chosen. As the number of objectives increases, convergence becomes more crucial for CSEA since the rate of non-dominated solutions in the population will become higher. In this case, a small K will classify more solutions into category I solutions, thereby increasing the selection pressure. One K is determined, the radial projection based selection in Algorithm 6 can then be applied for reference set selection.

Given the selected reference set, a surrogate can then be trained to learn the classification boundary.

8.4.4 Surrogate Management

Surrogate management in CSEA consists of the following steps:

- *Surrogate training* The structure of the MLP is defined, including the number of hidden layers and the number of nodes in each hidden layer. Since the number of

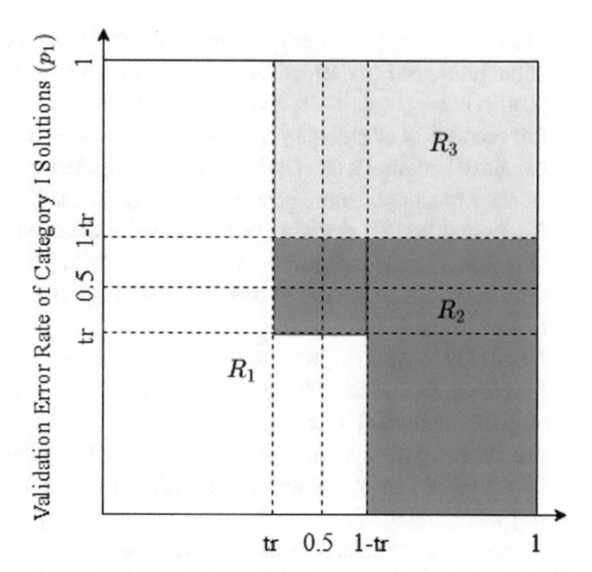

Fig. 8.11 Uncertainty estimation of the predictions by the MLP classifier according to the validation errors p_1 and p_2

data is usually limited in expensive optimization, one hidden layer is adequate. All weights are randomly initialized in [0; 1] and the sigmoid function is employed as the activation function. The MLP is trained using the gradient-based method or another supervised learning algorithm based on the training data. The maximum number of iterations is set to T.

- *Surrogate validation* The approximation error on the validation data is used for estimating the prediction uncertainty of the MLP. Instead of calculating the error on the entire validation data, the error of the category I solutions (denoted as p_1) and that of the category II solutions (denoted as p_2) in the validation set ($D_{\text{validation}}$) are calculated separately. Suppose Q_c is the set of category c solutions, and their predicted categories are $C_{p_1}, \ldots, C_{p_{|Q|}}$, where $|Q|$ is the number of solutions in Q_c. The mean absolute error is calculated by:

$$\text{MAE} = \frac{\sum_{i=1}^{|Q_c|} \text{abs}(c - C_{p_i})}{|Q_c|}, \tag{8.30}$$

where $|Q_c|$ is the number of solutions in Q_c, abs$(*)$ is the absolute value of $*$. MAE is employed as the validation error for category c solutions. Then p_1 and p_2 can be calculated using Eq. (8.30), which are employed to indicate the reliability (uncertainty) of the surrogate classifier.

- *Surrogate-Assisted Solution Selection* The reliability of the predictions is estimated by p_1 and p_2. Figure 8.11 presents the way in which the reliability is determined. In the figure, R_1 represents the region in which all category II solutions are reliable and therefore the MLP is expected to correctly select category II solutions to be evaluated using the expensive objective functions. R_1 consists of the following two subregions, one is defined by $p_2 < tr$, in which the classifier is able to reliably predict class II solutions, and the other by $p_1 < tr$ AND $p_2 < (1 - tr)$, in which the prediction of category I solutions is reliable and consequently the predicted category I solutions can be discarded. In the latter case, the boundary of p_2 can be relaxed to sample some promising category II solutions to promote exploration. If a poor classifier that always misclassifies a category I solution to be a category II solution, some category I solutions may be selected. Based on the predictions, only solutions predicted to belong to category II will be evaluated using the real objective functions.

None of the solutions predicted to be in R_2 will be evaluated using the real objective functions, in that the MLP is not able to reliably predict the dominance relationship of solutions. By contrast, region R_3 indicates that the MLP is very likely to predict category I solutions to be category II. Therefore, solutions predicted to be category I in R_3 might be promising and should be evaluated using the real objective functions.

A threshold much larger than 0.5 should be used to ensure that the predictions are reliable enough. Heuristically, the threshold tr can be set to $0.5 \times \min\{rr, 1 - rr\}$, where rr is the real rate of category II solutions given by:

$$rr = \frac{\sum_{i=1}^{|Q|} (Cr_i \text{ is category II})}{|Q|}, \tag{8.31}$$

where Q is the set of candidate solutions, $\{Cr_1, Cr_2, \ldots, Cr_{|Q|}\}$ are the true categories (classified according to the objective function values evaluated using the expensive objective functions. The predicted rate of category II solutions rp is defined by

$$rp = \frac{\sum_{i=1}^{|Q|} (Cp_i \text{ is category II})}{|Q|}, \tag{8.32}$$

where $\{Cp_1, Cp_2, \ldots, Cp_{|Q|}\}$ are the predicted categories of solutions in Q. Thus, the smaller the difference between rp and rr, the better accuracy of the MLP is.

8.4.5 Surrogate-Assisted Environmental Selection

Given the prediction of the surrogate as the reliability of the predictions, environmental selection can be performed. The offspring solutions can be generated by popular crossover and mutation operators, such as the simulated binary crossover and polynomial mutation. Then the location of the validation errors p_1 and p_2 on the reliability configuration is determined. For a solution in region R_1, those predicted to be category II are selected as parent individuals to generate offspring until the termination condition is met. If the solution is located in region R_3, the solutions predicted to be

Algorithm 8 SurrogateAssistedSelection(P, P_R, p_1, p_2, $g_{\max}$, tr)

Output:
 Q (offsprings).
1: $Q \leftarrow$ Variation($P \cup P_R$)
2: $L \leftarrow$ Prediction(MLP, Q)
3: $i \leftarrow 0$
4: **if** $p_2 < tr$ OR ($p_1 < tr$ AND $p_2 < 1 - tr$) **then**
5: **while** $i < g_{\max}$ **do**
6: $H \leftarrow$ Select $|P_R|$ solutions from Q with the largest L values
7: $Q \leftarrow$ Variation($H \cup P_R$)
8: $L \leftarrow$ Prediction(MLP, Q)
9: $i \leftarrow i + |Q|$
10: **end while**
11: $Q \leftarrow$ Select solutions from Q with $L > 0.9$
12: **else if** $p_1 > 1 - tr$ AND $p_2 > tr$ **then**
13: **while** $i < g_{\max}$ **do**
14: $H \leftarrow$ Select $|P_R|$ solutions from Q with the smallest L values
15: $Q \leftarrow$ Variation($H \cup P_R$)
16: $L \leftarrow$ Prediction(net, Q)
17: $i \leftarrow i + |Q|$
18: **end while**
19: $Q \leftarrow$ Select solutions from Q with $L < 0.1$
20: **else**
21: $Q \leftarrow \emptyset$
22: **end if**

category I are used to generate offspring solutions until the termination criterion is satisfied; otherwise, no solution will be selected. In surrogate-assisted selection in the second evolution loop, a maximum number of $g_{\max}$ solutions are predicted by the trained MLP, indicating the frequency that the surrogate has been used before the surrogate is updated. Note that only a small number of solutions are selected for evaluation using the expensive objective functions. Algorithm 8 presents the pseudo code of the surrogate-assisted environmental selection in the second loop of CSEA.

8.5 Dropout Neural Network Assisted Many-Objective Optimization

8.5.1 AR-MOEA

As discussed in Sect. 8.1, performance indicator based MOEAs are one class of evolutionary algorithms that can efficiently handle many-objective problems. However, some performance indicators, such as the hypervolume, are computationally intractable when the number of objectives increases. Here, we introduce one new performance indicator based algorithm based on an enhanced inverted generational distance with non-contributing solution detection, called IGD-NS (Tian et al., 2016).

The main motivation is to develop a computationally efficient performance indicator based MOEA that can perform efficiently regardless of the shape of the Pareto front. To this end, the reference points in the enhanced IGD-NS are adapted, and therefore the algorithm is called AR-MORA (Tian et al., 2018). The reference points (reference solutions) for calculating the IGD-NS consists of two sets, one containing a set of uniformly distributed solutions and the other is stored in an external archive.

Algorithm 9 provides the pseudo code of the overall AE-MOEA. In the following, we describe the IGD-NS indicator and the reference point adaptation method, which are two main components in AR-MOEA.

Algorithm 9 Overall Framework of AR-MOEA

Input: Population size N; Maximum number of generations $g_{\max}$;
Output: Final population P;
1: Create an initial population P_1 of size N;
2: Generate evenly distributed reference points W from a unit hyperplane;
3: $[A_1, R_1] \leftarrow UpdateRefPoint(P_1, W)$;
4: **for** $t = 1$ to $g_{\max}$ **do**
5: Use IGD-NS with respect to R_t to select solutions from P_t;
6: Generate offspring O_t using simulated binary crossover and polynomial mutation;
7: $[A_{t+1}, R_{t+1}] \leftarrow UpdateRefPoint([A_t; O_t], W)$;
8: Use IGD-NS with respect to R_{t+1} to select P_{t+1} from a combination of P_t and O_t for the next generation;
9: **end for**

8.5.1.1 IGD-NS Indicator

IGD-NS is based on IGD, which evaluates both convergence and diversity of a solution set. To more accurately assess the performance, IGD-NS considers the influence of the so-called the non-contributing solutions in the solution set under consideration. A solution is said to be non-contributing if it is not a nearest neighbor of any reference point. Mathematically, IGD-NS is defined as follows:

$$\text{IGD-NS}(Q, R) = \sum_{r \in R} \min_{q \in Q} \text{dis}(q, r) + \sum_{q' \in Q'} \min_{r \in R} \text{dis}(q', r) \tag{8.33}$$

where function $\text{dis}(\cdot)$ is the Euclidean distance between two solutions in the objective space, and Q, $Q'(Q' \in Q)$ and R denotes the non-dominated solution set, the non-contributing solutions, and reference points, respectively. The first term of IGD-NS sums up the minimum distance from each non-dominated solution to all points in R, while the second term sums up the minimum distance from each non-contributing solution to all solutions in R, which is meant to minimize the number of non-contributing solutions. The fitness of an individual p in terms of IGD-NS is calculated by IGD-NS$(P\backslash\{p\}, R)$, where $P\backslash\{p\}$ means the set of P after removing

p. Therefore, the larger the fitness value, the better. Similar to NSGA-II, the parent and offspring populations are combined, but then efficient non-dominated sorting in (Zhang et al., 2015b) is applied to sort the combined population for better computational efficiency. Then, parents are selected based on the sorted fronts until the critical front, where only subset of the solutions can be selected. Unlike NSGA-II that uses the crowding distance for selecting the solution, AR-MOEA relies on the individuals fitness as calculated in Eq. (8.33).

The objective values of solutions in P and Q are normalized based on the ideal point z^* and nadir point z^{nad}, which can be obtained by minimizing and maximizing each objective separately. Then, each reference point $r \in R$ is adjusted as follows. Each solution $q \in Q$ having the minimum perpendicular distance to vector $\overrightarrow{z^*r}$ is identified by minimizing $\|\mathbf{f}(q)\| \sin(\overrightarrow{z^*r}, \mathbf{f}(q))$, where $\mathbf{f}$ is the function that maps q from the decision space to the objective space. Then, each $r \in R$ is adjusted to the orthogonal projection of $\mathbf{f}(q)$ on vector $\overrightarrow{z^*r}$ in the following way:

$$r_i' = \frac{r_i}{\|r\|} \times \|\mathbf{f}(q)\| \cos(\overrightarrow{z^*r}, \mathbf{f}(q)), \quad i = 1, 2, \ldots, m \tag{8.34}$$

where m is the number of objectives.

8.5.1.2 Adaptation of Reference Points

One important modification of IGD-NS is the introduction of a mechanism for adapting the reference points R and an external archive A. The external archive A aims to capture the geometry of the Pareto front, which, together with a fixed evenly distributed reference points W, helps the adaptation of the reference set R to various shapes of the Pareto fronts. At first, the external archive A is adjusted based on a fixed reference set W and the current offspring population, as well as its contribution to IGD-NS. First, duplicate and dominated solutions in $A(t)$ are deleted. Then, the contributing solutions in $A(t)$, t is the generation index, with respect to W, denoted as $A'^c(t)$, are copied to $A(t+1)$. Finally, the solution p from $A(t) \setminus A(t+1)$ that maximizes $\min_{q \in A(t+1)} \arccos(\mathbf{f}(p), \mathbf{f}(q))$ is chosen one by one to fill $A(t+1)$ until the size of $A(t+1)$ reaches $\min(3|W|, |A(t)|)$. This ensures that solutions with the maximum angles to the selected solutions are added to the archive, thereby maintaining a good degree of diversity.

Once $A(t+1)$ is updated, the reference set R will be adjusted too. At first, the reference points in W closest to solutions in $A'^c(t)$, the contributing solutions in generation t, are identified and denoted as the valid reference point set W^v. W^v are copied to $R(t+1)$. Then, point q from $A(t+1) \setminus R(t+1)$ with the maximum value of $\min_{r \in R(t+1)} \arccos(\mathbf{f}(q), r)$ is chosen one by one to fill $R(t+1)$ until the size of $R(t+1)$ reaches $\min(|W|, |W^v| + |A(t+1)|)$ to promote the diversity of $R(t+1)$. The pseudo code for adapting the external archive and reference set is given in Algorithm 10.

Algorithm 10 Reference Point Adaptation

Input: $A(t)$; $R(t)$;P
Output: $A(t+1)$;$R_(t+1)$
1: //Archive adaptation//
2: Delete duplicate candidate solutions in $A(t)$;
3: Delete dominated candidate solutions in $A(t)$;
4: $A^c(t) \leftarrow \{p \in A \mid \exists r \in W : \text{dis}(r, F(p)) = \min_{q \in A(t)} \text{dis}(r, F(q))\}$
5: $A(t+1) \leftarrow A^c(t)$;
6: **while** $|A(T+1)| < \min(3|W|, |A(t)|)$ **do**
7: $\quad p \leftarrow \arg\max_{p \in A_t \setminus A(t+1)} \min_{q \in A(t+1)} \arccos(\mathbf{f}(p), \mathbf{f}(q))$
8: $\quad A(t+1) \leftarrow A(t+1) \cup p$;
9: **end while**
10: //Reference set adaptation//
11: $W^v \leftarrow \{r \in R \mid \exists p \in A'^c(t) : \text{dis}(r, F(p)) = \min_{s \in R} \text{dis}(s, F(p))\}$;
12: $R(t+1) \leftarrow W^v$;
13: **while** $|R(t+1)| < \min(|W|, |W^v| + |A(t+1)|)$ **do**
14: $\quad p \leftarrow \arg\max_{p \in A(t+1) \setminus R(t+1)} \min_{r \in R(t+1)} \arccos(r, \mathbf{f}(q))$
15: $\quad R(t+1) \leftarrow R(t+1) \cup p$;
16: **end while**

8.5.2 *Efficient Deep Dropout Neural Networks*

The dropout technique was meant for improving the generalization capability of deep neural networks (Srivastava et al., 2014). The theoretical results presented in (Gal & Ghahramani, 2016) implies that a dropout neural network can be seen as a Bayesian network, making it a good candidate to replace a Gaussian process model. In EDN-ARMORA (Guo et al., 2020), a dropout deep neural network assisted AR-MORA for many-objective optimization algorithm, a deep neural network containing a large number of hidden layers is too complex to serve as a surrogate in the presence of data paucity. Here, a feedforward neural network with two hidden layers is adopted to reduce the computational complexity, while making it still complex enough for properly performing dropout. The first hidden layer uses the ReLU activation function to achieve fast training, whereas the second hidden layer adopts the tanh function, making the neural network suited for regression. The stochastic gradient based method with mini-batches learning is used for training.

The output of EDN in the forward pass for the training stage can be expressed as follows:

$$\begin{aligned} \mathbf{y}_R &= f_R\left(\frac{1}{p_I}\left(\mathbf{x}_s \circ \mathbf{d}_I\right) W_1 + B_1'\right) \\ \mathbf{y}_T &= f_T\left(\frac{1}{p_R}\left(\mathbf{y}_R \circ \mathbf{d}_R\right) W_2 + B_2'\right) \\ \hat{\mathbf{y}}_s &= \mathbf{y}_T W_3 + B_3' \end{aligned} \tag{8.35}$$

where $\mathbf{x}_s$ and $\hat{\mathbf{y}}_s$ are the inputs and outputs of the efficient dropout network (EDN), W_1, W_2, W_3, and B_1, B_2, B_3, are the weight and bias matrices, respectively, $\mathbf{y}_R$ and

$\mathbf{y}_T$ are the outputs of the first and second hidden layer, respectively. The elements of $\mathbf{d}_I$ (for the input layer) and $\mathbf{d}_R$ (for the first hidden layer) are sampled from a Bernoulli distribution with parameter p_I and p_R, respectively. Matrices B_1', B_2' and B_3' are composed of *batchsize* copies of B_1, B_2 and B_3. f_R and f_T represent the ReLU and the tanh activation function, respectively. Note that the number of hidden nodes in the first and second hidden layers are J and K, respectively.

The loss function of the neural network is defined by:

$$\begin{aligned} \mathbf{e} &= \hat{\mathbf{y}}_s - \mathbf{y}_s \\ E &= \frac{1}{2}\left(\sum_{i=1}^{batchsize}\sum_{j=1}^{m}\mathbf{e}_{ij}^2 + \eta\, W_{\text{sum}}\right) \end{aligned} \tag{8.36}$$

where $\mathbf{y}_s$ is a matrix composed of the true objective values of the decision variables $\mathbf{x}_s$, η is the hyperparameter for weight decay, while W_{sum} is the quadratic sum of all elements in W_1, W_2 and W_3.

After training is complete, the EDN will be used to predict the fitness of the individuals of the population $\mathbf{x}_{\text{pop}}$ in AR-MOEA,

$$\begin{aligned} \hat{\mathbf{y}}_i = f_T&\left(\frac{1}{p_R}\left(f_R\left(\frac{1}{p_I}\left(\mathbf{x}_{\text{pop}} \circ \mathbf{d}_I\right)\times W_1 + B_1'\right)\circ \mathbf{d}_R\right)\times W_2 + B_2'\right) \\ &\times W_3 + B_3'. \end{aligned} \tag{8.37}$$

It should be stressed that, different from the original dropout technique, the dropout in EDN takes place at both the training and prediction stage, in order to estimate the uncertainty of the predictions. The output will be calculated for $Iter_{\text{pre}}$ times to estimate the fitness of new solutions, where $\mathbf{d}_I$ and $\mathbf{d}_R$ are regenerated anew every time. The output is mapped to its original range, and then the mean fitness estimations $\hat{\mathbf{y}}_{\text{pop}}$ of $\mathbf{x}_{\text{pop}}$ can be calculated, and the confidence level (degree of uncertainty) $\hat{\mathbf{s}}_{\text{pop}}$ of the predictions is estimated by

$$\begin{aligned} \hat{\mathbf{y}}_{\text{pop}} &= \frac{1}{Iter_{\text{pred}}}\sum_{i=1}^{Iter_{\text{pred}}}\hat{\mathbf{y}}_i \\ \hat{\mathbf{s}}_{\text{pop}} &= \sqrt{\frac{1}{Iter_{\text{pred}}}\sum_{i=1}^{Iter_{\text{pred}}}\hat{\mathbf{y}}_i^{\mathrm{T}}\hat{\mathbf{y}}_i - \hat{\mathbf{y}}_{\text{pop}}^{\mathrm{T}}\hat{\mathbf{y}}_{\text{pop}}}. \end{aligned} \tag{8.38}$$

Note that the number of iterations for updating the EDN during the optimization ($iter_r$) once new samples are added is typically much smaller than the number of iterations $iter_{train}$ for constructing the EDN the first time. This capability of efficient incremental learning is also very attractive compared to Gaussian process. When new samples are included in the training data during the optimization, there is only one single network model to be updated in EDN rather than multiple neural networks in the conventional dropout neural networks, making the EDN computationally more efficient for estimating the confidence level.

The pseudo code for training and prediction using the EDN is listed in Algorithm 11. Given a new candidate solution, the EDN can then predict its fitness and provide a confidence level of the predicted. Consequently, model management can be done based on these two types of information.

Algorithm 11 Efficient Dropout Network

Input: X_s and Y_s (training data pair), p_I and p_R (probability of retaining neurons), η (weight decay), α (learning rate), $iter_{\text{train}}$ (number of forward-backward passes), $batchsize$ (batch size), $iter_{\text{pred}}$ (number of calculations for the tested outputs), $\mathbf{x}_{\text{pop}}$ (population of AR-MOEA)

Output: $\hat{\mathbf{y}}_{\text{pop}}$ and $\hat{\mathbf{s}}_{\text{pop}}$ (the predicted fitness and the estimated confidence level for $\mathbf{x}_{\text{pop}}$);

1: Initialize the network Net. The weights $W = \{W_1, W_2, W_3\}$ and biases $B = \{B_1, B_2, B_3\}$ are randomly initialized within $[-0.5, 0.5]$;
2: **for** $i = 1$ to $iter_{\text{train}}$ **do**
3: A mini-batch of size $batchsize$ ($\mathbf{x}_s$ and $\mathbf{y}_s$) is created by randomly selecting a subset from X_s and Y_s;
4: $(Net, W, B) \leftarrow$TrainNet$(\mathbf{x}_s, \mathbf{y}_s, Net, W, B, p_I, p_R, \eta, \alpha)$;
5: **end for**
6: **for** $i = 1$ to $iter_{\text{pred}}$ **do**
7: $\hat{\mathbf{y}}_i \leftarrow$ PredictionNet$(\mathbf{x}_{\text{pop}}, Net, W, B, p_I, p_R)$;
8: **end for**
9: Calculate $\hat{\mathbf{y}}_{\text{pop}}$ and $\hat{\mathbf{s}}_{\text{pop}}$.

8.5.3 Model Management

The new samples to be evaluated by the expensive objective function are chosen from the solutions in the final generation of AR-MOEA, which optimizes the objective functions predicted by the EDN. To reduce the number of solutions to be sampled in each round, k-means clustering algorithm is applied to select a representative subset. The method for model management here is similar to that in K-RVEA as presented in 8.3.1. That is, when convergence of the solutions is prioritized, the solution having the minimum Euclidean distance to the origin in the objective space in each cluster will be selected for real function evaluations. Otherwise, the solution that has the maximum mean uncertainty in each cluster will be selected for evaluation using the expensive objective functions when diversity of the population must be promoted. The diversity of the population is measured by the ratio of $r = |W^v|/|W|$, where $|W^v|$ and $|W|$ are the number of active reference points and total number of fixed reference points. A reference point is called active if there is a contributing solution in the population associated with the reference point. If $r_{i-1} - r_i > \delta$, where δ is a predefined threshold, and r_{i-1} and r_i are the ratio in the i-th and $(i-1)$-th rounds of model update, then we consider the diversity of the population needs to be promoted.

All objective values should be translated by the minimum objective values before calculating the Euclidean distance.

Similar to K-RVEA, the number of the training data (X_1,Y_1) here is also limited to $11n - 1$, where n is the number of decision variables, to reduce the computation time in updating the EDN. At first, solutions that are most recently sampled, $(X_{\text{new}},Y_{\text{new}})$, are added to the empty (X_1,Y_1). The rest of the data in (X,Y) is selected one by one, and each time, the solution p in (X,Y) having the maximum angles to solutions in (X_1, Y_1) will be added promote the diversity of the training data.

8.5.4 Overall Framework of EDN-ARMOEA

The overall framework of the EDN assisted AR-MOEA, called EDN-ARMOEA is described in Algorithm 12. The general procedure of EDN-ARMOEA is similar to K-RVEA. Latin hypercude sampling (LHS) is adopted for sampling $11n - 1$ solutions based on the expensive objective functions. Then the EDN is initialized and trained using these samples. The AR-MOEA is then run on the EDN for $g_{\max}$ generations. Note that only one neural network is needed regardless the number of objectives, and the number of outputs of the END is m. Then, the mean fitness and uncertainty level for the solutions in the last generation of the population are calculated using the dropout predictions of the EDN to select k solutions to be sampled, i.e., to be evaluated using the real expensive objective functions. Then, the training data is updated and the EDN is re-trained using the updated training data. Then, the AR-MOEA is run again on the updated EDN, and this process repeats until the budget is exhausted, typically in terms of the maximum number of expensive function evaluations.

Algorithm 12 Overall Framework of EDN-ARMOEA

Input: $\mathbf{f}$ (expensive functions), $FE_{\max}$ (total function evaluations), n (search dimension), $\delta \in [0, 1]$ is a parameter, k (the number of real fitness evaluations in each generation), P is the population size, and $iter$ is the maximum generation)

Output: X and Y (the decision variables and objective values of final solutions);

1: $X \leftarrow$ **LHSDesign**$(n, 11n - 1)$, $Y \leftarrow \mathbf{f}(X)$;
2: $X_1 = X, Y_1 = Y, FE = 11n - 1$;
3: **while** $FE \leq FE_{\max}$ **do**
4: $(X_s, Y_s) \leftarrow$ **Normalize**(X_1, Y_1);
5: The EDN is initialized by **Net** $\leftarrow$ **EDN**(X_s, Y_s) or updated by **Net** $\leftarrow$ **EDN**$(X_s, Y_s, \mathbf{Net})$;
6: $(\mathbf{x}_{\text{pop}}, \hat{\mathbf{y}}_{\text{pop}}, \hat{\mathbf{s}}_{\text{pop}}, rf, rf^0) \leftarrow$ **AR-MOEA**$(N, g_{\max}, \mathbf{Net})$;
7: If there is no rb, rb is initialized by $rb = rf^0$;
8: $X_{\text{new}} \leftarrow$ **SelectIndividual**$(\mathbf{x}_{\text{pop}}, \hat{\mathbf{y}}_{\text{pop}}, \hat{\mathbf{s}}_{\text{pop}}, rf, rb, \delta, k)$, $Y_{\text{new}} \leftarrow \mathbf{f}(X_{\text{new}})$;
9: $X \leftarrow X \cup X_{\text{new}}, Y \leftarrow Y \cup Y_{\text{new}}$;
10: $(X_1, Y_1) \leftarrow$ **SelectTrainData**(X, Y);
11: $rb = rf, FE = FE + |X_{\text{new}}|$;
12: **end while**

Compared to a Gaussian process, the dropout neural network has the following attractive properties:

- The computational complexity of the EDN is more scalable to the increase in the number of training data. In addition, EDN is capable of incremental learning, and consequently the computation time in updating the END is much lower than that for initial construction of the EDN.
- Only one neural network is needed regardless the number of objective, unlike Gaussian processes, the computational complexity linearly increases with the number of objectives.

8.5.5 Operational Optimization in Crude Oil Distillation Units

EDN-ARMOEA is applied to solve a real-world operational optimization problem of crude oil distillation units, is computationally intensive because performance prediction of this process requires expensive numerical simulations. We compare the performance of EDN-AMOEAR with GP-ARMOEA, AR-MOEA assisted by a Gaussian process,HeE-ARMOEA, AR-MOEA assisted by a heterogeneous ensemble, and AR-MOEA without surrogates.

8.5.5.1 Problem Formulation

Crude oil distillation is very widely used separation technique in petroleum industry. By assigning operational conditions, we can improve the separation performance and the energy efficiency of distillation systems. According to (Ochoa-Estopier & Jobson, 2015), Table 8.1 lists 14 operational conditions to be optimized, which are the decision variables. The two objectives include the maximization of the net value of the products (f_1) and the minimization of energy consumption (f_2). The constraints include the product quality and distillation system specifications. Specifically, the optimization of the operational conditions in crude oil distillation units is defined as follows:

$$
\begin{aligned}
\max f_1 &= \sum_{i=1}^{4} P_{\text{prod},i} F_{\text{prod},i} - P_{\text{crude}} F_{\text{crude}} - P_{\text{steam}} \sum_{j=1}^{3} F_{\text{steam},j} \\
\min f_2 &= P_{\text{fuel}} U_{\text{fuel}} + P_{\text{cw}} U_{\text{cw}} \\
s.t.\ U_{\text{fuel}} &= gcc(F_{PA}, Q_{\text{PA}}, T_f, F_{\text{prod}}) \\
U_{\text{cw}} &= gcc(F_{PA}, Q_{\text{PA}}, T_f, F_{\text{prod}}) \\
& T95_i^{\text{lb}} \leq T95_i \leq T95_i^{\text{up}}, i = 1, 2, 3, 4 \\
& CC(F_{\text{prod}}, F_{\text{steam}}, F_{\text{PA}}, Q_{\text{PA}}, T_f) = 1
\end{aligned}
\tag{8.39}
$$

Table 8.1 Operational conditions in crude oil distillation units

No.	Symbol of operational conditions	Meaning (Units)
1	$F_{\text{prod},1}$	Naphtha flow rate (bbl/h)
2	$F_{\text{prod},2}$	Kerosene flow rate (bbl/h)
3	$F_{\text{prod},3}$	Light diesel flow rate (bbl/h)
4	$F_{\text{prod},4}$	Heavy diesel flow rate (bbl/h)
5	$F_{\text{steam},1}$	Main column steam flow rate (kmol/h)
6	$F_{\text{steam},2}$	1st stripper steam flow rate (kmol/h)
7	$F_{\text{steam},3}$	2nd stripper steam flow rate (kmol/h)
8	$F_{\text{PA},1}$	1st pump-around flowrate (bbl/h)
9	$F_{\text{PA},2}$	2nd pump-around flowrate (bbl/h)
10	$F_{\text{PA},3}$	3rd pump-around flowrate (bbl/h)
11	$Q_{\text{PA},1}$	1st pump-around duty (MW)
12	$Q_{\text{PA},2}$	2nd pump-around duty (MW)
13	$Q_{\text{PA},3}$	3rd pump-around duty (MW)
14	T_f	Furnace outlet temperature (°C)

where P and F represent prices and flow rates, respectively; subscripts $prod$, $crude$, $steam$, $fuel$ and cw represent products, crude oil, stripping steam, fuel and cooling water, respectively; U denotes the minimum requirement; U_{fuel} and U_{cw} are estimated using the grand composite curve (gcc) (Ochoa-Estopier & Jobson, 2015); $T95_i$ represents the 95% true boiling point temperature for product i, which is a product quality indicator; $CC(\cdot) = 1$ requires that the convergence criterion must be satisfied.

8.5.5.2 Parameter Settings

Before EDN-ARMOEA can be applied to the problem, one additional step needs to be done, mainly because the problem in Eq. (8.39) is a constrained problem. To handle the constraints, two classifiers to distinguish infeasible solutions are constructed, one determining whether the convergence criterion is satisfied, and the other verifying whether the operational conditions are feasible. The classifiers are built based on simulation data. Once they are built, these constraints model will no longer be updated. We randomly generate 150 sets of operational conditions to validate the accuracy of the classifiers, of which 118 out of 150 solutions converge, while 28 conditions are feasible based on the time-consuming numerical simulations. The classifier for judging convergence achieves an accuracy of 95.76% and 93.75%, respectively, for converging and non-converging solutions simulations. By contrast, the feasibility classifier achieves 75% accuracy for feasible solutions, and 92.22% for infeasible solutions. Overall, we consider the performance two classifier models are acceptable. These two classifiers are used for all four optimization algorithms under comparison.

Since the number of decision variables of this application example is 14, a total of 153 offline samples are generated for EDN-ARMOEA, GP-ARMOEA and HeE-ARMOEA, three surrogate assisted MOEAs. In addition, 145 real fitness evaluations are made during the optimization and therefore the surrogate can be updated for 29 times. The population size of AR-MOEA is set to 50, and consequently six generations can be run. This amounts to a total of 300 real fitness evaluations, which is closet to 298 fitness evaluations used by EDN-ARMOEA, GP-ARMOEA and HeE-ARMOEA.

8.5.5.3 Experimental Results

Hypervolume (HV) is adopted as the performance indicator to compare the three surrogate-assisted algorithms with respect to convergence and diversity. The reference point for calculating the HV is set to $r_j = \max_j + 0.01 \times (\max_j - \min_j)$, $j = 1, \ldots, m$, where $\max_j$ and $\min_j$ are the maximization and minimization of the non-dominated solutions obtained by the four compared algorithms. Additionally, HV is normalized by dividing $\prod_{j=1}^{m} r_j$. Figure 8.12 plots profiles of the HV values over the number of expensive fitness evaluations averaged over three independent runs. The final average HV value of the solution set obtained by EDN-ARMOEA, GP-

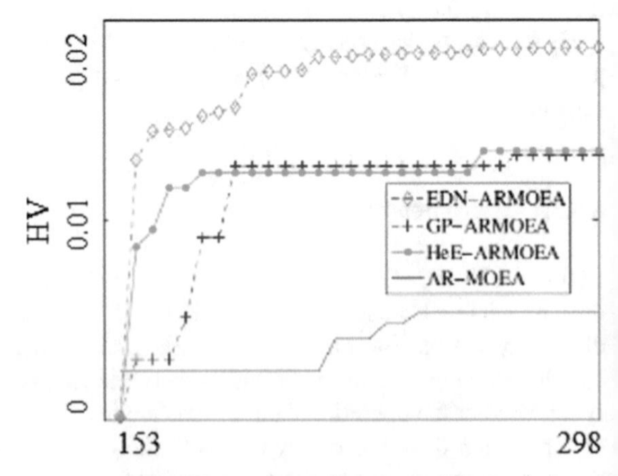

Fig. 8.12 The average HV values over the number of time-consuming simulations of EDN-ARMOEA (denoted by diamonds), GP-ARMOEA (crosses), HeE-ARMOEA (dots) and AR-MOEA

ARMOEA, HeE-ARMOEA and AR-MOEA are 0.0186, 0.0132, 0.0134 and 0.0054, respectively. The average run time of EDN-ARMOEA, GP-ARMOEA and HeE-ARMOEA is 139 h, 192 h and 182 h, respectively.

From these results, it can be observed that all three surrogate assisted MOEAs performs better than the one without a surrogate, indicating that the surrogate in all these algorithms is successful in improving the optimization performance. In addition, we can see that HeE-MOEA converges faster in the early stage of optimization compared to GP-ARMOEA, although finally the two algorithms have achieved similar results, confirming again that HeE-MOEA is a valuable replacement of Gaussian process assisted MOEA. Finally, the results confirm that EDN-ARMOEA is the most competitive in terms of both optimization performance and computation efficiency.

8.6 Summary

Many-objective optimization is challenging due to the high-dimensional objective space, leading to many additional complexities in the number of Pareto optimal solutions and the shape of the Pareto front. As a result, solving many-objective optimization is much harder than multi-objective optimization. Thus, surrogate-assisted evolutionary optimization of many-objective optimization problems remains a challenging topic with respect to both the performance and computational efficiency, in particular when the number of decision variables is also large.

In the next chapter, we will present methods to address the data paucity issue in data-driven optimization with the help of semi-supervised learning, transfer learning and transfer optimization.

References

Aggarwal, C. C., Hinneburg, A., & Keim, D. A. (2001). On the surprising behavior of distance metrics in high dimensional space. In *International conference on database theory* (pp. 420–434). Springer.

Bhattacharjee, K. S., & Ray, T. (2010). A novel constraint handling strategy for expensive optimization problems. In *11th world congress on structural and multidisciplinary optimization*.

Branke, J., Deb, K., Dierolf, H., & Osswald, M. (2004). Finding knees in multi-objective optimization. In *International conference on parallel problem solving from nature* (pp. 722–731). Springer.

Cheng, R., Jin, Y., Narukawa, K., & Sendhoff, B. (2015). A multiobjective evolutionary algorithm using Gaussian processbased inverse modeling. *IEEE Transactions on Evolutionary Computation, 19*(6), 838–856.

Cheng, R., Jin, Y., Olhofer, M., & Sendhoff, B. (2016). A reference vector guided evolutionary algorithm for many objective optimization. *IEEE Transactions on Evolutionary Computation, 20*(5), 773–791.

Chugh, T., Jin, Y., Miettinen, K., Hakanen, J., & Sindhya, K. (2018). A surrogate-assisted reference vector guided evolutionary algorithm for computationally expensive many-objective optimization. *IEEE Transactions on Evolutionary Computation, 22*, 129–142.

Cornell, J. A. (2011). *Experiments with mixtures: designs, models, and the analysis of mixture data*, (Vol. 403). John Wiley & Sons.

Deb, K., Pratap, A., Agarwal, S., & Meyarivan, T. (2002). A fast and elitist multiobjective genetic algorithm: NSGA-II. *IEEE Transactions on Evolutionary Computation*, *6*(2), 182–197.

Fleming, P. J., Purshouse, R. C., & Lygoe, R. J. (2005). Many-objective optimization: An engineering design perspective. In *International conference on evolutionary multi-criterion optimization* (pp. 14–32). Springer.

Gal, Y., & Ghahramani, Z. (2016). Dropout as a bayesian approximation: Representing model uncertainty in deep learning. *International conference on machine learning*, pp. 1050–1059.

Guo, D., Wang, X., Gao, K., Jin, Y., Ding, J., & Chai, T. (2020). Evolutionary optimization of high-dimensional multi- and many-objective expensive problems assisted by a dropout neural network. *IEEE Transactions on Systems, Man and Cybernetics: Systems*.

He, C., Tian, Y., Jin, Y., Zhang, X., & Pan, L. (2017). A radial space division based evolutionary algorithm formany-objective optimization. *Applied Soft Computing*, *61*, 603–621.

Hua, Y., Jin, Y., & Hao, K. (2019). A clustering based adaptive evolutionary algorithm for multi-objective optimization with irregular pareto fronts. *IEEE Transactions on Cybernetics*, *49*(7), 2758–2770.

Hua, Y., Jin, Y., Hao, K., & Cao, Y. (2020). Generating multiple reference vectors for a class of many-objective optimization problems with degenerate pareto fronts. *Complex & Intelligent Systems*, *6*, 275–285.

Hua, Y., Liu, Q., Hao, K., & Jin, Y. (2021). A survey of evolutionary algorithms for multi-objective optimization problems with irregular pareto fronts. *IEEE/CAA Journal of Automatica Sinica*, *8*(2), 303–318.

Ishibuchi, H., Tsukamoto, N., & Nojima, Y. (2008). Evolutionary many-objective optimization: A short review. In *Proceedings of the IEEE congress on evolutionary computation (CEC)*. IEEE.

Jones, D. R., Schonlau, M., & Welch, W. J. (1998). Efficient global optimization of expensive black-box functions. *Journal of Global Optimization*, *13*(4), 455–492.

Knowles, J. (2006). ParEGO: A hybrid algorithm with on-line landscape approximation for expensive multiobjective optimization problems. *IEEE Transactions on Evolutionary Computation*, *10*(1), 50–66.

Li, B., Li, J., , Tang, K., & Yao, X. (2015a). Many-objective evolutionary algorithms: A survey. *ACM Computing Surveys*, *48*(1), 13.

Li, K., Deb, K., Zhang, Q., & Kwong, S. (2015b). Combining dominance and decomposition in evolutionary many-objective optimization. *IEEE Transactions on Evolutionary Computation*, *19*(5), 694–716.

Liu, Q., Jin, Y., Heiderich, M., Rodemann, T., & Yu, G. (2020). An adaptive reference vector guided evolutionary algorithm using growing neural gas for many-objective optimization of irregular problems. *IEEE Transactions on Cybernetics*.

Loshchilov, I., Schoenauer, M., & Sebag, M. (2010). A mono surrogate for multiobjective optimization. In *Proceedings of the 12th annual conference on genetic and evolutionary computation* (pp. 471–478). ACM.

Lu, X.-F., & Tang, K. (2012). Classification-and regression-assisted differential evolution for computationally expensive problems. *Journal of Computer Science and Technology*, *27*(5), 1024–1034.

Morgan, R., & Gallagher, M. (2013). Sampling techniques and distance metrics in high dimensional continuous landscape analysis: Limitations and improvements. *IEEE Transactions on Evolutionary Computation*, *18*(3), 456–461.

Ochoa-Estopier, L. M., & Jobson, M. (2015). Optimization of heat-integrated crude oil distillation systems. Part I: The distillation model. *Industrial & Engineering Chemistry Research*, *54*(18), 4988–5000.

Pan, L., He, C., Tian, Y., Wang, H., Zhang, X., & Jin, Y. (2018). A classification-based surrogate-assisted evolutionary algorithm for expensive many-objective optimization. *IEEE Transactions on Evolutionary Computation*, *23*(1), 74–88.

Praditwong, K., & Yao, X. (2006). A new multi-objective evolutionary optimisation algorithm: The two-archive algorithm. In *2006 International conference on computational intelligence and security* (Vol. 1, pp. 286–291). IEEE.

Purshouse, R., & Fleming, P. (2003). Evolutionary many-objective optimisation: an exploratory analysis. In *Proceedings of the IEEE congress on evolutionary computation (CEC)*. IEEE.

Purshouse, R. C., & Fleming, P. J. (2007). On the evolutionary optimization of many conflicting objectives. *IEEE Transactions on Evolutionary Computation, 11*(6), 770–784.

Rachmawati, L., & Srinivasan, D. (2009). Multiobjective evolutionary algorithm with controllable focus on the knees of the pareto front. *IEEE Transactions on Evolutionary Computation, 13*(4), 810–824.

Srivastava, N., Hinton, G., Krizhevsky, A., Sutskever, I., & Salakhutdinov, R. (2014). Dropout: A simple way to prevent neural networks from overfitting. *Journal of Machine Learning Research, 15*(1), 1929–1958.

Tian, Y., Zhang, X., Cheng, R., & Jin, Y. (2016). A multi-objective evolutionary algorithm based on an enhanced inverted generational distance metric. In *Congress on evolutionary computation*. IEEE.

Tian, Y., Cheng, R., Zhang, X., Cheng, F., & Jin, Y. (2018). An indicator based multi-objective evolutionary algorithm with reference point adaptation for better versatility. *IEEE Transactions on Evolutionary Computation, 3*(4), 609–622.

Tian, Y., Cheng, R., Zhang, X., Su, Y., & Jin, Y. (2019). A strengthened dominance relation considering convergence and diversity for evolutionary many-objective optimization. *IEEE Transactions on Evolutionary Computation, 23*(2), 331–345.

Tian, Y., Wang, H., Zhang, X., & Jin, Y. (2017). Effectiveness and efficiency of non-dominated sorting for evolutionary multi-and many-objective optimization. *Complex & Intelligent Systems, 3*(4), 247–263.

Wang, H., Jiao, L., & Yao, X. (2015). Two_arch2: An improved two-archive algorithm for many-objective optimization. *IEEE Transactions on Evolutionary Computation, 19*(4), 524–541.

Wang, H., & Yao, X. (2014). Corner sort for Pareto-based many-objective optimization. *IEEE Transactions on Cybernetics, 44*(1), 92–102.

Yu, G., Jin, Y., & Olhofer, M. (2019). References or preferences–rethinking many-objective evolutionary optimization. In *2019 IEEE congress on evolutionary computation (CEC)* (pp. 2410–2417). IEEE.

Yu, G., Jin, Y., & Olhofer, M. (2020). A multi-objective evolutionary algorithm for finding knee regions using two localized dominance relationships. *IEEE Transactions on Cybernetics*.

Yuan, Y., Xu, H., Wang, B., & Yao, X. (2016). A new dominance relation-based evolutionary algorithm for many-objective optimization. *IEEE Transactions on Systems, Man, and Cybernetics-Part B, 20*(1), 16–37.

Yu, G., Jin, Y., & Olhofer, M. (2020). Benchmark problems and performance indicators for search of knee points in multiobjective optimization. *IEEE Transactions on Cybernetics, 50*(8), 3531–3544.

Zhang, J., Zhou, A., & Zhang, G. (2015a). A classification and pareto domination based multiobjective evolutionary algorithm. In *2015 IEEE congress on evolutionary computation (CEC)* (pp. 2883– 2890). IEEE.

Zhang, X., Tian, Y., Cheng, R., & Jin, Y. (2015b). An efficient approach to non-dominated sorting for evolutionary multi-objective optimization. *IEEE Transactions on Evolutionary Computation, 19*(6), 761–776.

Zhang, X., Tian, Y., & Jin, Y. (2015c). A knee point driven evolutionary algorithm for many-objective optimization. *IEEE Transactions on Evolutionary Computation, 19*(5), 761–776.

Zitzler, E. & Künzli, S. (2004). Indicator-based selection in multiobjective search. In *Proceedings of the parallel problem solving from nature-PPSN* (pp. 832–842). Springer.

Zou, X., Chen, Y., Liu, M., & Kang, L. (2008). A new evolutionary algorithm for solving many-objective optimization problems. *IEEE Transactions on Systems, Man, and Cybernetics-Part B, 38*(5), 1402–1412.

Chapter 9
Knowledge Transfer in Data-Driven Evolutionary Optimization

9.1 Introduction

One big difference between human intelligence and artificial intelligence is that the human being is able to perform multiple tasks by learning from a small amount of data, while artificial intelligence systems such as deep learning models can learn a single task very well if a huge amount of training data is made available.

To address the above issues, several advanced machine learning techniques, as briefly accounted for in Sect. 4.3.4, have been proposed. Recently, semi-supervised learning, multi-tasking learning, and transfer learning have attracted vast attention in machine learning, and found increasing applications in evolutionary computation, in particular in data-driven evolutionary optimization. Meanwhile, multi-tasking evolutionary optimization has also shown to be effective in accelerating evolutionary optimization by simultaneously evolving multiple tasks in a population.

Semi-supervised learning, such as co-training, is particularly helpful for surrogate-assisted evolutionary optimization, since in each generation, only a small number of individuals are labeled (evaluated using the real objective function), while the fitness of the majority of the individuals are estimated by the surrogate, which can be seen as unlabeled data. Thus, it will be very helpful to exploit the information in the unlabeled data to improve the quality of the surrogates, thereby more effectively assisting the evolutionary search. Meanwhile, transfer learning is well suited for knowledge transfer in data-driven optimization, since machine learning models are always involved.

In data driven optimization, knowledge may come from different sources, such as previous optimization tasks, different objectives, different fidelity representations of the same problem, and different scenarios of the same problem. Multi-fidelity representations are widely seen in many simulation based optimization, e.g., partial simulations instead of full simulation, 2D simulations instead of 3D simulations, among many others. In addition, many real-world problems must be optimized in multiple scenarios (operating conditions). For example, a vehicle may be driving

Y. Jin et al., *Data-Driven Evolutionary Optimization*,
Studies in Computational Intelligence 975,
https://doi.org/10.1007/978-3-030-74640-7_9

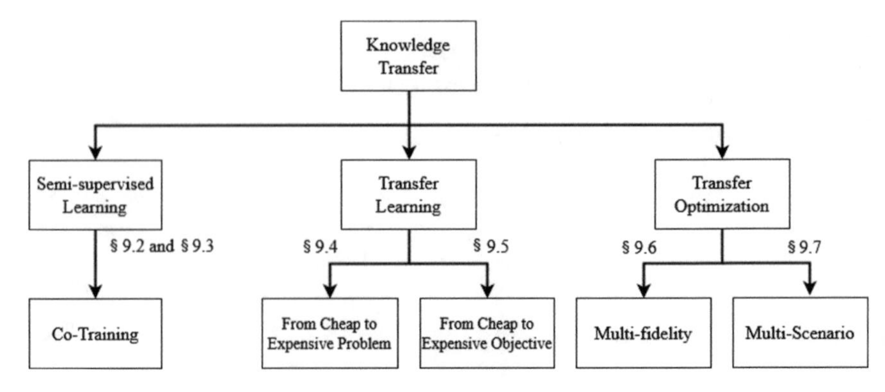

Fig. 9.1 Three major approaches to knowledge transfer in data-driven evolutionary optimization: Semi-supervised learning, transfer learning and transfer optimization

on a high way, or passing a bridge, on a curvature or in a city. The designs for optimizing the vehicle dynamics in these different scenarios may be slightly different, but are closely related, and knowledge can be transferred between these scenarios to accelerate the optimization.

In the following sections, we are going to present some six selected data-driven evolutionary optimization algorithms in which knowledge transfer is considered. A summary of the knowledge transfer approaches to knowledge transfer introduced in this chapter are summarized in Fig. 9.1.

9.2 Co-Training for Surrogate-Assisted Interactive Optimization

Many real-world design optimization problems, such as product design, art design, and music composition, are not able to be described with an analytic or numeric fitness function to evaluate the performance of a solution. Therefore, a human user may often participate to evaluate the quality of the candidate solutions by assigning a fitness value based on her or his subjective preferences. These evolutionary algorithms are called interactive evolutionary algorithms. Interactive genetic algorithms (IGAs) are genetic algorithms in which the fitness values are evaluated by a human user. Over the past decade, IGAs have successfully been applied to many real-world applications such as color design, hearing aid fitting and design of robotic systems. However, the fitness value given by the human user often has a large degree of uncertainty, especially in the beginning of the optimization, because of the limited cognitive ability or a very unclear preference of the human user. Furthermore, not many fitness evaluations can be allowed for optimization due to the human fatigue. To address these challenges, surrogate models are proposed to replace, at least in part, human evaluations. On the other hand, fitness assignment using intervals, i.e.,

$f(\mathbf{x}) = [\underline{f}(\mathbf{x}), \overline{f}(\mathbf{x})]$, is easier for human users than using accurate values for problems without an explicit objective function, where $\underline{f}(\mathbf{x})$ and $\overline{f}(\mathbf{x})$ are the lower and upper bounds of the fitness of a solution $\mathbf{x}$. Therefore, in this section, we will present an improved co-training semi-supervised learning(iCSSL) algorithm for model training in the surrogate-assisted IGAs with interval-based fitness (IGA-IF) (Sun et al., 2013b).

9.2.1 *Overall Framework*

Generally, the surrogate models used in IGA are trained with a supervised learning algorithm, whose success heavily depends on the availability of sufficient training samples. However, not so many labeled data are available because of the human fatigue. Therefore, some unlabeled data are proposed to be used to train the surrogate models. The pseudocode of the surrogate-assisted IGA-IF with iCSSL is given in Algorithm 1.

Algorithm 1 Pseudocode of the surrogate-assisted IGA-IF with iCSSL

Input: Number of the clusters k;
Output: The optimal solution found so far.
1: Initialization of a population $P(t)$, $t = 0$;
2: **repeat**
3: Cluster the population into k subpopulations;
4: Evaluate the center of each cluster in intervals by the user;
5: Train two surrogate models using the evaluated data;
6: Construct or update the surrogate models using the improved co-training semi-supervised learning algorithm; **(Refer to Algorithm 2)**
7: Estimate the fitness values of the rest solutions using the surrogate models;
8: Generate a candidate population $P(t)$, $t = t + 1$ using the reproduction operators of GA;
9: Approximate the individuals in $P(t)$ using the surrogate models;
10: Determine and evaluate the most informative individuals;
11: **until** The stopping criterion is not satisfied

In Algorithm 1, a new population $P(t)$, $t = 0$ will be initialized in the decision space. The following procedure will be repeated until the stopping criterion is satisfied. The population will be clustered into a number of subproblems using k-mean clustering method. The center of each cluster will be evaluated in intervals by the users. All center points are used to train two surrogate models M_1 and M_2, which will be described in Sect. 9.2.2. After that, some unlabeled data are selected for updating M_1 and M_2, respectively, which will be described in Sect. 9.2.4. Then the rest individuals in each cluster will be approximated using the final updated surrogate models using Eq. (9.11). After that, a new population will be generated using the GA reproduction operators. All of individuals in the new population will be approximated using the current surrogate models. Finally, the most informative information in the current population will also be selected for user evaluation. In the following, we will give a detailed description on the selection of unlabeled data, the reliability estimation, the surrogate model training, and the infill criterion.

9.2.2 Surrogate for Interval Prediction

The interval-based fitness value is proposed to be used in IGA to effectively reduce the user fatigue. Let $\mathbf{x}_i(t)$, $i = 1, 2, ..., N$ be the i-th individual in the current population $P(t)$ at generation t, N is the population size. The interval-based fitness of $\mathbf{x}_i(t)$ is $f(\mathbf{x}_i(t)) = [\underline{f}(\mathbf{x}_i(t)), \overline{f}(\mathbf{x}_i(t))]$. Thus, an intuitively way for training a surrogate model is that it provides two outputs that approximate the upper and lower bounds of an interval. Therefore, the method to measure the difference between the values the user gives and those approximated by the surrogate models should be re-defined, which will be used to optimize the weights of the surrogate model. The mathematical equation of the error function used in IGA-IF is defined as:

$$E = \frac{1}{2} \sum_{l=1}^{|L(t)|} e_l \tag{9.1}$$

where

$$e_l = \left(\hat{\underline{f}}(\mathbf{x}_l(t)) - \underline{f}(\mathbf{x}_l(t))\right)^2 + \left(\hat{\overline{f}}(\mathbf{x}_l(t)) - \overline{f}(\mathbf{x}_l(t))\right)^2 \tag{9.2}$$

In Eq. (9.1), $|L(t)|$ represents the number of data in the labeled data set $L(t)$. $\hat{\underline{f}}(\mathbf{x}_l(t))$ and $\hat{\overline{f}}(\mathbf{x}_l(t))$ in Eq. (9.2) are lower and upper approximated outputs of the surrogate model on the solution $\mathbf{x}_l(t)$.

It is easy to understand that, for the most or least interested individuals, the uncertainty degree of the evaluations of the users is often small. In these cases, the evaluations of the users are more reliable. That is, smaller the uncertainty degree of its evaluation is, more reliable the evaluation of the user is. However, when the user preference on the solution is neutral or uncertain, the fitness the user evaluate may be rough and correspondingly, the value of the uncertainty will be large and the reliability of the evaluation will be low. For any solution $\mathbf{x}_i(t)$, the center and width of its interval fitness are calculated as follows:

$$c\left(f(\mathbf{x}_i(t))\right) = \frac{\overline{f}(\mathbf{x}_i(t)) + \underline{f}(\mathbf{x}_i(t))}{2}, \tag{9.3}$$

$$w\left(f(\mathbf{x}_i(t))\right) = \overline{f}(\mathbf{x}_i(t)) - \underline{f}(\mathbf{x}_i(t)), \tag{9.4}$$

From Eqs. (9.3) and (9.4), we can easily understand that the value of $c\left(f\left(\mathbf{x}_i(t)\right)\right)$ reflects the preferred degree of user to the solution $\mathbf{x}_i(t)$ and the value of $w(f\left(\mathbf{x}_i(t)\right))$ depicts the uncertainty of the user when she/he evaluates the solution $\mathbf{x}_i(t)$. From optimization point of view, we expect to search for a solution who has a higher value on fitness $c\left(f\left(\mathbf{x}_i(t)\right)\right)$ and a smaller value of uncertainty $w\left(f\left(\mathbf{x}_i(t)\right)\right)$, thus, the evaluated reliability is proposed to be evaluated as follows:

$$r(\mathbf{x}_i(t)) = \frac{c\left(f\left(\mathbf{x}_i(t)\right)\right)}{w\left(f\left(\mathbf{x}_i(t)\right)\right) + \epsilon} \tag{9.5}$$

In Eq. (9.5), ϵ is an extremely small positive number, which is used to ensure that the denominator will not be zero.

As training samples having a higher reliability are more important, thus, a higher priority should be given to these samples in learning. The reason is that these samples represent the current preference of the user, and they should be approximated with a higher accuracy. To this end, the evaluation reliability, which is used as a penalty coefficient, is proposed to be integrated in the error function for adapting the weights of the RBFN. So the modified error function for training the RBFN can be written as follows:

$$E = \frac{1}{2}\sum_{l=1}^{|L(t)|} r'(\mathbf{x}_l(t))e_l, \tag{9.6}$$

where $r'(\mathbf{x}_l(t)) = r(\mathbf{x}_l(t))/\max_{l=1,2,\ldots,|L(t)|} r(\mathbf{x}_l(t))$ is a normalized value of $r(\mathbf{x}_l(t))$ between 0 and 1.

Without loss of generality, a Gaussian kernel is adopted to be the radial function and is expressed as $K(\mathbf{x}_l, \mathbf{C}_i) = \exp(-\|\mathbf{x}_l - \mathbf{C}_i\|^2/2\sigma_i^2)$, where $\mathbf{C}_i$ is the center of i-th node and σ is the width of the function. Suppose the weights between i-th node and the output nodes are W_{i1} and W_{i2}, respectively, then the outputs of RBFN will be

$$\hat{\bar{f}}(\mathbf{x}_l) = \sum_{i=1}^{N_h} W_{i1}K(\mathbf{x}_l, \mathbf{C}_i), \tag{9.7}$$

$$\hat{\underline{f}}(\mathbf{x}_l) = \sum_{i=1}^{N_h} W_{i2}K(\mathbf{x}_l, \mathbf{C}_i). \tag{9.8}$$

In Eqs. (9.7) and (9.8), N_h represents the number of nodes in the hidden layer.

Then, the learning rules will be modified as follows:

$$\Delta w_{i1} = \eta \sum_{l=1}^{|L|} r^{'}(\mathbf{x}_l(t))\overline{e}(\mathbf{x}_l)K(\mathbf{x}_l, \mathbf{c}_i),$$
$$\Delta w_{i2} = \eta \sum_{l=1}^{|L|} r^{'}(\mathbf{x}_l(t))\underline{e}(\mathbf{x}_l)K(\mathbf{x}_l, \mathbf{c}_i),$$
$$\Delta \mathbf{c}_i = \sum_{l=1}^{|L|} r^{'}(\mathbf{x}_l(t))(\overline{e}(\mathbf{x}_l)w_{i1} + \underline{e}(\mathbf{x}_l)w_{i2})K(\mathbf{x}_l, \mathbf{c}_i)\frac{\mathbf{x}_l - \mathbf{c}_i}{\delta_i^2},$$
$$\Delta \sigma_i^2 = \eta \sum_{l=1}^{|L|} r^{'}(\mathbf{x}_l(t))(\overline{e}(\mathbf{x}_l)w_{i1} + \underline{e}(\mathbf{x}_l)w_{i2})K(\mathbf{x}_l, \mathbf{c}_i)\frac{\|\mathbf{x}_l - \mathbf{c}_i\|^2}{\delta_i^3}, \tag{9.9}$$

where $\overline{e}(\mathbf{x}_l) = f(\mathbf{x}_l - \overline{\hat{f}}(\mathbf{x}_l))$ and $\underline{e}(\mathbf{x}_l) = f(\mathbf{x}_l - \underline{\hat{f}}(\mathbf{x}_l))$. Based on Eq. (9.9), we can see that the approximation errors on the more reliable training samples have a higher priority to be reduced. Therefore, the approximation accuracy is higher for these data than those less reliable training samples. Correspondingly, the less reliable training samples will not be selected in the evolution and thus, will not influence the search.

9.2.3 Fitness Estimation

Two approximation values will be obtained from two surrogate models for each solution, therefore, it needs to be aggregated appropriately to be the final output of the approximated fitness value of a solution. Different aggregate operators, such as a weighted aggregation, a minimization, a maximization, or a product operator, can be used. Note, however, that in IGA-IF, the fitness of a solution is an interval. Thus, the approximated fitness values are also intervals for each surrogate model, although the confidence levels of the approximated fitness values of two models may be different. Therefore, a new weighting method is introduced, where the weights relies on the approximation confidence, which is given by

$$\hat{r}_j(\mathbf{x}_i(t)) = \frac{c\left(\hat{f}_j(\mathbf{x}_i(t))\right)}{w\left(\hat{f}_j(\mathbf{x}_i(t))\right) + \epsilon} \tag{9.10}$$

where $\hat{r}_j(\mathbf{x}_i(t)), i = 1, 2, ..., N, j = 1, 2$ is the approximation confidence of the surrogate model $M_j, j = 1, 2$ in approximating the fitness value of $\mathbf{x}_i$, ϵ is a very small positive number as is given in Eq. (9.5). Therefore, the final output of the approximation on solution $\mathbf{x}_i$ will be given as follows:

$$\hat{f}(\mathbf{x}_i(t)) = \frac{1}{\hat{r}_1(\mathbf{x}_i(t)) + \hat{r}_2(\mathbf{x}_i(t))}(\hat{r}_1(\mathbf{x}_i(t))\hat{f}_1(\mathbf{x}_i(t)) + \hat{r}_2(\mathbf{x}_i(t))\hat{f}_2(\mathbf{x}_i(t))) \quad (9.11)$$

9.2.4 An Improved CSSL

The contribution of the improved CSSL (iCSSL) is to train two surrogate models using both the labeled and unlabeled solutions in the current population. Algorithm 2 gives the pseudo code of the improved CSSL method. As the number of labeled data is very limited because of the user fatigue, some unlabeled data will be selected from the current population to be used together with the labeled data to train the surrogate model. At each generation, the population will be classified into k subpopulations. The center of each cluster is evaluated using the users and saved to the data set $L(t)$. All unlabeled data in the population are saved in the data set $U(t)$. The strategy to select unlabeled data for updating the surrogate models are given as follows: Given two initial models with different structures and parameters, denoted as M_1 and M_2, are trained using all labeled data $L(t)$. Then model $M_1(M_2)$ estimates the labels of those unlabeled solutions in each subpopulation. The model $M_1(M_2)$ will be retrained using these labeled data and tested the approximation error on the samples in $L(t)$. The solutions in the subpopulation with the lowest test error will be selected to augment the training set of the other model, i.e., $M_2(M_1)$, and this subpopulation will be deleted from $U(t)$. The process will be repeated until there is no solution in $U(t)$, i.e., $U(t) = \emptyset$.

Algorithm 2 Pseudocode of iCSSL

Input: M_1 and M_2: two surrogate models trained on the labeled data of the current population;
 $L(t)$: sets of labeled data selected from $P(t)$;
 $U(t)$: sets of unlabeled data selected from $P(t)$;
Output: The updated surrogate models M_1 and M_2.
1: **while** $U(t) \neq \emptyset$ **do**
2: Estimate the interval fitness values of the unlabeled individuals using M_1 and M_2, respectively, on each cluster;
3: A surrogate model for each cluster is trained using the solutions estimated by $M_1(M_2)$, and used to approximate the data in $L(t)$.
4: Augment the training data of $M_2(M_1)$ using the solutions in the cluster who has the minimum test error on the data in $L(t)$ and delete these solutions from $U(t)$;
5: Retrain the surrogate models M_1 and M_2 using the updated training data set;
6: **end while**

9.2.5 *Surrogate Management*

The surrogate model can be used to replace the human user to estimate the fitness value of a solution and thus reduce the user fatigue. However, the fitness evaluation should also be conducted by the user in the optimization to ensure that the optimization may not converge to a false optimum. That means, the surrogate model should be updated by additional solutions evaluated by the user so that the evolutionary search can converge to a correct optimum. However, which solutions should be selected to be evaluated by the user to speed up finding the good solution is very important. In the surrogate-assisted IGA-IF with iCSSL, the solutions with a high estimated fitness and a higher estimation confidence will be selected for reproduction, while those solutions with a high estimated fitness but with a lower estimation confidence should be chosen to be evaluated by the user to improve the approximation quality of the surrogate model in a promising search space. Furthermore, the solutions that can enhance the population diversity will also be chosen to improve the exploration capability of the evolutionary search. As the estimation confidence of a solution $\mathbf{x}_i(t)$ can indicate the estimated quality, it is adopted to identify the potentially good solutions. Note that the approximated fitness value of each solution is an interval. So the final output of an individual will be of high estimation accuracy if two estimated interval values are sufficiently similar, and on the contrary, the approximated fitness value may deviate from the user's evaluation if the outputs of the two surrogate models have a big discrepancy. So the similarity of the two intervals estimated by two surrogate models will be used to measure the quality of the final approximated value, which is given as follows:

$$S_{\hat{f}_1,\hat{f}_2}(\mathbf{x}_i(t)) = \frac{w(\hat{f}_1(\mathbf{x}_i(t)) \cap \hat{f}_2(\mathbf{x}_i(t)))}{w(\hat{f}_1(\mathbf{x}_i(t)) \cup \hat{f}_2(\mathbf{x}_i(t)))} \tag{9.12}$$

where $S_{\hat{f}_1,\hat{f}_2}(\mathbf{x}_i(t)) \in [0, 1]$. The larger the value of $S_{\hat{f}_1,\hat{f}_2}(\mathbf{x}_i(t))$, the greater the similarity between approximated values, and the more accurate the estimate of $\hat{f}(\mathbf{x})$. The similarity of the two approximated values will be used to determine whether the surrogate model needs to be updated or not. If the approximation quality of the surrogate model is smaller than a predefined threshold, the surrogate model will be updated. In addition, the surrogate model can also be updated if the user feels that the approximated fitness considerably deviates from his/her evaluation.

9.3 Semi-Supervised Learning Assisted Particle Swarm Optimization

For the computationally expensive optimization problems, despite the success of various surrogate techniques reported in the literature, most surrogate modeling approaches rely only on a small number of labeled data (i.e., on solutions that have been evaluated using the computationally expensive objective function). However, a majority of unlabeled data can be got and used to enhance the learning performance of the surrogate model. In this section, a semi-supervised learning assisted particle swarm optimization (SSL-assisted PSO), in which a number of unlabeled solutions are selected to train an RBF surrogate model together with those labeled data.

9.3.1 Algorithm Framework

For computationally expensive problems, only a limited amount of labeled data are available, however, there is usually a large amount of unlabeled data. Therefore, borrowing the idea from semi-supervised learning, a new strategy is proposed to select unlabeled data to be used to train a surrogate model together with the labeled solutions. Furthermore, some solutions are also required to be evaluated and used to update the surrogate model so that the search will not deviate from the correct direction. Fig. 9.2 gives the framework of the semi-supervised learning assisted PSO.

An initial population $pop(t)$, $t = 0$ will be generated and evaluated using the exact objective function. The positional information, including the position and its fitness value of each solution, will be saved to both *DB* and *EDB* archives, where the archive *DB* is used to save all labeled solutions who have been evaluated using the exact objective function, and the archive *EDB* is provided for saving all labeled solutions and some representative unlabeled solutions. The personal best position of each solution and the global best position of the population found so far will be determined. Then the following procedure will be repeated before the stopping criterion is satisfied. Two surrogate models, M_1 and M_2, will be trained using the data in *DB* and *EDB*, respectively. Both of them are used to approximate the fitness value of each individual in the new population. After that, a final approximated fitness value of each solution will be determined for providing information to be used in model management. In the model management stage, two main issues should be addressed. First, which solutions should be selected to be evaluated using the exact objective function and saved to both archives *DB* and *EDB*. Second, which solutions can be chosen to be saved to archive *EDB* and train the model together with the labeled data. In the following, we will give a detailed description on important parts of SSL-assisted PSO.

9.3.2 Social Learning Particle Swarm Optimization

Many particle swarm optimization variances have been proposed for either speeding up the convergence (Clerc and Kennedy, 2002) or improving the diversity of the population to prevent the search from getting stuck in a local optimum (Liang et al., 2006b). In SSL-assisted PSO, a modification of the social learning particle swarm optimization, which is also a PSO variance proposed in (Cheng and Jin, 2015),

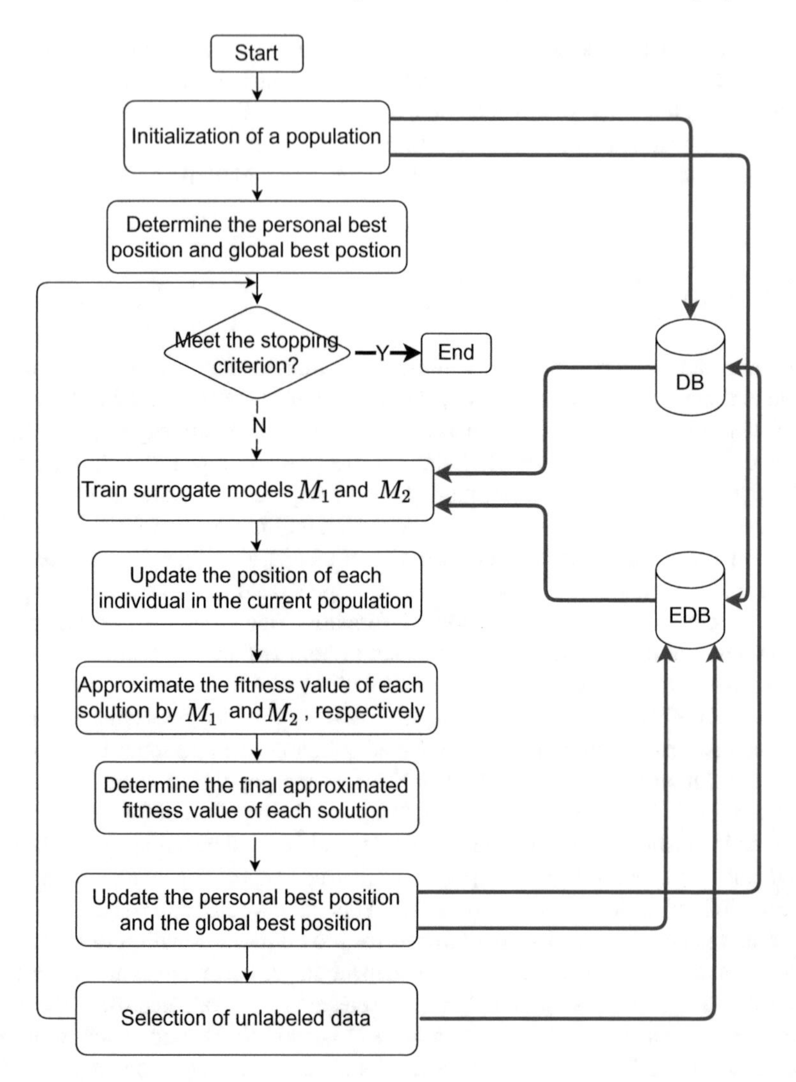

Fig. 9.2 The framework of SSL-assisted PSO

is adopted to be the optimization algorithm. The SL-PSO has been shown to have better performance for solving large-scale optimization problems, however, its convergence speed is slow especially in the beginning of the search, which, therefore, is limited to solve the expensive problems as not many fitness evaluations are allowed. Therefore, the position of individual i will be updated as follows:

$$x_{ij}(t+1) = \begin{cases} x_{ij}(t) + \Delta x_{ij}(t+1), & \text{if } pr_j(t) \leq pr_j^L \\ x_{ij}(t), & \text{otherwise} \end{cases} \tag{9.13}$$

where

$$\Delta x_{ij}(t+1) = r_1 \cdot \Delta x_{ij}(t) + r_1 \cdot (p_{kj}(t) - x_{ij}(t)) + r_3 \cdot \epsilon \cdot (\bar{x}_d(t) - x_{ij}(t)) \quad (9.14)$$

In Eqs. (9.13) and (9.14) , $pr_j, 0 \leq pr_j \leq 1$ is a randomly generated probability and pr_j^L is the probability threshold for particle j to update its position. r_1, r_2 and r_3 are three random numbers uniformly generated in the range [0, 1]. p_{kj} is the j-th dimension of personal best position of particle k whose fitness is better than that of particle i. $\bar{x}_j(t) = \frac{\sum_{i=1}^{n} x_{ij}(t)}{N}$ is the mean position on j-th dimension of the swarm. ϵ is a parameter called the social influence factor that controls the influence of $\bar{x}_j(t)$. From Eq. (9.13), we can see that any individual i updates its position by learning from the better personal best position of other individuals as well as the mean position of the swarm. Therefore, the search is expected to be accelerated on the convergence while still preserve an adequate degree of population diversity in the beginning of the search.

9.3.3 Surrogate Management Strategy

In the semi-supervised learning, a number of labeled data are also required. Thus, sampling strategy is very important for SSL-assisted PSO to find a good solution in a limited fitness evaluations. From Fig. 9.2, we can see that the labeled data in DB comes from the following parts. All solutions of the initial population, which are the first source of the labeled data, are evaluated using the expensive fitness function and saved in both *DB* and *EDB*. From the second generation onward, all solutions are approximated by the two surrogate models, M_1 and M_2. Let $\hat{f}_{M_1}(\mathbf{x}_i)$ and $\hat{f}_{M_2}(\mathbf{x}_i(t))$ represent the fitness values approximated by M_1 and M_2 for solution $\mathbf{x}_i$, respectively. Then its final fitness will be set as follows: If both conditions, $\hat{f}_{M_1}(\mathbf{x}_i) < f(\mathbf{p}_i)$ and $\hat{f}_{M_2}(\mathbf{x}_i) < f(\mathbf{p}_i)$, are satisfied, then particle $\mathbf{x}_i$ will be evaluated using the exact objective function, and saved to both *DB* and *EDB*. Otherwise, the maximum approximated value will be the fitness of the solution $\mathbf{x}_i$, i.e., $f(\mathbf{x}_i) = \max\{\hat{f}_{M_1}(\mathbf{x}_i), \hat{f}_{M_2}(\mathbf{x}_i)\}$. Note that after a solution is evaluate using the exact objective function, its personal best position should also be updated if the fitness value of this solution is better than that of its personal best position. All evaluated solutions will be used to update the global best position if they are better than their personal best position. However, the situation may not always be true, that is, no solution in the current population is able to meet the condition to be evaluated using the real objective function. In this case, the particle having the best approximated fitness value will be evaluated using the exact expensive fitness function, and used to update its personal best position and the best position found so far as well. This mechanism is expected to reduce the possibility of the algorithm to get stuck in a local optimum. Algorithm 3 gives the pseudocode of the process to update the global best position.

Algorithm 3 Update of the global best position

Input: $\mathbf{g}$: the best position found so far;
if at least one solution has been evaluated using the exact objective function **then**
  Suppose particle k in the current population has the best personal best fitness $f(\mathbf{p}_k)$;
  if $f(\mathbf{p}_k) < f(\mathbf{g})$ **then**
    $\mathbf{g} = \mathbf{p}_k$ and $f(\mathbf{g}) = f(\mathbf{p}_k)$;
  end if
else
  Find the particle i that has the minimum approximated value;
  Evaluate the particle i using the real objective function, save to both DB and EDB;
  Update the personal best position if $f(\mathbf{x}_i) < f(\mathbf{p}_i)$;
  Update the global best position if $f(\mathbf{p}_i) < f(\mathbf{g})$;
end if
Output the updated best position found so far $\mathbf{g}$;

9.3.4 Selection of Unlabeled Data

In semi-supervised learning, it is critical to select appropriate unlabeled data to train a surrogate model together with those labeled data. Fig. 9.3 gives two examples to show which unlabeled data are helpful to train a good required surrogate model. In Fig. 9.3, the solid line denotes the real objective function, the dashed line represents a surrogate model, denoted as M_2, before any new unlabeled data is added in the training set. Assume $\mathbf{x}_2$ is a solution in the current population that has been selected to be evaluated using the real objective function and its fitness value is given as a circle. $\mathbf{x}_1$ is a solution in the current population and its value approximated by the model M_2 ($\hat{f}_{M_2}(\mathbf{x}_1)$) is denoted by a diamond. Now suppose $\mathbf{x}_1$ is the selected unlabeled data which will be added to the training data set to train a new surrogate model $M_2^{'}$, denoted by the dotted line. The fitness value approximated by $M_2^{'}$ on solution $\mathbf{X_2}$ ($\hat{f}_{M_2^{'}}(\mathbf{x}_2)$) is denoted as a four-point star. From Fig. 9.3a, we can see that when $\mathbf{x}_1$ is added to train the surrogate model $M_2^{'}$, the fitness value of solution $\mathbf{x}_2$ will much closer to its real objective value, thus it is beneficial to include $\mathbf{x}_1$ in the training data set. However, in Fig. 9.3b, it can be found easily that the approximated fitness value will much far away from the real objective value, which shows that the approximation error of the surrogate model $M_2^{'}$ will be increased if the unlabeled solution $\mathbf{x}_1$ is included. So a new strategy is proposed to select beneficial solutions to be included in the training data set in order to train an effective surrogate model to speed up the search for the global optimum.

Algorithm 4 gives the pseudocode of selecting the unlabeled data. The current population is separated into two subsets, one including all solutions that have been evaluated using the exact objective function, denoted as S_l, and the other containing those having not been evaluated using the real objective function, given as S_u. Each solution i in S_u will be used to temporary train a surrogate model $M_2^{'}$, and used to approximate the fitness value of all solutions in S_l. The following equation is proposed to measure the difference in the approximation errors before and after an unlabeled solution $\mathbf{x}_i$ is added to train a new surrogate model.

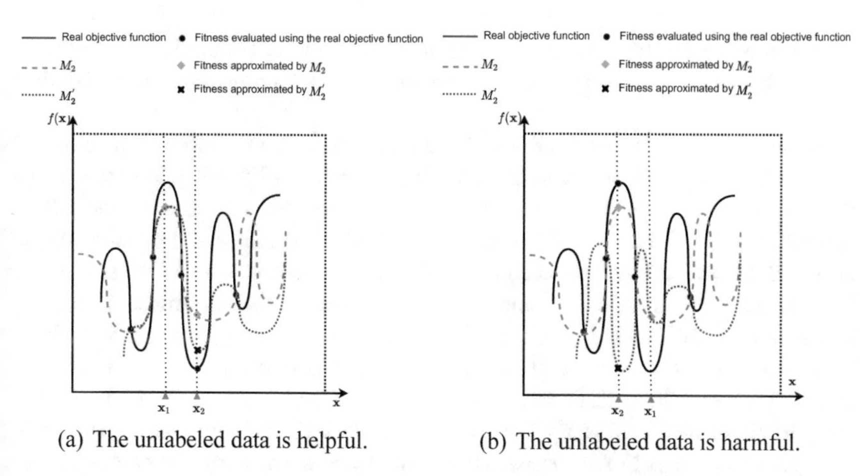

(a) The unlabeled data is helpful. (b) The unlabeled data is harmful.

Fig. 9.3 Two examples to show the influence of adding an unlabeled data into the training data set

$$\phi(\mathbf{x}_i) = \min_{j \in S_l}\{|\hat{f}_{M_2}(\mathbf{x}_j) - f(\mathbf{x}_j)| - |\hat{f}_{M_2'}(\mathbf{x}_j) - f(\mathbf{x}_j)|\} \tag{9.15}$$

Finally solution i that achieves the maximum $\phi(\mathbf{x}_i), i \in S_u$ will be eventually added to the archive *EDB* and the surrogate model will be update.

Algorithm 4 The strategy for selecting unlabeled data

Input: The labeled set of current population S_l; The unlabeled set of current population S_u
1: **for** $i \in S_u$ **do**
2: Training a surrogate model using all data in *EDB* and solution i;
3: Approximate the fitness value of each solution in S_l;
4: Calculate $\phi(\mathbf{x}_i)$ using Eq. (9.15);
5: **end for**
6: Find the solution that has the best $\phi(\mathbf{x}_i)$ value and save it in the archive *EDB*;

9.3.5 Experimental Results and Discussions

To examine the performance of SSL-assisted PSO, a set of empirical studies by comparing its performance with a state-of-the-art surrogate-assisted PSO on five widely utilized uni-modal and multi-modal benchmark problems, including Ellipsoid, Rosenbrock, Ackley, Griewank, and Rastrigin. The parameter settings used in the experiments are given as follows: 20 independent runs are performed on each test problem. The size of the population of all compared algorithms is set to 50, with the probability threshold $pr_j^L, j = 1, 2, ..., n$ being set to 1 and ϵ set to 0.001. The maximum number of fitness evaluation is set to $11 \times n$. The size of the training set for M_1 is set to $2 \times (n + 1)$ and the size of the training set for M_2 is set to twice of that for M_1, i.e., $4 \times (n + 1)$, since we can obtain more training data from *EDB* than from *DB*.

Table 9.1 gives the mean results of the compared algorithms as well as the standard deviations on the five test problems with 30 dimensions. In Table 9.1, PSO represents the PSO given in Sect. 9.3.2 without the surrogate model assistance. SL-only PSO and SSL-only PSO are the PSO assisted by an RBF network trained by labeled data only and that trained by unlabeled data only, respectively. CAL-SAPSO is inspired by committee-based active learning proposed by Wang et al. in (Wang et al., 2017a), in which only the labeled data are utilized to train the surrogate ensemble. From Table 9.1, we can observe that SSL-assisted PSO has obtained better results on all five problems than PSO, SL-only PSO and SSL-only PSO. Furthermore, the comparison results to SL-only PSO and SSL-only PSO indicate that the surrogate management strategy in SSL-assisted PSO making use of two surrogates are effective. In comparison with CAL-SAPSO, SSL-assisted PSO also obtains better results on three out of five test instances, which show that the strategy to use the unlabeled data together with the labeled data is helpful.

9.4 Knowledge Transfer between Problems in Multi-objective Optimization

Evolutionary multi-objective Bayesian optimization aims to solve an expensive multi-objective optimization problem (MOP) by optimizing the corresponding acquisition functions, as described in Sect. 5.7. The question here is, can we transfer knowledge from cheap problems to enhance the performance of evolutionary Bayesian optimization when the computational budget for the expensive MOP is very limited? In this section, we describe an evolutionary multi-objective Bayesian optimization algorithm in which knowledge is transferred from a cheap MOP to the expensive MOP by means of online online domain adaptation (Li et al., 2021).

9.4.1 Domain Adaptation for Transfer Learning

Transfer learning is able to transfer knowledge from a source task to a target task (Zhuang et al., 2020). One challenge in achieving knowledge transfer is that the data distributions of the source and target tasks may not be the same, meaning that that the target and source tasks have different sample spaces and different marginal or conditional distributions. To overcome this hurdle, techniques have been proposed to adapt and match the marginal and conditional distributions between source and target domains (Tahmoresnezhad and Hashemi, 2017). Among these techniques, balanced distribution adaptation (BDA) (Wang et al., 2017d) that can adjust the relative

Table 9.1 Comparisons of the statistical results on 30-dimensional problems, shown as the mean value±the standard deviations

Prob.	PSO	SL-only PSO	SSL-only PSO	CAL-SAPSO	SSL-assisted PSO
Ellipsoid	1.02e+03±1.46e+02	1.02e+02±5.87e+01	4.84e+02±1.17e+02	4.02e+00±1.08e+00	**3.56e+00±1.52e+00**
Rosenbrock	6.15e+02±1.13e+02	7.99e+01±1.63e+01	2.79e+02±4.76e+01	5.10e+01±1.15e+01	**4.76e+01±1.02e+01**
Ackley	1.85e+01±4.71e−01	1.57e+01±1.94e+00	1.71e+01±5.43e−01	1.62e+01±4.13e−01	**7.19e+00±2.94e+01**
Griewank	2.37e±3.99e+01	1.73e+01±8.43e+00	1.29e+02±2.03e+01	**9.95e-01±3.99e−02**	1.04e+00±7.66e−02
Rastrigin	3.19e+02±2.05e+01	2.86e+02±2.49e+01	2.85e+02±2.10e+01	**8.78e+01±1.65e+01**	2.76e+02±2.99e+01

importance of the marginal and conditional distributions has been shown effective for domain adaptation. The main idea here is to use BDA to transfer knowledge from cheap optimization problems, for which a large number of data (fitness evaluations) can be collected, to facilitate the solution of expensive ones where only a very limited fitness evaluations are allowed.

Let $\Phi = (Z, Y)$ denote a domain, P a probability distribution, and Q a conditional probability distribution. For simplicity, subscripts Φ_c and Φ_e are used to denote the domains of the source (the cheap problem) and target (the expensive problem). In each domain, $Z = \{z_i\}_{i=1}^{n}$, represents the decision or objective space, Y is the label vector $Y = \{y_i\}_{i=1}^{n}$, where n equals either n_c or n_e, is the number of samples for the cheap or expensive problem, respectively. $P(z)$ is then the corresponding marginal distribution and $Q(y\,|z)$ is the conditional distribution. Let χ and γ denote the feature and category space, respectively. Consequently, the feature space of the cheap problem is denoted as χ_c, and feature space of the expensive problem χ_e, which is the same as χ_c. It is easy to see that the category space of the cheap problem γ_c (dominated or non-dominated) and that of the expensive space γ_e are also the same.

Usually, neither the marginal nor the conditional distributions of the cheap problem and target problem domains are the same. To transfer knowledge from the cheap problem to expensive problem, we need to reduce the distance between the probability distributions of the cheap and the expensive problem domains. However, the marginal and conditional distributions may not be equally important, and the source and target data may be unbalanced. To address this issue, BDA is thus proposed, mainly by adjusting the importance of the marginal and conditional distributions according to the given tasks. To this end, a balance factor (Wang et al., 2017d) by minimizing the following weighted distance:

$$\begin{aligned} D(\Phi_c, \Phi_e) \approx (1-\mu)D(P(z_c), P(z_e)) \\ +\beta D(P(y_c\,|z_c), P(y_e\,|z_e)) \end{aligned} \tag{9.16}$$

where $D(\Phi_c, \Phi_e)$ is the distance between the source and target domains, and $\beta \in [0, 1]$ is the balance factor. If β is close to zero, the marginal distribution is more important in minimizing the distance, while if $\beta = 1$, the conditional distribution plays a more critical role.

If the target data is unlabeled, it is difficult to compute $P(y_e\,|z_e)$. One idea is to use $P(z_e\,|y_e)$ to approximate $P(y_e\,|z_e)$, provided that we have sufficient samples. To calculate $P(z_e\,|y_e)$, we can train a classifier on Φ_c and use the classifier to predict Φ_e, which is called the soft label, must be iteratively refined since the prediction may be inaccurate in the beginning.

Let us denote the expensive problem data as $\left\{z_{e_i}, y_{e_i}\right\}_{i=1}^{n_e}$, where y_{e_i} is a given class label based on the dominance relationship of the solutions. Assume we have a large set of data for the cheap problem denoted as $\left\{z_{c_j}\right\}_{j=1}^{n_c}$. According to the maximum mean difference (Pan et al., 2010), Eq. (9.16) can be rewritten as:

$$D(\Phi_c, \Phi_e) \approx (1-\beta) \left\| \frac{1}{n_c} \sum_{i=1}^{n_c} z_{c_i} - \frac{1}{n_e} \sum_{j=1}^{n_e} z_{e_j} \right\|_{\mathrm{H}}^2 + \beta \sum_{k=1}^{K} \left\| \frac{1}{n_{c_k}} \sum_{z_{c_i} \in \Phi_c^{(k)}} z_{c_i} + \frac{1}{n_{e_k}} \sum_{z_{e_j} \in \Phi_e^{(k)}} z_{e_j} \right\|_{\mathrm{H}}^2 \tag{9.17}$$

where H denotes the reproducing kernel Hilbert space (RKHS), k is the class label, $n_{c_k} = \left|\Phi_c^{(k)}\right|$ and $n_{e_k} = \left|\Phi_e^{(k)}\right|$ represents the number of samples belonging to this class in the source and target domains, respectively. The first term in the above equation represents the distance between the marginal distributions of the two domains, and the second term is the distance between the conditional distributions. The minimization problem of Eq. (9.16) can be rewritten as follows by including a regularization term:

$$\begin{aligned} \min tr(\Theta^T Z((1-\beta)M_0 + \beta \sum_{k=1}^{K} M_k) Z^T \Theta) + \lambda \left\|\Theta\right\|_F^2 \\ \text{s.t. } \Theta^T Z H Z^T \Theta = I. \end{aligned} \tag{9.18}$$

where $0 \leq \beta \leq 1$. The first term is the marginal and conditional distribution adaptation with the balance factor, while the second term is the regularization term, and λ is hyperparameter. In the equation, $\|\bullet\|_F^2$ is the Frobenius norm, and the equality constraint makes sure that $(\Theta^T Z)$ keeps the properties of the original data. Additionally, $Z = [z_c; z_e]$ is the input data, $\Theta \in R^{(n_c+n_e)\times(n_c+n_e)}$ is the transformation matrix, $H = I - (1/n)I$ is the centering matrix, and M_0 and M_k are MMD matrices:

$$(M_0)_{i,j} = \begin{cases} \frac{1}{n_c^2}, z_i, z_j \in \Phi_c \\ \frac{1}{n_e^2}, z_i, z_j \in \Phi_e \\ -\frac{1}{n_c n_e}, otherwise \end{cases} \tag{9.19}$$

$$(M_k)_{i,j} = \begin{cases} \frac{1}{n_{c_k}^2}, z_i, z_j \in \Phi_s^{(k)} \\ \frac{1}{n_{e_k}^2}, z_i, z_j \in \Phi_t^{(k)} \\ -\frac{1}{n_c n_e}, \begin{cases} z_i \in \Phi_c^{(k)}, z_j \in \Phi_e^{(k)} \\ z_i \in \Phi_e^{(k)}, z_j \in \Phi_c^{(k)} \end{cases} \\ 0, otherwise \end{cases} \tag{9.20}$$

The constrained minimization problem defined in (9.18) can be solved defining the following Lagrangian function:

$$L = tr(\Theta^T Z((1-\beta)M_0 + \beta \sum_{k=1}^{K} M_k)Z^T \Theta) + \lambda \|\Theta\|_F^2, \quad +tr((I - \Theta^T ZHZ^T \Theta)\phi) \tag{9.21}$$

where $\phi = (\varphi_1, \varphi_2, \cdots, \varphi_l)$ is Lagrangian multipliers. If $\frac{\partial L}{\partial \Theta} = 0$, the optimization problem treated as a generalized eigendecomposition problem:

$$(Z((1-\beta)M_0 + \beta \sum_{k=1}^{K} M_k)Z^T + \lambda I)\Theta = ZHZ^T \Theta \phi \tag{9.22}$$

The l smallest eigenvectors can be obtained by solving the above optimization problem, consequently, the optimal transformation matrix Θ can be obtained.

BDA aims to learn the transferable components, also known as the latent variables, between the target and source domains. This is done by projecting data distributions in different domains onto a common sub-domain (known as the latent space) based on feature mapping so that the distribution of the mapped data is approximately the same in this latent space. The dimension of the latent space *dim* is a parameter to be specified by the user.

9.4.2 Knowledge Transfer from Cheap to Expensive Problems

In this algorithm, a subset of the non-dominated solutions according to the acquisition functions of the objectives are sampled and used for training the GP models of the expensive problem. Thus, knowledge about the non-dominated solutions of the cheap problem can be used to guide the search for the non-dominated solutions of the expensive problem. This is done by generating additional synthetic training data for the expensive problem based on the relevant (transferable) non-dominated solutions of the cheap problem. We call the BDA based method for transferring knowledge from the cheap to expensive problem CE-BDA.

CE-BDA consists of two knowledge transfer processes, one in the decision space, $Z_X = [X_c, X_e]$, and the other in the objective space, $Z_Y = [F_c, F_e]$, where X_c and X_e are the decision vectors of the cheap and expensive multi-objective optimization problems (MOPs), and F_c and F_e are the objective vectors of the cheap and expensive MOPs, respectively. It should be noted that the two processes contain same the category space, i.e., whether a solution is dominated (labeled '0') or non-dominated (labeled '1'). For the sake of efficient knowledge transfer, all objective values of the cheap and expensive MOPs are normalized between [0, 1].

Given the initial data samples of the cheap and expensive MOPs $D_c(0)$ and $D_e(0)$, which are sampled using the Latin hypercube sampling method, the first knowledge transfer process calculates the optimal transformation matrix (Θ_X) and the balance factor (β_X) according to Eq. (9.22) using the decision variables of the solutions in

$D_c(0)$ (source) and $D_e(0)$ (target). Similarly, the optimal transformation matrix Θ_F and the balance factor β_F for the objective functions can be obtained.

The pseudo code of CE-BDA is given in Algorithm 5. Specifically, all solutions in $D_c(t)$ and $D_e(t)$ are sorted into dominated or non-dominated solutions and a label ('0' or '1') is assigned to each solution, denoted by L_c, and L_e. Construct the feature space for the two knowledge transfer process $Z_X = [X_c, X_e]$ and $Z_F = [F_c, F_e]$ (line 1 of Algorithm 5), initialize M_0^X and M_0^F. M_k^X, M_k^F can then be calculated according to Eqs. (9.19) and (9.20), and K^X, K^F can be constructed (lines 2-4). By solving the problem in Eq. (9.22) and using the l smallest eigenvectors, we can then calculate Θ_X,Θ_F (lines 6-7).

Algorithm 5 CE-BDA algorithm.

Input: $D_c(0)$,$D_e(0)$ are the initial data samples of the cheap and expensive MOPs, L_c,L_e are their class labels, *dim*, μ, λ, and $T1$ is the number of iterations.
Output: $\hat{D}(t)$.
1: $Z_X \leftarrow [X_c, X_e]$, $Z_F \leftarrow [F_c, F_e]$;
2: Initialize M_0^X and M_0^F;
3: Evaluate M_k^X, M_k^F using (9.19) and (9.20);
4: $K^X, K^F \leftarrow$ the Kernel Matrix of BDA;
5: **for** $i = 1$ to $T1$ **do**
6: Solve (9.22);
7: Build $\Theta_X, \Theta_F \leftarrow l$ smallest eigenvectors;
8: Train the classifiers C_X, C_F. $C_X \leftarrow \{\Theta_X^T X_c, L_c\}$, $C_F \leftarrow \{\Theta_F^T F_c, L_c\}$;
9: Update the soft labels. $\hat{L}_e^X = C_X(\Theta_X^T X_e)$, $\hat{L}_e^F = C_F(\Theta_F^T F_e)$;
10: Update matrix M_k^X, M_k^F using (9.20);
11: **end for**
12: Obtain the final transformation matrix Θ_X,Θ_F;
13: Adaptive balance factor $\beta \leftarrow$ Estimation of adaptive balance factor β;
14: Train the classifiers for cheap problem. $C_{X_c} \leftarrow \{\Theta_X^T X_c^1, L_c^1, \beta_X\}$, $C_{F_c} \leftarrow \{\Theta_F^T F_c^1, L_c^1, \beta_F\}$;
15: $\hat{L}_{c_2}^X \leftarrow C_{X_c}(\Theta_X^T X_c^2)$, $\hat{L}_{c_2}^F \leftarrow C_{F_c}(\Theta_F^T F_c^2)$;
16: Select the data points $\hat{L}_{c_2}^X$=1 and $\hat{L}_{c_2}^F$=1, denote it as $\hat{D}(t)$;

To determine the balance factors according to the distribution of the data for the two knowledge transfer processes, we can tune the balance parameters based on the *A*-distance, which is defined as the error of building a linear classifier to distinguish two domains (Wang et al., 2020b). To this end, a classifier C_l is trained to distinguish the cheap and expensive problem domain, denoted by Φ_c, Φ_e, respectively. Then, *A*-distance can be calculated by:

$$Dis_A(\Phi_c, \Phi_e) = 2(1 - 2e(C_l)) \tag{9.23}$$

where $e(C_l) = E_{Z \sim \Phi_e} |Y_e \neq C_l(Z)|$. The *A*-distance of the marginal distributions, denoted by Dis_A^M, can be calculated by (9.23). For calculating the *A*-distance between the conditional distributions, let $\Phi_c^{(k)}$ and $\Phi_e^{(k)}$ be the samples from the k-th class, $k \in \{0, 1\}$ for dominated and non-dominated solutions. Thus, we have:

$$Dis_A^C(k) = Dis_A(\Phi_c^{(k)}, \Phi_e^{(k)}) \tag{9.24}$$

Thus, β can be estimated as follows:

$$\hat{\beta} = 1 - \frac{Dis_A^M}{Dis_A^M + \sum_{k=0}^{1} Dis_A^C(k)} \tag{9.25}$$

where $k = 0$ means that a solution is a dominated solution, and $k = 1$ means it is non-dominated.

Algorithm 6 presents the main steps for estimating β_X for the decision variables and β_F for the objectives. In this algorithm, the k-nearest neighbor is adopted as

Algorithm 6 Estimation of adaptive balance factor β

Input: Cheap problem domain Φ_c, expensive problem domain Φ_e, number of iterations $T2$.
Output: adaptive balance factor β.
1: **for** $i = 1$ to $T2$ **do**
2: Calculate the A-distance of the marginal distribution Dis_A^M by (9.23);
3: Calculate $Dis_A^C(k)$ corresponding to category k by (9.24);
4: Estimation of adaptive balance factor μ by (9.25);
5: **end for**

the classifiers. In the initial stage, the classifiers may be inaccurate and therefore soft label method is adopted to make the classification more accurate through an iterative training iterations. In each training iteration, classifiers C_X, C_F are trained, the soft labels are updated, matrices M_k^X, M_k^F are calculated and Eq. (9.22) is solved according to the updated matrix to update the transformation matrix. The above process is repeated for $T1$ iterations to obtain the final transformation matrix Θ_X,Θ_F.

9.4.3 *CE-BDA for Data Augmentation*

Data generated for the cheap MOP can be selected for augmenting the training data of the expensive MOP, once the transformation matrices and the balance factors are calculated. m_1 solutions can be randomly sampled and evaluated on the cheap MOP. One classifier C_{X_c}, is trained using the data pair $\{X_c^1, L_c^1\}$ based on the transformation matrix Θ_X, and β_X, and the other, C_{F_c}, is trained using the data pair $\{F_c^1, L_c^1\}$. Then, another m_2 solutions are randomly sampled, denoted by X_c^2, which are also evaluated using the objective functions of the cheap MOP. Then, the dominance label of X_c^2 will be predicted using C_{X_c} and C_{F_c}. If a data point is predicted to be non-dominated by both classifiers, it will be added as a synthetic training data for building the GP models of the expensive MOP, which is denoted as $\hat{D}(t)$.

9.4.4 *Evolutionary Multi-Objective Bayesian Optimization*

Before the optimization starts, a set of data evaluated on the cheap MOP $D_c(0)$ and a set of data evaluated on the expensive MOP ($D_e(0)$) (usually of size $11d-1$ as recommended in (Jones et al., 1998)) are generated. Before optimization starts, a set of synthetic data $\hat{D}(1)$ is generated based on dynamic distribution adaptation and data augmentation. Then a GP model is built for each objective function using the augmented training data $D_e(1) \cup \hat{D}(1)$, where $D_e(1) = D_e(0)$. Given the m GP models (m is the number of objectives), the following expensive MOP will be solved using NSGA-II:

$$\min\{u_1(x), \cdots, u_m(x)\} \tag{9.26}$$

where $m = 2$ or $m = 3$,, $u_i(x) = \mu_i(x) - \alpha\sigma_i(x), i = 1, 2, ..., m$, are the LCB of the i-th objective.

Algorithm 7 Framework of EMBO-OTL

Input: $D_c(0)$, $D_e(0)$, and maximum iterations $t_{\max}$.
Output: Solutions X_e and Y_e.
1: **while** $t \le t_{\max}$ **do**
2: $\hat{D}(t) \leftarrow$ CE-BDA algorithm;
3: Build GPs based on $D_e(t) \cup \hat{D}(t)$;
4: Solutions $\leftarrow$ **NSGA-II**(GPs);
5: $D_e'(t)$ is selected from solutions by k-means;
6: Set $D_e(t+1) = D_e(t) \cup D_e'(t)$, $t = t+1$;
7: **end while**;

NSGA-II will be run for a predefined number of generations to solve the MOP in (9.26). Once the optimization is complete, a number of non-dominated solutions is obtained. Then, k-means clustering algorithm is applied to group the non-dominated solutions into k_s clusters, where k_s is a hyperparameter. k_s solutions closest to the cluster centers are chosen and evaluated using the expensive objective functions and the resulting data is denoted as $D_e'(t)$. Finally, set $D_e(t+1) = D_e(t) \cup D_e'(t)$ and $t = t+1$ before starting the next iteration of Bayesian optimization.

The pseudo code of the overall evolutionary multi-objective Bayesian optimization assisted with online transfer learning, EMBO-OTL for short, is given in Algorithm 7.

9.5 Knowledge Transfer between Objectives in Multi-objective Optimization

9.5.1 Motivation

Most multi-objective evolutionary algorithms assume that all objective functions of the same problem have similar time complexity. Thus, the numbers of evaluations for different objectives are always the same. This assumption may fail to hold in many real-world applications, where different objective functions require highly different computation times, which is also known as latencies (Allmendinger et al., 2015). For example in car design, computation times for evaluating aerodynamic performance and for evaluating the cost may be very different, just as in evolutionary search of deep neural network architectures, the computation of model complexity is very fast, while evaluating the classification performance may take hours for a large model trained on a huge dataset. In the following, we consider a bi-objective optimization problem, of which one objective is computationally cheap, and the other is expensive (Allmendinger et al., 2015; Chugh et al., 2018a). The main motivation is to perform more search on the cheap objective while evaluating the expensive objective, and transfer the knowledge from the cheap objective to the expensive one to enhance the overall performance (Wang et al., 2020c). Here, the basic assumption is that the two objectives are evaluated in parallel and surrogates, here Gaussian processes, must be built for both objective functions.

9.5.2 Parameter Based Transfer Learning

Let us denote the slow and fast objective functions as $f^s(\mathbf{x})$ and $f^f(\mathbf{x})$, respectively, and the corresponding surrogates are GP^s and GP^f. $\mathbf{Y}^s$ and $\mathbf{Y}^f = \{y^1, y^2, ...y^N\}^T$ are the vectors of the slow and fast objective functions for a given solution set X. We assume that the time for evaluating the expensive objective $f^s(\mathbf{x})$ once is τ times of that for evaluating the fast objective $f^f(\mathbf{x})$, where τ is a positive integer larger than 1.

For Gaussian processes, their (hyper-) parameters are supposed to represent the knowledge learned, and therefore, can be used for transferring knowledge from a source task to a target task by sharing the model parameters (Zhuang et al., 2019). Here, parameter-based transfer learning is adopted to build GP_s by sharing some of its parameters with those of GP^f. For more efficient knowledge transfer, a feature selection method can be employed to identify a compact set of most relevant parameters to be shared to improve the performance of GP^s. Let $\theta = [\theta_1, \theta_2, ..., \theta_n]$ denote the parameter vector indicating the importance of the corresponding k-th decision variable and $\boldsymbol{\theta}^s(t)$ and $\boldsymbol{\theta}^f(t)$ represent the importance of the GP^s and GP^f of the selected features at the t-th iteration. $\boldsymbol{\theta}^s(t)$ can then be calculated by

$$\boldsymbol{\theta}^s(t+1) = (1-\alpha)\,\boldsymbol{\theta}^f(t) + \alpha\,\boldsymbol{\theta}^s(t) \tag{9.27}$$

where

$$\alpha = -0.5 \cdot \cos\left(\frac{FE}{FE^s_{\max} \cdot \pi}\right) + 0.5 \quad (9.28)$$

where FE and $FE^s_{\max}$ denote the current number of the function evaluations using the real objective function. α is an adaptation parameter defined by a cosine function, aiming to assign adaptive weights to $\boldsymbol{\theta}^s$ and $\boldsymbol{\theta}^f$ in the aggregation function. Note that parameter sharing as described in Eq. (9.27) allows GP^s to learn knowledge from GP^f.

In the early stage of the optimization, we allocate a larger weight to $\boldsymbol{\theta}^f(t)$ to make sure that mainly the high-quality GP^f is shared since it has been trained on sufficient data. As the optimization proceeds, the number of training samples for GP^s increases, and the weight of $\boldsymbol{\theta}^s(t)$ will gradually be made larger.

9.5.3 Overall Framework

Here, we also adopt the Bayesian approach to solve the bi-objective optimization problem that used an acquisition function for model management. The pseudocode of the overall transfer learning based bi-objective optimization algorithm, called T-SABOEA is listed in Algorithm 8.

Like most SAEAs, Latin hypercube sampling (LHS) is used to generate the initial population, and objectives are evaluated using the real objective function, $f^s(\mathbf{x})$ and $f^f(\mathbf{x})$, respectively. A single-objective search is conducted on $f^f(\mathbf{x})$ while waiting for the evaluations of $f^s(\mathbf{x})$. The evaluated solutions of the expensive objective are stored in D^s for constructing the surrogate GP^s. Solutions evaluated only on the cheap objective function are stored in D^f used to train GP^f. Different from conventional SAEAs in which we always have the same number of training data for all objectives, we have here much more data in D^f than in D^s, because more function evaluations can be made on $f^f(\mathbf{x})$. This means that GP^f is usually more accurate than GP^s. However, if there is a correlation between the two objectives, and if we are able to transfer knowledge of $f^f(\mathbf{x})$ to GP^s, we can improve the quality of GP^s. Similar to other SAEAs, an MOEA, e.g., the RVEA (Cheng et al., 2016) is employed to search on two GP surrogates. After the optimization on the surrogate is complete, an acquisition function, for example, LCB, can be adopted to evaluate the optimized solutions and APD in RVEA is employed to determine the samples x^s_{new} to be sampled using $f^s(\mathbf{x})$.

During evaluating the selected x^s_{new}, additional samples x^f_{new} can be collected by sampling around x^f_{new} using the LHS method. While GP^f is updated on D^f, GP^s is updated using two different methods alternately. Specifically, for τ iterations, GP^s is updated once using the data D^s. Then, the parameters in GP^s corresponding to the decision variables identified by the feature selection method are updated for $\tau - 1$ iterations using the parameter sharing method in (9.27).

Algorithm 8 Framework of T-SABOEA

Input: FE^s_{max}: the maximum number of fitness evaluations for the slow objective; τ: the ratio of the evaluation times of the two objectives; u: the number of new samples to update the GPs; w_{max}: the maximum number of generations before updating GPs
Output: The final solution population in the archive D^s

1: Initialization: Sample the initial population D_0 using LHS and set $D^f_0 = D_0$; Run a single-objective EA to optimize $f^f(\mathbf{x})$ and save solutions in D^f_0; set $FE^s = size(D_0)$ and $Iter = 1$
2: **While** $FE^s \leqslant FE^s_{max}$ **do**
3: Train GP^f with the training data D^f
4: **If** ($Iter \mod \tau == 0$)
5: Train GP^s with the training data D^s;
6: **Else**
7: Use a feature selection method to select the relevant features;
8: Use the Eq.(9.27) to update GP^s;
9: **End If**
10: **While** $w \leqslant w_{max}$ **do**
11: Run an EA to find samples for updating GPs;
12: $w = w + 1$;
13: **End While**
14: Calculate the LCB of the optimized individuals ;
15: Use APD to determine u points x^s_{new} to be evaluated by $f^s(\mathbf{x})$ and add x^s_{new} to D^s;
16: Sample $\tau * u$ points x^f_{new} around x^s_{new} to be evaluated by $f^f(\mathbf{x})$ and add them to D^f;
17: Update $FE^s = FE^s + u$, $Iter = Iter + 1$;
18: **End While**
19: Return the final solutions D^s;

9.6 Data-Driven Multi-fidelity Transfer Optimization

Most existing algorithms optimize each problem separately without considering the similarity between them, which might waste computation resources. Recently, transfer learning (Pan and Yang, 2010) has successfully been applied to evolutionary optimization (Gupta et al., 2018) to address to make use of knowledge from previously solved problems. For example, the algorithm in (Luo et al., 2018a) applies the knowledge transfer among sub-problems which are decomposed from the original MOP.

Many data-driven optimization problems are driven by computationally very intensive simulations (Branke et al., 2017; Jin and Sendhoff, 2009), where the fidelity level and complexity can be controlled (Wang et al., 2018d). In fact, the useful information can be transferred among different fidelity levels to benefit the optimization process (Yang et al., 2019; Wang et al., 2020a).

9.6.1 *Transfer Learning for Bi-Fidelity Optimization*

When the data-driven optimization problems have two available fidelity levels (high- and low-fidelity levels) of function evaluations, they can be called bi-fidelity optimization problems (Wang et al., 2018d). Their mathematical expression is shown below.

$$\min f_h(\mathbf{x}) = f_l(\mathbf{x}) - e(\mathbf{x}), \tag{9.29}$$

where $f_l(\mathbf{x})$ and $f_h(\mathbf{x})$ are the low- and high-fidelity fitness evaluations for a solution $\mathbf{x}$, $e(\mathbf{x})$ is the difference from $f_l(\mathbf{x})$ to $f_h(\mathbf{x})$. The low-fidelity fitness function $f_l(\mathbf{x})$ is easy to be calculated but inaccurate, whereas the high-fidelity fitness function $f_h(\mathbf{x})$ is accurate but computationally expensive. Generally, both functions are correlated to each other, but their accuracy and complexity are different. Therefore, how to make use the advantages of both fidelity level is key to the effectiveness and efficiency of the optimization algorithm.

So far, the existing bi-fidelity EAs can be divided into two categories:

- **Fidelity adjustment**: A large number of bi-fidelity EAs change the fidelity levels to achieve both high accuracy and low computation cost. Those algorithms can control when to increase the fidelity level or which solution to be assign the high-fidelity function evaluation.
- **Surrogate assistance**: In fact, surrogate models that approximate the high- or low-fidelity function evaluations are additional low-fidelity function evaluations. In most existing surrogate-assisted bi-fidelity EAs, both fidelity levels are used separately. Co-kriging models (Myers, 1984) provide a different idea, in which the low-fidelity fitness evaluations are usually used together with the high-fidelity fitness evaluations for the prediction(Le Gratiet and Garnier, 2014).

In fact, the problem in Eq. (9.29) can be re-formulated in the view of transfer learning. The source task T_S aims to find the optimum of the low-fidelity fitness function $f_l(\mathbf{x})$, and the source data $\mathcal{D}_S$ is the data by evaluating solutions using $f_l(\mathbf{x})$. The target task T_T is to find the optimum of the high-fidelity fitness function $f_h(\mathbf{x})$, the target data $\mathcal{D}_T$ is collected by evaluating solutions using $f_h(\mathbf{x})$. As we know, $\mathcal{D}_S$ is cheaper to be obtained than $\mathcal{D}_T$, so the size of $\mathcal{D}_S$ is much larger than that of $\mathcal{D}_T$. Thus, with proper transfer learning techniques, T_T can be improved by T_S with its rich $\mathcal{D}_S$.

9.6.2 *Transfer Stacking*

Transfer stacking (Pardoe and Stone, 2010) is a transfer learning algorithm for regression, which uses multiple regression models $h_{S1}, h_{S2}, \ldots, h_{SB}$ trained from corresponding source data $\mathcal{D}_{S1}, \mathcal{D}_{S2}, \ldots, \mathcal{D}_{SB}$ to assist the learning process of the target model h_T as below:

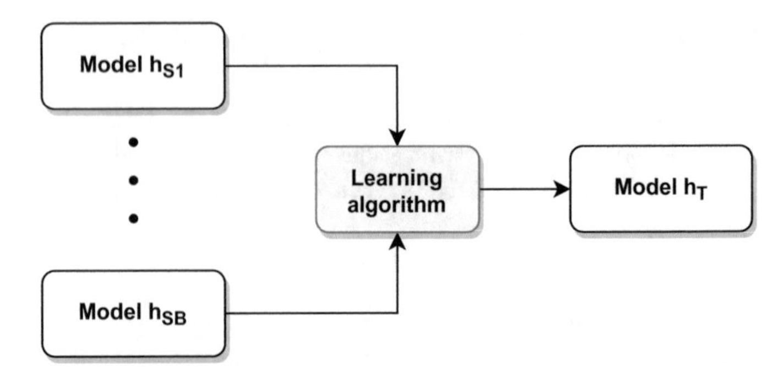

Fig. 9.4 A diagram of transfer stacking

$$h_T(\mathbf{x}) = \sum_{i=1}^{B} a_i h_{Si}(\mathbf{x}) + a_{B+1}, \tag{9.30}$$

where h_T is a linear combination of h_{S1},h_{S2},...,h_{SB} by weights a_1,a_2,...,a_B, a_{B+1}. Those weights can be estimated by minimizing MSE on $\mathcal{D}_T$ as shown in Fig.9.4.

9.6.3 Surrogate-Assisted Bi-Fidelity Evolutionary Optimization

In fact, both the low-fidelity fitness function and surrogate models are cheap and able to generate a large number of source data to approximate the high-fidelity fitness function. In this section, we will introduce an example of transferring the low-fidelity level to the the high-fidelity level for data-driven optimization.

A transfer surrogate-assisted bi-fidelity evolutionary algorithm (TSA-BFEA) is reported in (Wang et al., 2020a), which applies the transfer stacking algorithm to transfer knowledge from low-fidelity $f_l(\mathbf{x})$ to high-fidelity $f_h(\mathbf{x})$ in a generic EA.

In TSA-BFEA, an RBFN model $\hat{f}_{RBF}(\mathbf{x})$ is employed as another source regression model in addition to $f_l(\mathbf{x})$. Thus, $f_h(\mathbf{x})$ can be replaced by a transfer surrogate model $\hat{f}_T(\mathbf{x})$, which is trained from $f_l(\mathbf{x})$ and $\hat{f}_{RBF}(\mathbf{x})$ by using the transfer stacking algorithm. The main framework is shown in in Fig. 9.5, which follows the main steps of existing data-driven EAs except the black components. The differences are building surrogate model and choosing the online samples.

To build the transfer surrogate model locally, the target data $\mathcal{D}_T$ in each generation is selected from the whole training data D_S using the high-fidelity fitness evaluations, which is the subset near the current population. The selection process is shown in Algorithm 9. In the t-th generation, TSA-BFEA chooses the nearest neighbors in D_S to P^t for D_T^t.

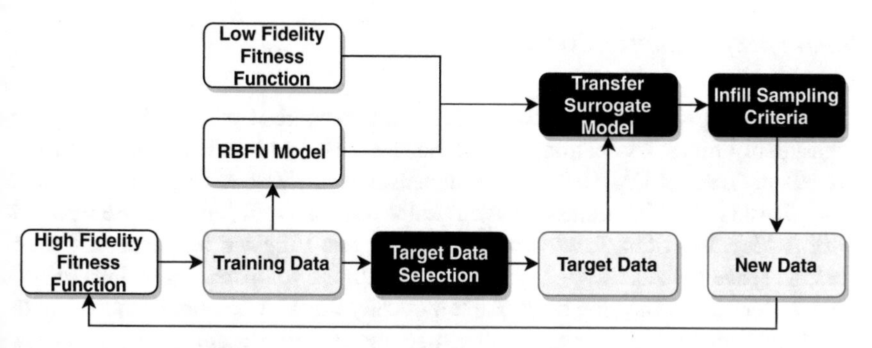

Fig. 9.5 A diagram of the model management strategy in TSA-BFEA

Algorithm 9 Pseudo code of data selection for target data D_T^t in the t-th generation.

Input: D_S^t-the source data in the t-th generation, P^t-the current population.
1: Set the target data D_T^t empty.
2: **for** i=1:$|P^t|$ **do**
3: Find the nearest neighbor of P_i^t (NB) in D_S^t.
4: **if** $NB \notin D_T^t$ **then**
5: Add NB to D_T^t.
6: **end if**
7: **end for**
Output: D_T^t.

As another source regression model, $\hat{f}_{RBF}(\mathbf{x})$ is trained using the whole training data. Different with Eq. (9.30), the transfer surrogate model combines $\hat{f}_{RBF}(\mathbf{x})$ and $f_l(\mathbf{x})$ as below:

$$\hat{f}_T(\mathbf{x}) = a_1 f_l(\mathbf{x}) + a_2 \hat{f}_{RBF}(\mathbf{x}). \tag{9.31}$$

As $\mathcal{D}_T$ is in a small area, we exclude the bias in Eq. (9.30) from Eq. (9.31). The least square method is employed to obtain a weight $\{a_1, a_2\}$ by minimizing MSE of $\hat{f}_T(\mathbf{x})$ for D_T^t. Thus, the transfer surrogate model $\hat{f}_T(\mathbf{x})$ can enhance the approximation accuracy in the local area of P^t.

Using $\hat{f}_T(\mathbf{x})$ to evaluate solutions in the population, selection operation can be executed. To further enhance the quality of $\hat{f}_T(\mathbf{x})$ in each generation, two solutions ($\mathbf{x}^p$ and $\mathbf{x}^u$) in the population P are selected as the infilling samples as the following two equations:

$$\mathbf{x}^p = \arg\min_{\mathbf{x}\in P} \hat{f}_T(\mathbf{x}), \tag{9.32}$$

$$\mathbf{x}^u = \arg\max_{\mathbf{x}\in P} \left| f_l(\mathbf{x}) - \hat{f}_{RBF}(\mathbf{x}) \right|, \tag{9.33}$$

where $\mathbf{x}^p$ is the current predicted best solutions and $\mathbf{x}^u$ is the most uncertain solutions. As (Wang et al., 2017a) shows, both samples can improve the exploration and exploitation ability of TSA-BFEA.

9.6.4 Experimental Results

To verify the performance of TSA-BFEA, we compare it with an EA using the high-fidelity fitness evaluations only (termed EA-HF) and an EA using an RBFN model only (termed EA-RBFN) on 30-dimensional modified MFB problems (Wang et al., 2018d). The high fitness functions in the original MFB problems are replaced with Ackley, Rastrigin, Rosenbrock, Griewank, and Ellipsoid functions, and their feasible space is scaled to $[-1, 1]^n$. To be fair, all the compared algorithms run 30 independent times and stop by 10000 low-fidelity fitness evaluations (equivalent to 1000 high-fidelity fitness evaluations). Also, those three algorithms use simulated binary crossover (SBX) with $\eta = 15$, polynomial mutation with $\eta = 15$ as their variation operators. For the RBFN models in EA-RBFN and TSA-BFEA, the initial training data is of 100 samples, and the hidden layer consists n Gaussian radial basis functions.

Table 9.2 is the optimal solutions obtained by EA-HF, EA-RBFN, and TSA-BFEA on bi-fidelity Ackley, Rastrigin, Rosenbrock, Griewank, and Ellipsoid with resolution, stochastic, and instability errors. TSA-BFEA outperforms other two algorithms on the problems with resolution and stochastic errors, but fails on the problems with instability errors. The instability errors of MFB12 problems is large, the transfer surrogate model cannot be accurate without a bias in Eq. (9.31). The reasons why TSA-BFEA is effective on the problems with resolution and stochastic errors are:

Table 9.2 Optimal solutions obtained by EA-HF, EA-RBFN, and TSA-BFEA on bi-fidelity Ackley, Rastrigin, Rosenbrock, Griewank, and Ellipsoid with resolution, stochastic, and instability errors. The results are shown in the form of mean ± standard deviation. The best fitness values among all the compared algorithms for each problem are highlighted

Problem	EA-HF	EA-RBFM	TSA-BFEA
MFB1-Ackley	18.3±0.4	10.6±1.3	**7.7±0.8**
MFB1-Rastrigin	259.4±21.9	106.7±20.9	**63.1±9.1**
MFB1-Rosenbrock	1584.7±244.9	198.0±42.1	**57.2±9.1**
MFB1-Griewank	188.8±31.7	5.7±1.1	**4.0±0.8**
MFB1-Ellipsoid	773.5±106.9	15.2±6.2	**8.2±1.7**
MFB8-Ackley	18.5±0.5	11.1±1.1	**7.6±0.9**
MFB8-Rastrigin	252.8±16.9	97.7±15.5	**54.1±10.3**
MFB8-Rosenbrock	1622.7±345.8	164.1±27.7	**60.1±11.1**
MFB8-Griewank	192.9±30.6	4.7±0.8	**3.9±0.7**
MFB8-Ellipsoid	750.4±105.1	15.6±4.4	**6.4±1.3**
MFB12-Ackley	18.5±0.4	**7.6±0.7**	12.3±0.9
MFB12-Rastrigin	251.9±17.0	**104.9±16.2**	192.6±19.3
MFB12-Rosenbrock	1566.0±308.2	**61.1±11.6**	123.2±21.9
MFB12-Griewank	196.0±22.3	**3.9±1.0**	12.0±2.7
MFB12-Ellipsoid	756.7±120.3	**15.9±4.8**	30.8±8.9

- Although $f_l(\mathbf{x})$ is inaccurate, it still can economically help the optimizer.
- Surrogate models are cheaper than both $f_l(\mathbf{x})$ and $f_h(\mathbf{x})$, they are good for the evolutionary search.
- TSA-BFEA uses both fidelity levels to effectively solve bi-fidelity optimization problems.

9.7 Surrogate-Assisted Multitasking Multi-Scenario Optimization

Robustness is one key factor considered in engineer design (Beyer and Sendhoff, 2007), because the operation scenario of an design has uncertainty. In fact, multi-scenario optimization is a practical formulation, but it requires multiple evaluations in different scenarios, which makes the optimization process expensive. Therefore, a multitasking optimization technique (Ong and Gupta, 2016) is employed to accelerate the multi-scenario optimization process.

9.7.1 Multi-Scenario Minimax Optimization

Different from generic optimization problems, multi-scenario optimization problems have an additional scenario space $\mathbf{s} \in \pi$ to describe multiple scenarios. Optimizing the worst-case scenario performance as a minimax optimization problem in Eq. (9.34) is one formulation of multi-Scenario optimization.

$$\min_{\mathbf{x} \in \Omega} \max_{\mathbf{s} \in \pi} f(\mathbf{x}, \mathbf{s}), \tag{9.34}$$

where $\max_{\mathbf{s} \in \pi} f(\mathbf{x}, \mathbf{s})$ is the worst-scenario performance of $\mathbf{x}$ for all the possible scenarios π. In other words, the worst-case scenario for each maximization sub-problem should be found for the outer minimization problem in Eq. (9.34), which makes the optimization for $\mathbf{x}$ is a hierarchical search process, because the worst scenario $\mathbf{s}$ needs to be searched to evaluate a candidate $\mathbf{x}$. When the scenario space π is infinite, a multi-scenario optimization process will be very expensive.

Coevolutionary algorithms are widely used for minimax problems, where two populations P_x and P_s are involved to search the design and scenario spaces separately. For P_x, $h(\mathbf{x}, P_s)$ in Eq. (9.35) of a solution $\mathbf{x}$ is its worst objective value among the scenarios in P_s, which works as the fitness function.

$$h(\mathbf{x}, P_s) = \max_{\mathbf{s} \in P_s} f(\mathbf{x}, \mathbf{s}). \tag{9.35}$$

For P_s, $g(\mathbf{s}, P_x)$ in Eq. (9.36) of a scenario $\mathbf{s}$ is its best objective value among the solutions in P_x as the fitness function.

$$g(\mathbf{s}, P_x) = \min_{\mathbf{x} \in P_x} f(\mathbf{x}, \mathbf{s}). \tag{9.36}$$

Based on $h(\mathbf{x}, P_s)$ and $g(\mathbf{s}, P_x)$, two separate selection operations are used to choose good solutions and scenarios.

In fact, as a parallel process, coevolutionary algorithms reduce the search space of the original problem. Therefore, those coevolutionary algorithms might get stuck in an endless optimization cycle of finding the optima in the decision and scenario spaces when the symmetrical condition (Cramer et al., 2009) is not met.

To address this issue, a recent minimax differential evolution (MMDE) algorithm (Qiu et al., 2018) used a min heap to allocate computational costs between the decision and scenario spaces. Due to the large search space, MMDE can solve problems with only two decision variables with 100,000 function evaluations.

To reduce the required number of function evaluations, surrogate models have been employed in two minimax EAs (Ong et al., 2006; Zhou et al., 2007), where the uncertainty in the decision space is addressed or the worst-scenario performance is approximated.

9.7.2 *Surrogate-Assisted Minimax Multifactorial Evolutionary Optimization*

To further reduce the required high computational cost, a multifactorial evolutionary algorithm (MFEA) (Ong and Gupta, 2016) is employed to parallelize the multiple worst scenario search in the current population and an RBFN to approximate $f(\mathbf{x}, \mathbf{s})$. The proposed algorithm is called surrogate-assisted minimax multifactorial evolutionary algorithm (SA-MM-MFEA).

As shown in Fig. 9.6, SA-MM-MFEA follows the very basic EA framework, where the scenario and decision variables are encoded in one individual, but scenarios and solutions are selected separately. In addition, it uses a trained RBFN as the approximated function evaluations, which is updated in every N_g generations.

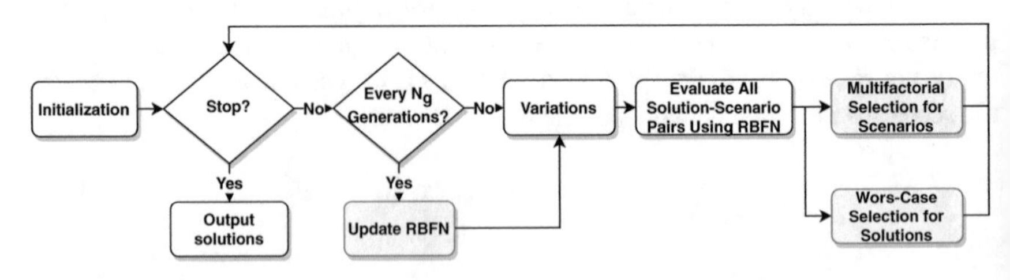

Fig. 9.6 A diagram of SA-MM-MFEA

For a population P with N individuals, each individual contains both $\mathbf{x}$ and $\mathbf{s}$. Before selecting solutions based on their worst-case fitness, N search processes for worst scenarios of N candidate solutions need to be done. Instead of repeatedly optimizing N problems, they can be modeled as a multitasking problems:

$$\begin{aligned}&\{\mathbf{s}^1, ..., \mathbf{s}^j, ..., \mathbf{s}^N\} = \\ &\{\arg\max_{\mathbf{s}\in\pi} f(\mathbf{x}^1, \mathbf{s}), ..., \arg\max_{\mathbf{s}\in\pi} f(\mathbf{x}^j, \mathbf{s}), ..., \arg\max_{\mathbf{s}\in\pi} f(\mathbf{x}^N, \mathbf{s})\} \cdot\end{aligned} \tag{9.37}$$

SA-MM-MFEA employs the scalar fitness in MFEA (Ong and Gupta, 2016) for the scenario selection. The scalar fitness of multiple tasks is defined as below:

$$\begin{aligned}\varphi^i &= 1/\min R_d(\mathbf{s}^i, \varsigma^j), 1 \le j \le N, \\ \varsigma^j &= \{f(\mathbf{x}^j, \mathbf{s}^1), ..., f(\mathbf{x}^j, \mathbf{s}^N)\},\end{aligned} \tag{9.38}$$

where ς^j is the set of fitness of N scenarios for $\mathbf{x}^j$ and $R_d(\mathbf{s}^i, \varsigma^j)$ is rank of $\mathbf{s}^i$ in ς^j. Thus, φ^i shows the best performance of $\mathbf{s}^i$ on N tasks in Eq. (9.37). SA-MM-MFEA selects candidate scenarios based on the scalar fitness φ. Thus, good solutions for each tasks can be kept in one single population.

During the calculation of φ, the worst scenario $\mathbf{sw}^j$ for $\mathbf{x}^j$ can be estimated as:

$$\mathbf{sw}^j = \arg\max_{\mathbf{s}^i} f(\mathbf{x}^j, \mathbf{s}^i). \tag{9.39}$$

Thus, the worst-case fitness of $\mathbf{x}^j$ can be estimated as $f(\mathbf{x}^j, \mathbf{sw}^j)$, and solutions are selected according to the estimated worst-case fitness.

SA-MM-MFEA adopts a generation-based model management strategy, as described in Sect. 5.3.2, to update its surrogate model (RBFN). The initial model is built based on the initial data $\mathcal{D}_T$ sampled by LHS in the whole joint scenario-decision space. In every N_g generations, the optimization state is detected and different update strategies are applied according to the detected state. Through a small number of samples around the predicted optimum (written as $\mathcal{D}_N$), the state can be divided into following three cases:

- **Case 1**: $\mathcal{D}_N$ is worse than $\mathcal{D}_T$. The average objective value of D_N is worse than that of D_T, or p-value PV of a t-test is larger than the best previous p-value PV_{best}. This state means the current RBFN guides a very wrong direction.
- **Case 2**: $\mathcal{D}_N$ is slightly better than $\mathcal{D}_T$ based on a t-test, whose p-value PV is larger than 0.05. Such state shows that the current RBFN provides a right searching direction.
- **Case 3**: $\mathcal{D}_N$ is significantly better than $\mathcal{D}_T$ based on a t-test, whose p-value PV is smaller than 0.05. This state shows that a very promising local area has been found.

SA-MM-MFEA updates the RBFN and optimization region using different strategies according to those above cases.

- **Case 1**: To correct the current RBFN, the complexity of the RBFN is rebuilt based on all the sampled data. Since the population has been guided to a wrong area, the population requires a restart by an re-initialization.
- **Case 2**: The RBFN is rebuilt based on all the sampled data. The optimization region is reduced to the neighbor area of the current optimum.
- **Case 3**: A local RBFN with the smallest complexity is built based on $\mathcal{D}_N$. The optimization region is reduced to the neighbor area of the current optimum.

9.7.3 Experiment Results

To further study the performance of SA-MM-MFEA, we compare it with two algorithms on six minimax optimization benchmark problems in (Qiu et al., 2018) as shown in Table 9.3. The two compared algorithms are minimax evolutionary algorithm (MMEA) and minimax differential evolution algorithm (MMDE) (Qiu et al., 2018), MMEA is very basic minimax EA with separate worst scenario search, and MMDE is the well-performing minimax EA.

Table 9.3 Test problems for minimax optimization

F1	Objective	$f(x,s)=(x-5)^2-(s-5)^2$
	Domain	$x\in[0,10], s\in[0,10]$
	Optimum	$(x^*,s^*)=(5,5)$
F2	Objective	$f(x,s)=\min\{3-0.2x+0.3s, 3+0.2x-0.1s\}$
	Domain	$x\in[0,10], s\in[0,10]$
	Optimum	$(x^*,s^*)=(0,0)$
F3	Objective	$f(x,s)=\frac{\sin(x-s)}{\sqrt{x^2+s^2}}$
	Domain	$x\in(0,10], s\in(0,10]$
	Optimum	$(x^*,s^*)=(10,2.125683)$
F4	Objective	$f(x,s)=\frac{cos\sqrt{x^2+s^2}}{\sqrt{x^2+s^2}+10}$
	Domain	$x\in[0,10], s\in[0,10]$
	Optimum	$(x^*,s^*)=(7.044146333751212, 10\ or\ 0)$
F5	Objective	$f(\mathbf{x},\mathbf{s})=100(x_2-x_1^2)^2+(1-x_1)^2 - s_1(x_1+x_2^2)-s_2(x_1^2+x_2)$
	Domain	$\mathbf{x}\in[-0.5,0.5]\times[0,1], \mathbf{s}\in[0,10]^2$
	Objective	$(\mathbf{x}^*,\mathbf{s}^*)=(0.5,0.25,0,0)$
F6	Objective	$f(\mathbf{x},\mathbf{s})=100(x_1-2)^2+(x_2-2)^2 + s_1(x_1^2-x_2)+s_2(x_1+x_2-2)$
	Domain	$\mathbf{x}\in[-1,3]^2, \mathbf{s}\in[0,10]^2$
	Objective	$(\mathbf{x}^*,\mathbf{s}^*)=(1,1,any,any)$

In the comparative experiment, we apply the following settings. The population size of SA-MM-MFEA is 100, while the population size of MMEA and MMDE is 10. Two parameters K_s and T controlling the number of real function evaluations in the bottom-boosting scheme and partial-regeneration strategy of MMDE are reduced to 19 and 1, whose original settings are 190 and 10 for inexpensive problems in (Qiu et al., 2018). N_g in the proposed algorithm is set as 100. All the compared EAs use simulated binary crossover with $\eta = 15$ and polynomial mutation with $\eta = 15$, and all the compared DE use binomial recombination with $Cr = 0.5$ and mutation with $F = 0.7$, which follows the settings of MMDE. The number of hidden nodes of the initial RBF network is set as the total dimension of the decision and scenario space, where the centers of hidden nodes are determined by applying the k-means algorithm, the widths of radial basis functions are the largest distances between those centers, and the weights of radial basis functions are calculated using the pseudo-inverse method.

All the compared algorithms stop by 50 real function evaluations. The results are analyzed by the Friedman test with the Bergmann-Hommel post-hoc test (Derrac et al., 2011b), which are shown in Table 9.4.

The results show that MMEA is worst and SA-MM-MFEA is the best. The good performance of SA-MM-MFEA comes from the benefits of surrogate models and the parallelism of multitasking. Both techniques accelerate the convergence speed. Firstly, finding the worst-case scenario for different candidate solutions is considered as the optimization of similar problems which can be accelerated by using the evolutionary multi-tasking optimization approach. Secondly, a surrogate model is built to replace part of the expensive function evaluations in SA-MM-MFEA, which saves the high computational cost of function evaluations.

Table 9.4 MSE of compared algorithms on F1-6. The results are shown in the form of mean ± standard deviation. The results are analyzed by the Friedman test with the Bergmann-Hommel post-hoc test (SA-MM-MFEA is the control method and the significance level is 0.05). The best fitness values among all the compared algorithms for each problem are highlighted

	MMEA	MMDE	SA-MM-MFEA
F1	25.4±8.7	15.1±12.1	**0.5±0.6**
F2	57.6±49.1	35.1±24.6	**15.5±33.9**
F3	63.0±34.8	50.4±46.4	**43.7±30.7**
F4	21.8±20.5	11.0±13.1	**6.9±10.9**
F5	43.1±25.0	29.8±24.2	**21.2±37.0**
F6	95.0±50.3	45.2±36.5	**18.9±28.8**

9.8 Summary

Knowledge transfer is a powerful approach to addressing data paucity widely encountered in data-driven expensive optimization. However, many practical issues remain open for more efficient knowledge transfer so that negative transfer (Weiss et al., 2016), i.e., performance deterioration resulting from knowledge transfer can be avoided. The most important question to answer is how to properly determine if a source task contains transferable knowledge to the target optimization task and how to achieve effective knowledge transfer. To the former question, similarity metrics have been adopted to measure whether an existing optimization task can provide positive knowledge transfer, and to the latter, domain adaptation and transformation techniques have been shown to be valuable.

References

Allmendinger, R., Handl, J., & Knowles, J. (2015). Multiobjective optimization: When objectives exhibit non-uniform latencies. *European Journal of Operational Research*, *243*(2), 497–513.

Beyer, H.-G., & Sendhoff, B. (2007). Robust optimization-a comprehensive survey. *Computer Methods in Applied Mechanics and Engineering*, *196*(33), 3190–3218.

Branke, J., Asafuddoula, M., Bhattacharjee, K. S., & Ray, T. (2017). Efficient use of partially converged simulations in evolutionary optimization. *IEEE Transactions on Evolutionary Computation*, *21*(1), 52–64.

Cheng, R., & Jin, Y. (2015). A social learning particle swarm optimization algorithm for scalable optimization. *Information Sciences*, *291*, 43–60.

Cheng, R., Jin, Y., Olhofer, M., & Sendhoff, B. (2016). A reference vector guided evolutionary algorithm for many objective optimization. *IEEE Transactions on Evolutionary Computation*, *20*(5), 773–791.

Chugh, T., Allmendinger, R., Ojalehto, V., & Miettinen, K. (2018a). Surrogate-assisted evolutionary biobjective optimization for objectives with non-uniform latencies. In *Proceedings of the Genetic and Evolutionary Computation Conference* (pp. 609–616). ACM.

Clerc, M., & Kennedy, J. (2002). The particle swarm—explosion, stability, and convergence in a multidimensional complex space. *IEEE Transactions on Evolutionary Computation*, *6*(1), 58–73.

Cramer, A. M., Sudhoff, S. D., & Zivi, E. L. (2009). Evolutionary algorithms for minimax problems in robust design. *IEEE Transactions on Evolutionary Computation*, *13*(2), 444–453.

Derrac, J., García, S., Molina, D., & Herrera, F. (2011b). A practical tutorial on the use of nonparametric statistical tests as a methodology for comparing evolutionary and swarm intelligence algorithms. *Swarm and Evolutionary Computation*, *1*(1), 3–18.

Gupta, A., Ong, Y.-S., & Feng, L. (2018). Insights on transfer optimization: Because experience is the best teacher. *IEEE Transactions on Emerging Topics in Computational Intelligence*, *2*(1), 51–64.

Jin, Y., & Sendhoff, B. (2009). A systems approach to evolutionary multiobjective structural optimization and beyond. *IEEE Computational Intelligence Magazine*, *4*(3), 62–76.

Jones, D. R., Schonlau, M., & Welch, W. J. (1998). Efficient global optimization of expensive black-box functions. *Journal of Global Optimization*, *13*(4), 455–492.

Le Gratiet, L., & Garnier, J. (2014). Recursive co-kriging model for design of computer experiments with multiple levels of fidelity. *International Journal for Uncertainty Quantification*, *4*(5),

Li, H., Jin, Y., & Chai, T. (2021). Evolutionary multi-objective bayesian optimization based on online transfer learning. *IEEE Transactions on Cybernetics*.

Liang, J. J., Qin, A. K., Suganthan, P. N., & Baskar, S. (2006b). Comprehensive learning particle swarm optimizer for global optimization of multimodal functions. *IEEE Transactions on Evolutionary Computation*, *10*(3), 281–295.

Luo, J., Gupta, A., Ong, Y.-S., & Wang, Z. (2018a). Evolutionary optimization of expensive multi-objective problems with co-sub-pareto front Gaussian process surrogates. *IEEE Transactions on Cybernetics*, *49*(5), 1708–1721.

Myers, D. E. (1984). Co-kriging-new developments. In *Geostatistics for natural resources characterization* (pp. 295–305). Springer.

Ong, Y.-S., & Gupta, A. (2016). Evolutionary multitasking: a computer science view of cognitive multitasking. *Cognitive Computation*, *8*(2), 125–142.

Ong, Y.-S., Nair, P. B., & Lum, K. (2006). Max-min surrogate-assisted evolutionary algorithm for robust design. *IEEE Transactions on Evolutionary Computation*, *10*(4), 392–404.

Pan, S. J., Tsang, I. W., Kwok, J. T., & Yang, Q. (2010). Domain adaptation via transfer component analysis. *IEEE Transactions on Neural Networks*, *22*(2), 199–210.

Pan, S. J., & Yang, Q. (2010). A survey on transfer learning. *IEEE Transactions on Knowledge and Data Engineering*, *22*(10), 1345–1359.

Pardoe, D. and Stone, P. (2010). Boosting for regression transfer. In *Proceedings of the 27th International Conference on International Conference on Machine Learning* (pp. 863–870). Omnipress.

Qiu, X., Xu, J.-X., Xu, Y., & Tan, K. C. (2018). A new differential evolution algorithm for minimax optimization in robust design. *IEEE Transactions on Cybernetics*, *48*(5), 1355–1368.

Sun, X., Gong, D., Jin, Y., & Chen, S. (2013b). A new surrogate-assisted interactive genetic algorithm with weighted semisupervised learning. *IEEE Transactions on Cybernetics*, *43*(2), 685–698.

Tahmoresnezhad, J., & Hashemi, S. (2017). Visual domain adaptation via transfer feature learning. *Knowledge and Information Systems*, *50*(2), 585–605.

Wang, J., Chen, Y., Feng, W., Yu, H., Huang, M., and Yang, Q. (2020b). Transfer learning with dynamic distribution adaptation. *ACM Transactions on Intelligent Systems and Technology*, *11*(1: Article No. 6).

Wang, J., Chen, Y., Hao, S., Feng, W., & Shen, Z. (2017d). Balanced distribution adaptation for transfer learning. In *2017 IEEE International Conference on Data Mining (ICDM)* (pp. 1129–1134). IEEE.

Wang, H., Jin, Y., & Doherty, J. (2018d). A generic test suite for evolutionary multi-fidelity optimization. *IEEE Transactions on Evolutionary Computation*. to appear.

Wang, X., Jin, Y., Schmitt, S., & Olhofer, M. (2020c). Transfer learning for gaussian process assisted evolutionary bi-objective optimization for objectives with different evaluation times. In *Proceedings of the Genetic and Evolutionary Computation Conference* (pp. 587–594). ACM.

Wang, H., Jin, Y., Yang, C., & Jiao, L. (2020a). Transfer stacking from low-to high-fidelity: A surrogate-assisted bi-fidelity evolutionary algorithm. *Applied Soft Computing* (pp. 106276).

Wang, H., Jin, Y., & Doherty, J. (2017a). Committee-based active learning for surrogate-assisted particle swarm optimization of expensive problems. *IEEE Transactions on Cybernetics*, *47*(9), 2664–2677.

Weiss, K., Khoshgoftaar, T. M., & Wang, D. (2016). A survey of transfer learning. *Journal of Big Data*, *3*(9), 40.

Yang, C., Ding, J., Jin, Y., Wang, C., & Chai, T. (2019). Multitasking multiobjective evolutionary operational indices optimization of beneficiation processes. *IEEE Transactions on Automation Science and Engineering*, *16*(3), 1046–1057.

Zhou, Z., Ong, Y. S., Nair, P. B., Keane, A. J., & Lum, K. Y. (2007). Combining global and local surrogate models to accelerate evolutionary optimization. *IEEE Transactions on Systems, Man, and Cybernetics, Part C: Applications and Reviews*, *37*(1), 66–76.

Zhuang, F., Qi, Z., Duan, K., Xi, D., Zhu, Y., Zhu, H., Xiong, H., & He, Q. (2019). A comprehensive survey on transfer learning. arXiv preprint arXiv:1911.02685.

Zhuang, F., Qi, Z., Duan, K., Xi, D., Zhu, Y., Zhu, H., et al. (2020). A comprehensive survey on transfer learning. *Proceedings of IEEE*, *109*(1), 43–76.

Chapter 10
Surrogate-Assisted High-Dimensional Evolutionary Optimization

10.1 Surrogate-Assisted Cooperative Optimization for High-Dimensional Optimization

Most surrogate-assisted evolutionary algorithms we discussed are not meant for solving high-dimensional problems because a large number of training data are required to train a sufficiently accurate surrogate model, the the computational cost for constructing the surrogates will be prohibitive. In this section, a surrogate-assisted cooperative swarm optimization algorithm (SA-COSO for short) for solving high-dimensional time-consuming optimization problems up to a dimension of 200 will be introduced. Two PSO variants, PSO with a constriction factor (Clerc, 1999) and SL-PSO (Cheng and Jin, 2015), are cooperatively utilized. The PSO learns not only from its personal and global best individuals, but also from the best position found so far by SL-PSO, while the SL-PSO also learn from promising solutions provided by the PSO. On the other hand, the SL-PSO algorithm performs the global search for the optimal solution of the RBF surrogate model and the PSO algorithm conducts a local search assisted by the fitness estimation strategy. Fig. 10.1 gives a simple example to show the coupling between the FES-assisted PSO and RBF-assisted SL-PSO in the proposed SA-COSO. From Fig. 10.1, we can see that all solutions selected to be evaluated using the exact objective function are saved in an archive *DB*, which will be used to train a surrogate model and provide demonstrators for RBF-assisted SL-PSO as well. Both PSO and SL-PSO utilized the surrogate modal to approximate the fitness of an individual to reduce the number of fitness evaluation at each generation. Note that the best position found so far by RBF-assisted SL-PSO is also provided for individual learning in PSO, which will be

Y. Jin et al., *Data-Driven Evolutionary Optimization*,
Studies in Computational Intelligence 975,
https://doi.org/10.1007/978-3-030-74640-7_10

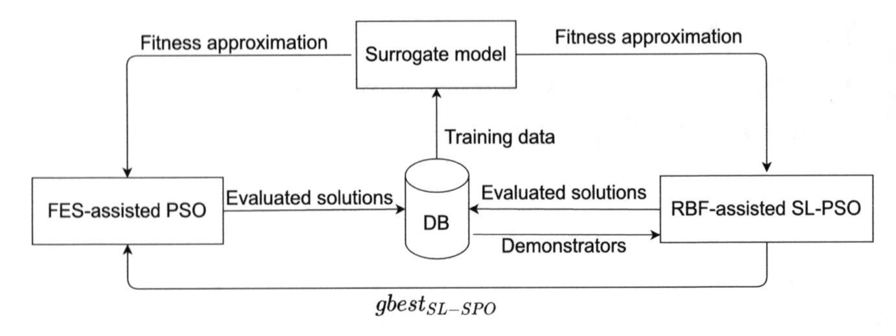

Fig. 10.1 Coupling between FES-assisted PSO and RBF-assisted SL-PSO in SA-COSO

$$
\begin{aligned}
v_{ij}(t+1) = \ & \chi(v_{ij}(t) + c_1 r_1 (p_{ij}(t) - x_{ij}(t)) + c_2 r_2 (p_{gj}(t) - x_{ij}(t)) \\
& + c_3 r_3 (p_{rg,j}(t) - x_{ij}(t))),
\end{aligned} \tag{10.1}
$$

$$
x_{ij}(t+1) = x_{ij}(t) + v_{ij}(t+1), \tag{10.2}
$$

where r_1, r_2 and r_3 are random number generated uniformly in the range of [0, 1], c_1, c_2 and c_3 are positive constant, where c_1 is called the cognitive parameter and both c_2 and c_3 are known as the social learning parameter. $\mathbf{p}_i = (p_{i1}, p_{i2}, \ldots, p_{in})$ and $\mathbf{p}_g = (p_{g1}, p_{g2}, \ldots, p_{gn})$ (i.e., $gbest_{PSO}$) are the personal best position of individual i and the global best position of the swarm of the FES-assisted PSO. $\mathbf{p}_{rg} = (p_{rg,1}, p_{rg,2}, \ldots, p_{rg,n})$ is the global best position found so far by the RBF-assisted SL-PSO (i.e., $gbest_{SL-PSO}$).

Algorithm 1 presents the pseudo code of the SA-COSO. Two populations, pop_{PSO} and pop_{SL-PSO}, are initialized and evaluated using the real objective function. All of these evaluated solutions will be saved in the archive *DB* for the model training. The personal best position of each individual and the global best position found so far by PSO (denoted as $gbest_{PSO}$) will be determined. Meanwhile, the global best position found so far by SL-PSO, denoted as $gbest_{Sl-PSO}$, will also be determined. Thus, the best position found so far by PSO and SL-PSO, labeled as *gbest*, can be correspondingly identified. The FES-assisted PSO and RBF-assisted SL-PSO algorithms are then run simultaneously and some information will be exchanged at each generation. The process will be repeated until the stopping criterion is met. N_{PSO} and N_{SL-PSO} are the population size of PSO and SL-PSO, respectively. In the following, the details of RBF-assisted SL-PSO, FES-assisted PSO, and the model updating will be presented.

Algorithm 1 The pseudocode of SA-COSO

1: Initialize a population pop_{PSO} for PSO, including velocity and position initialization, fitness evaluation for each individual. Determine the personal best position for each solution and the global best position for the population pop_{PSO};
2: Initialize a population pop_{SL-PSO} for SL-PSO, including position initialization, fitness evaluation for each individual and global best position determination;
3: Determine the global best position of two populations found so far, $gbest = \min\{gbest_{PSO}, gbest_{SL-PSO}\}$;
4: Save all solutions that have been evaluated using the exact fitness function into DB;
5: $N_{FE} = |N_{PSO}| + |N_{SL-PSO}|$;
6: **while** the stopping criterion is not met **do**
7: Training/updating an RBF surrogate model;
8: $DB_t = \emptyset$;
9: Run RBF-assisted SL-PSO; (**Refer to Algorithm 2**)
10: Run FES-assisted PSO; (**Refer to Algorithm 3**)
11: Determine the global best position of two populations found so far, $gbest = \min\{gbest_{PSO}, gbest_{SL-PSO}\}$;
12: **end while**
13: Output the global best position found so far $gbest$ and its fitness value;

10.1.1 RBF-Assisted SL-PSO

Generally, each individual (named imitator) learns from its demonstrators who are in the current population and have better fitness values than this individual. However, in RBF-assisted SL-PSO, the fitness values of individuals are approximated using the surrogate model. Therefore, the individuals marked as the demonstrators are probably not actually better than the imitator. Therefore, to avoid the population searching in a wrong direction, N_{SL-PSO} solutions in the archive DB will also be chosen as potential demonstrators for individual learning to update its position. Thus, the demonstrators of an individual i will be selected from the subset of the union of N_{SL-PSO} solutions in the current population of SL-PSO and N_{SL-PSO} solutions randomly selected from the archive DB whose fitness values are better than that of individual i. Note that in SL-PSO, the individual will be kept to the next generation if no solution has better fitness value than this individual.

Once the demonstrators of an individual are identified, the position of individual i can then be updated. After that, the fitness values of all updated solutions will be approximated by the RBF surrogate model. The best position of all current population will be compared to the global best position found so far by the SL-PSO algorithm. It will be evaluated using the real expensive objective function if this position has better fitness value than the global best position, and will be used to update the global best position found so far. Note that in each generation of SL-PSO, at most one solution will be evaluated using the real-objective function. Algorithm 2 gives the pseudocode of RBF-assisted SL-PSO.

Algorithm 2 The pseudocode of the RBF-assisted SL-PSO

Input: M_{RBF}: RBF surrogate model;
  DB: the archive to save all solutions that have been evaluated using the exact objective function;
  P_{SL-PSO}: current population of SL-PSO;
  N_{FE}: the number of fitness evaluation;
  $gbest_{SL-PSO}$: the best position found so far by SL-PSO;
Output: N_{FE}, $gbest_{SL-PSO}$
  for each individual i in P_{SL-PSO} **do**
    Identify the demonstrators from current population and those drawn from the archive DB;
    Update the position using Eq. (3.44);
    Estimate the fitness value using M_{RBF}
  end for
  Identify the best position $best_{SL-PSO}$ from the current population;
  if $\hat{f}_{RBF}(best_{SL-PSO}) < f(gbest_{SL-PSO})$ **then**
    Evaluate the fitness of $best_{SL-PSO}$ using the exact objective function;
    Save to the temporary archive DB_t
    $N_{FE} = N_{FE} + 1$;
    if $f_{RBF}(best_{SL-PSO}) < f(gbest_{SL-PSO})$ **then**
      Update the global best position of SL-PSO found so far $gbest_{SL-PSO}$;
    end if
  end if

10.1.2 FES-Assisted PSO

Algorithm 3 gives the pseudocode of FES-assisted PSO.

Algorithm 3 Pseudocode of the FES-assisted PSO

Input: P_{PSO}: current population of PSO;
  N_{FE}:the number of fitness evaluation;
  M_{RBF}: RBF surrogate model;
Output: N_{FE}, $gbest_{PSO}$
  while the stopping criterion is not met **do**
    Update the velocity and position of each individual in P_{PSO} using Eqs. (10.1) and (10.2), respectively;
    Fitness evaluation/approximation for each individual in the current population; (**Refer to Algorithm 4**)
  end while

In FES-assisted PSO, after the positions of individuals have been updated, the fitness values of individuals will be evaluated using the expensive objective function (at the first two generations) or be approximated using the fitness estimation strategy (in the following generations. Fitness estimation strategy (Sun et al., 2013a) is a computationally simple yet effective fitness approximation technique based on the positional relationship between the particles of PSO. According to Eqs. (10.1) and (10.2), the new positions of individuals i and j in the population can be updated, respectively, as follows:

$$\begin{aligned}\mathbf{x}_i(t+1) = & \mathbf{x}_i(t) + \chi((\mathbf{x}_i(t) - \mathbf{x}_i(t-1)) + c_1\mathbf{r}_{i1}(\mathbf{p}_i(t) - \mathbf{x}_i(t)) + \\ & c_2\mathbf{r}_{i2}(\mathbf{p}_g(t) - \mathbf{x}_i(t)) + c_3\mathbf{r}_{i3}(\mathbf{p}_{rg}(t) - \mathbf{x}_i(t))) \\ = & (1 + \chi(1 - c_1\mathbf{r}_{i1} - c_2\mathbf{r}_{i2} - c_3\mathbf{r}_{i3}))\mathbf{x}_i(t) - \chi\mathbf{x}_i(t-1) + \\ & \chi c_1\mathbf{r}_{i1}\mathbf{p}_i(t) + \chi c_2\mathbf{r}_{i2}\mathbf{p}_g(t) + \chi c_3\mathbf{r}_{i3}\mathbf{p}_{rg}(t),\end{aligned} \tag{10.3}$$

$$\begin{aligned}\mathbf{x}_j(t+1) = & \mathbf{x}_j(t) + \chi((\mathbf{x}_j(t) - \mathbf{x}_j(t-1)) + c_1\mathbf{r}_{j1}(\mathbf{p}_j(t) - \mathbf{x}_j(t)) + \\ & c_2\mathbf{r}_{j2}(\mathbf{p}_g(t) - \mathbf{x}_j(t)) + c_3\mathbf{r}_{j3}(\mathbf{p}_{rg}(t) - \mathbf{x}_j(t))) \\ = & (1 + \chi(1 - c_1\mathbf{r}_{j1} - c_2\mathbf{r}_{j2} - c_3\mathbf{r}_{j3}))\mathbf{x}_j(t) - \chi\mathbf{x}_j(t-1) + \\ & \chi c_1\mathbf{r}_{j1}\mathbf{p}_j(t) + \chi c_2\mathbf{r}_{j2}\mathbf{p}_g(t) + \chi c_3\mathbf{r}_{j3}\mathbf{p}_{rg}(t),\end{aligned} \tag{10.4}$$

Thus, we can get a virtual position $\mathbf{x}_v$ given in the following by combining and rearranging Eqs. (10.3) and (10.4):

$$\begin{aligned}\mathbf{x}_v = & \mathbf{x}_i(t+1) + \chi\mathbf{x}_i(t-1) + (1 + \chi(1 - c_1\mathbf{r}_{j1} - c_2\mathbf{r}_{j2} - c_3\mathbf{r}_{j3}))\mathbf{x}_j(t) + \\ & \chi c_1\mathbf{r}_{j1}\mathbf{p}_j(t) + \chi c_2\mathbf{r}_{j2}\mathbf{p}_g(t) + \chi c_3\mathbf{r}_{j3}\mathbf{p}_{rg}(t) \\ = & \mathbf{x}_j(t+1) + \chi\mathbf{x}_j(t-1) + (1 + \chi(1 - c_1\mathbf{r}_{i1} - c_2\mathbf{r}_{i2} - c_3\mathbf{r}_{i3}))\mathbf{x}_i(t) + \\ & \chi c_1\mathbf{r}_{i1}\mathbf{p}_i(t) + \chi c_2\mathbf{r}_{i2}\mathbf{p}_g(t) + \chi c_3\mathbf{r}_{i3}\mathbf{p}_{rg}(t),\end{aligned} \tag{10.5}$$

Consequently, the fitness of the virtual position can be approximated using the weighted average of $f(\mathbf{x}_i(t+1))$, $f(\mathbf{x}_i(t-1))$, $f(\mathbf{x}_j(t))$, $f(\mathbf{p}_j(t))$, $f(\mathbf{p}_g(t))$ and $f(\mathbf{p}_{rg}(t))$ or of $f(\mathbf{x}_j(t+1))$, $f(\mathbf{x}_j(t-1))$, $f(\mathbf{x}_i(t))$, $f(\mathbf{p}_i(t))$, $f(\mathbf{p}_g(t))$ and $f(\mathbf{p}_{rg}(t))$ in the following form:

$$f(\mathbf{x}_v) = \frac{WS_1}{WD_1} = \frac{WS_2}{WD_2}, \tag{10.6}$$

where

$$WS_1 = \frac{f(\mathbf{x}_i(t+1))}{D_i(t+1)} + \frac{\mathbf{x}_i(t-1)}{D_i(t-1)} + \frac{\mathbf{x}_j(t)}{D_j(t)} + \frac{f(\mathbf{p}_j(t))}{D_{pj}(t)} + \frac{f(\mathbf{p}_g(t))}{D_g(t)} + \frac{\mathbf{p}_{rg}(t)}{D_{rg}(t)}, \tag{10.7}$$

$$WS_2 = \frac{f(\mathbf{x}_j(t+1))}{D_j(t+1)} + \frac{\mathbf{x}_j(t-1)}{D_j(t-1)} + \frac{\mathbf{x}_i(t)}{D_i(t)} + \frac{f(\mathbf{p}_i(t))}{D_{pi}(t)} + \frac{f(\mathbf{p}_g(t))}{D_g(t)} + \frac{\mathbf{p}_{rg}(t)}{D_{rg}(t)}, \tag{10.8}$$

$$WD_1 = \frac{1}{D_i(t+1)} + \frac{1}{D_i(t-1)} + \frac{1}{D_j(t)} + \frac{1}{D_{pj}(t)} + \frac{1}{D_g(t)} + \frac{1}{D_{rg}(t)}, \tag{10.9}$$

$$WD_1 = \frac{1}{D_j(t+1)} + \frac{1}{D_j(t-1)} + \frac{1}{D_i(t)} + \frac{1}{D_{pi}(t)} + \frac{1}{D_g(t)} + \frac{1}{D_{rg}(t)}, \tag{10.10}$$

where $D_i(t+1)$, $D_i(t-1)$, $D_j(t)$, $D_{pj}(t)$, $D_j(t+1)$, $D_j(t-1)$, $D_i(t)$, $D_{pi}(t)$, $D_g(t)$ and $D_{rg}(t)$ represent the Euclidean distances between the virtual position $\mathbf{x}_v$ and

$\mathbf{x}_i(t+1)$, $\mathbf{x}_i(t-1)$, $\mathbf{p}_j(t)$, $\mathbf{x}_j(t+1)$, $\mathbf{x}_j(t-1)$, $\mathbf{x}_i(t)$, $\mathbf{p}_i(t)$, $\mathbf{p}_g(t)$ and $\mathbf{p}_{rg}(t)$, respectively. From Eq. (10.6), we can see that only the fitness values of individual i and j at generation $t+1$ are not known. Thus, if one of them have a fitness value, suppose the fitness value of $\mathbf{x}_i(t+1)$ is provided, the fitness value of the other individual j can be estimated as follows:

$$\hat{f}_{FES}(\mathbf{x}_j(t+1)) = D_j(t+1) \cdot WF_{new}, \tag{10.11}$$

where

$$WF_{new} = \frac{WD_1 \times WS_1}{WD_2} - \frac{f(\mathbf{x}_j(t-1))}{D_j(t-1)} - \frac{f(\mathbf{x}_i(t))}{D_i(t)} - \frac{f(\mathbf{p}_i(t))}{D_{pi}(t)} - \frac{f(\mathbf{p}_g(t))}{D_g(t)} - \frac{\mathbf{p}_{rg}(t)}{D_{rg}(t)}. \tag{10.12}$$

In (Sun et al., 2013b), the individual will be evaluated using the real objective function if it has not been approximated by other individuals. To save the number of fitness evaluations, in FES-assisted PSO, each individual will be approximated using the RBF surrogate model at first. Then, the approximated fitness values of individuals will be updated by the fitness estimation strategy. Algorithm 4 gives the pseudocode of the determination of the fitness value of each individual in FES-assisted PSO. From Algorithm 4, we can see that the fitness value of each individual i is determined sequentially and its nearest neighbor who falls behind individual i will be updated on the approximated fitness value. The minimum value of all approximated values for an individual i will be set to the fitness value of this individual. In FES-assisted PSO, three parts consider to select individuals to be evaluated using the real-objective function. In the first two generations (lines 2–3), all individuals in P_{PSO} are evaluated using the real-objective function. Then in the procedure to update the personal best position and the global best position of the population (lines 10–24), if both approximated fitness values of an individual are better that the fitness value of its personal best position, this individual will be evaluated using the real expensive objective function. In order to ensure that the global best position must be evaluated using the exact expensive objective function, the personal best position that has the minimum fitness value among the current population will be evaluated using the real objective function and used to update the global best position. Note that there may be a situation that no solution in P_{PSO} satisfies to have two better approximated fitness values than its personal best position, and there is no personal best position having better approximated fitness than the global best position. In such a case, the solutions having larger approximation uncertainty than the mean approximation uncertainty will be selected to be evaluated using the real objective function. Eq. (10.13) gives the method to calculate the average degree of approximation uncertainty between the fitness values approximated by the RBF network and the fitness estimation strategy. Algorithm 5 shows the method to select individuals to be evaluated using the real objective function.

$$DF = \frac{1}{N_{PSO}} \sum_{i=1}^{N_{PSO}} |\hat{f}(\mathbf{x}_i) - \hat{f}_{RBF}(\mathbf{x}_i)| \tag{10.13}$$

Algorithm 4 Pseudocode of the fitness determination for each solution in FES-assisted PSO

Input: t: the current generation;
P_{PSO}: the current population of FES-assisted PSO;
DB_t: the temporary archive to save the solutions that have been evaluated using the exact objective function;
M_{RBF}: the RBF surrogate model;
N_{FE}: the number of fitness evaluations;
1: **if** $t < 2$ **then**
2: Evaluate each individual in P_{PSO} using the exact objective function, and save in the temporary archive DB_t;
3: Update the personal best position of each individual and the global best position $gbest_{PSO}$ found so far of PSO.
4: **else**
5: Approximate the fitness value of each individual $i, i = 1, 2, \ldots, N_{PSO}$ in P_{PSO} using the surrogate model M_{RBF}, denoted as $\hat{f}(\mathbf{x}_i(t+1)) = \hat{f}_{RBF}(\mathbf{x}_i(t+1))$;
6: **for** $i = 1$ to N_{PSO} **do**
7: Find the nearest neighbor j of particle i, $\mathbf{x}_j(t+1) \neq \mathbf{x}_i(t+1)$ and $j > i$;
8: Approximate the fitness value of $\mathbf{x}_j(t+1)$ using the fitness estimation strategy, denoted as $\hat{f}_{FES}(\mathbf{x}_j(t+1))$
9: $\hat{f}(\mathbf{x}_j(t+1)) = \min\{\hat{f}(\mathbf{x}_j(t+1)), \hat{f}_{FES}(\mathbf{x}_j(t+1))\}$
10: **if** $\hat{f}(\mathbf{x}_i(t+1)) < f(\mathbf{p}_i(t))$ **then**
11: **if** $\hat{f}_{RBF}(\mathbf{x}_i(t+1)) < f(\mathbf{p}_i(t))$ **then**
12: Evaluate the solution $\mathbf{x}_i(t+1)$ using the real objective function and save to the temporary archive DB_t;
13: Update the personal best position of individual i, $\mathbf{p}_i(t)$;
14: Update the global best position if $f(\mathbf{p}_i(t+1)) < f(gbest_{PSO})$;
15: **else**
16: Update the personal best position of individual i;
17: **end if**
18: **end if**
19: **end for**
20: Find the best position in personal best solutions of all individuals in P_{PSO}, denoted as $best_{PSO}$;
21: **if** $best_{PSO}$ has better value than $gbest_{PSO}$ **then**
22: Evaluate the solution $best_{PSO}$ using the real objective function and save to the temporary archive DB_t;
23: Update $gbest_{PSO}$ using $best_{PSO}$;
24: **end if**
25: **if** no solution in P_{PSO} has been evaluated **then**
26: Select solutions with high approximation uncertainty to be evaluated using the exact objective function; (Refer to Algorithm 5);
27: **end if**
28: **end if**

Algorithm 5 Individual selection based on the approximation uncertainty for exact evaluation

1: Calculate the mean difference of the values approximated by the RBF model and the fitness estimation strategy using Eq. 10.13;
2: **for** $i = 1$ to N_{PSO} **do**
3: **if** $|\hat{f}(\mathbf{x}_i) - \hat{f}_{RBF}(\mathbf{x}_i)| > DF$ **then**
4: Evaluate individual i using the real objective function and save to the temporary archive DB_t;
5: **if** $f(\mathbf{x}_i) < f(\mathbf{p}_i)$ **then**
6: Update the personal best position $\mathbf{p}_i$;
7: **end if**
8: **if** $f(\mathbf{x}_i) < f(\mathbf{g})$ **then**
9: Update the best position found so far of PSO;
10: **end if**
11: **end if**
12: **end for**

10.1.3 Archive Update

In SA-COSO, the RBF network is mainly utilized as a global surrogate model for SL-PSO to perform global search. To this end, the samples for training the RBF model should be properly selected. All data in the archive *DB* will be used to train the surrogate model, in order to reduce the computational time, there is no need for using all data that have been evaluated to train the surrogate model. Thus, we limited the number of data saved in the archive *DB*. When the number of data in the archive *DB* does not reach the maximum capacity, the solutions will be saved to *DB* directly. However, when the archive *DB* reaches the maximum number of data, some of them should be removed from *DB* and some of exact evaluated solutions saved in the temporary archive DB_t will be selected to add in *DB*. Fig. 10.2 gives an example to illustrate the strategy for updating the archive *DB*.

In Fig. 10.2, the horizontal axis represents the decision space and the vertical axis is the fitness value. The black circles denote solutions (data pairs) in DB, the blue diamonds represent solutions that have been evaluated in current generation, and the green triangles are individuals of P_{SL-PSO}. Suppose the maximum number of data that are allowed to be saved in *DB* is 5. From Fig. 10.2, we can see that there are three newly evaluated solutions in the current generation, denoted as ‘a’, ‘b’, and ‘c’. Among them, we can see that solution ‘a’ is far away from the current location of the population P_{SL-PSO}. If this solution is added to the archive *DB*, it will have little influence on the surrogate that covers the region of the current population. By contrast, if the solution ‘c’ is included in the archive *DB*, the quality of the surrogate model will be improved more effectively on the region where the current population locate. Similarly, if the data in *DB* is far away from the current population, for example, ‘1’ in *DB*, it will contribute little on the accuracy of the surrogate model to approximate the fitness value of the individuals in the current population. Therefore, if a solution must be discarded when the archive is already full, the solution that is

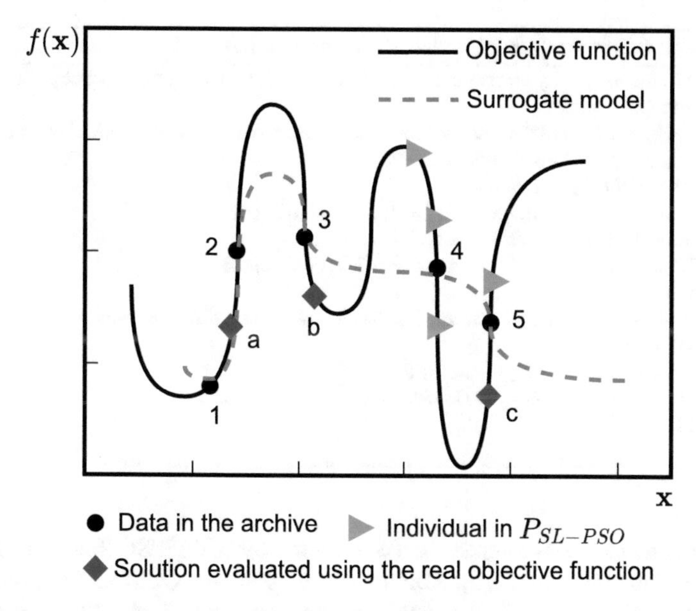

Fig. 10.2 An example to show the strategy for updating the archive *DB*

far away from the local of the current population should be selected to be replaced by a solution evaluated using the real objective function which is much closer to the current population. Algorithm 6 gives the pseudo code to update the data in the archive. For each solution $\mathbf{x}_k$ in the temporary archive DB_t, its minimum distance to the current population P_{SL-PSO} will be calculated at first (line 2). $D(\cdot)$ represents the Euclidean distance between two solutions. Also, the minimum distance of each solution in *DB* to the current population will be calculated (lines 3-5), and the solution that has the maximum distance to the current population will be found (line 6). if the solution $\mathbf{x}_k$ in the temporary archive DB_t is closer than the solution in *DB* that has the maximum distance to P_{SL-PSO}, it will replace this solution in *DB* (lines 7-9). Note that the solutions in DB_t will be added to *DB* one by one until the archive *DB* is full.

10.1.4 Experimental Results and Discussions

To investigate the effectiveness of SA-COSO, an empirical study on six widely used unimodal and multimodal benchmark problems (Liu et al., 2014a), including Ellipsoid (F1), Rosenbrock (F2), Ackley (F3), Griewank (F4), Shifted Rotated Rastrigin (F5) and Rotated hybrid Composition Function (F6), are conducted. In order to evaluate the effectiveness of SA-COSO for solving high-dimensional expensive optimization problems, a set of experiments on 100- and 200-dimensional test prob-

Algorithm 6 Pseudo code of updating the archive *DB*

Input: DB_t: the temporary archive that saves the data evaluated using the exact objective function at the current generation;
DB: the archive to save the data that have been evaluated using the real objective function;
P_{SL-PSO}: the current population of SL-PSO
1: **for** each solution $\mathbf{x}_k$ in DB_t **do**
2: $ind\_pop_{min}(k) = \min\{D(\mathbf{x}_k, \mathbf{x}_j), j = 1, 2, \ldots, N_{SL-PSO}\}$
3: **for** each data $\mathbf{x}_i$ in the archive DB **do**
4: $DB\_pop_{min}(i) = \min\{D(\mathbf{x}_i, \mathbf{x}_j), j = 1, 2, \ldots, N_{SL-PSO}\}$
5: **end for**
6: Find the solution, denoted as $\mathbf{x}_b$, that has the maximum distance in all minimum distance $\max(DB\_pop_{min})$
7: **if** $ind\_pop_{min}(k) < DB\_pop_{min}(b)$ **then**
8: Replace the solution $\mathbf{x}_b$ in DB using the solution $\mathbf{x}_k$ in DB_t
9: **end if**
10: **end for**

lems are conducted and compared to PSO (Clerc, 1999), FESPSO (Sun et al., 2013b), SL-PSO (Cheng and Jin, 2015), RBF-assisted SL-PSO and COSO. RBF-assisted SL-PSO is SL-PSO assisted by a radial basis function network, and COSO represents the cooperative swarm optimization without surrogate model assistance. Among them, FESPSO and RBF-assisted SL-PSO are two surrogate assisted particle swarm algorithms, while the rest are not. All experimental results are conducted for 20 independent runs.

The sizes of P_{PSO} and P_{SL-PSO} are set to 30 and 200, respectively. The cognitive c_1 is set 2.05, two social coefficients, c_2 and c_3, which aim to cooperatively guide the particles toward the global best position, are both set to 1.025. The maximum number of fitness evaluations is set to 1, 000. In order to make fair comparisons, the population size of both PSO and FESPSO is set to 30, the size of SL-PSO and RBF-assisted SL-PSO is set to 200, and the size of COSO is set to 230. The parameters of the RBF-assisted SL-PSO and in COSO are set the same as those in SA-COSO.

Table 10.1 lists the statistical results of all algorithms under comparison on 100- and 200-dimensional problems averaged 20 independent runs. It can be seen from Table 10.1 that SA-COSO has achieved significantly better results on all of the benchmark functions, confirming that SA-COSO performs well on high-dimensional problems. Furthermore, from the results of RBF-assisted SL-PSO, FESPSO and SA-COSO, we can get that SA-COSO integrating the advantages of global surrogate model and fitness estimation strategy is able to get better results than RBF-assisted SL-PSO using the global model only and FESPSO utilizing fitness estimation strategy only.

Table 10.1 Comparative results on 50- and 100-dimensional problems, shown as mean value (standard deviation)

Prob.	Dim.	PSO	FESPSO	SL-PSO	RBF-assisted SL-PSO	COSO	SA-COSO
F1	100	1.5309e+04 (1.7685e+03)	1.9912e+03 (1.8004e+03)	1.6935e+04 (1.2746e+03)	4.2610e+03 (3.0987e+04)	3.2459e+03 (1.2135e+03)	**5.1475e+01** (3.1718e+02)
	200	8.7570e+04 (5.8963e+03)	9.2915e+04 (5.4782e+03)	8.3447e+04 (3.0600e+03)	5.3455e+04 (1.5658e+04)	8.3989e+04 (4.5144e+03)	**1.6382e+04** (2.9811e+03)
F2	100	1.2160e+04 (2.0188e+03)	1.2991e+04 (1.8186e+03)	1.4755e+04 (1.5183e+03)	1.6895e+04 (6.1860e+03)	1.4405e+04 (1.4055e+03)	**2.7142e+03** (1.1702e+02)
	200	3.9989e+04 (3.0511e+03)	4.1495e+04 (4.4760e+03)	3.8801e+04 (2.4071e+03)	5.3149e+04 (5.5807e+03)	3.9679e+04 (2.3388e+03)	**1.6411e+04** (4.0965e+03)
F3	100	2.0239e+01 (1.8744e−01)	2.0178e+01 (3.5469e−01)	1.9981e+01 (1.9843e−01)	2.0876e+01 (1.7703e−01)	1.9949e+01 (1.5436e−01)	**1.5756e+01** (5.0245e−01)
	200	2.0647e+01 (1.4147e−01)	2.0632e+01 (1.2730e−01)	2.0328e+01 (6.9280e−02)	2.1022e+01 (3.6218e−02)	2.0356e+01 (1.1736e−01)	**1.7868e+01** (2.2319e−02)
F4	100	1.2162e+03 (9.2716e+01)	1.2305e+03 (1.0561e+02)	1.2232e+03 (9.7340e+01)	2.4023e+02 (3.1646e+02)	1.2898e+03 (9.6918e+01)	**6.3353e+01** (1.9021e+01)
	200	3.3073e+03 (2.2346e+02)	3.3245e+03 (2.8726e+02)	2.9726e+03 (1.6242e+02)	1.9394e+03 (3.5615e+02)	3.0148e+03 (1.6510e+02)	**5.7776e+02** (1.0140e+02)
F5	100	1.8946e+03 (1.5227e+02)	1.8636e+03 (1.9079e+02)	1.8604e+03 (1.3078e+02)	1.5629e+03 (1.3868e+02)	2.1028e+03 (5.6521e+01)	**1.2731e+03** (1.1719e+02)
	200	5.5872e+03 (3.4360e+02)	5.4966e+03 (3.4179e+02)	5.2454e+03 (1.6168e+02)	4.7900e+03 (2.7671e+02)	5.2272e+03 (1.5075e+02)	**3.9275e+03** (2.7254e+02)
F6	100	1.4083e+03 (5.2538e+01)	1.3810e+03 (3.9465e+01)	1.5407e+03 (2.4168e+01)	1.5721e+03 (7.5160e+01)	1.4852e+03 (2.6082e+01)	**1.3657e+03** (3.0867e+01)
	200	1.4162e+03 (3.4170e+01)	1.5035e+03 (3.6865e+01)	1.5045e+03 (1.5512e+01)	1.4228e+03 (2.7529e+01)	1.4972e+03 (1.6516e+01)	**1.3473e+03** (2.4665e+01)

The best solutions are highlighted in bold

10.2 A Multi-objective Infill Criterion for High-Dimensional Optimization

The main benefit of Gaussian process model to be used to assist evolutionary algorithms is that it can provide an estimate of the fitness approximation uncertainty together with the approximated fitness value. However, the strategy to select individuals to be evaluated using the real objective function, which is often called infill criterion, remains the most important issue to be studied. Generally, there are three main criteria to determine which individuals to be evaluated using the real objective function in the GP-assisted evolutionary algorithms. One commonly used criterion is to select individuals that have the global or local best approximated fitness to be evaluated using the real objective function. The second widely used criterion is to select the individual who has the approximated fitness with a large amount of uncertainty for fitness evaluation using the real fitness function. There are two reasons to select individuals with a large degree of approximation uncertainty for expensive fitness evaluation. First, a large degree of uncertainty in fitness approximation of the individuals indicates that the fitness landscape around these individuals has not been well explored and therefore evaluation of these individuals are very likely to find a good solution. Second, the accuracy of the surrogate model can be most effectively improved if individuals with a large degree of uncertainty in fitness approximation are evaluated using the real objective function. The third criterion proposed for selecting individuals is called the acquisition function which considers the approximation fitness and the approximation uncertainty simultaneously. The commonly utilized acquisition functions include the lower confidence bound (LCB), the expected improvement (EI), and the probability of improvement (PoI). Different to the infill criteria introduced above, this section will introduce a multiobjective infill criterion (MIC), which is used to select individuals to be evaluated using the real objective function. There are two main advantages to use the MIC as the model management strategy, especially for optimization of high-dimensional problems. First, the current population is nondominated sorted in terms of the approximated fitness and the degree of approximation uncertainty. The completely dominated individuals will be selected for fitness evaluation, which takes into account both exploration of the search space and enhancing the model quality. Second, no hyper-parameter is needed to be specified as done in LCB and EI to linearly or nonlinearly combine the estimated fitness and the uncertainty, which is able to strike a good balance between performance and uncertainty. In the following, the Gaussian process assisted social learning particle swarm optimization with multiobjective infill criterion, denoted as MGP-SLPSO, will be introduced.

10.2.1 Main Framework

The framework of MGP-SLPSO is shown in Fig. 10.3, in which the SL-PSO is adopted as the base optimizer and the GP model is used as the surrogate. Note that other evolutionary algorithms can also be used as the base optimizer. The SL-PSO algorithm (Cheng and Jin, 2015) has been shown to perform well in surrogate-assisted optimization of high-dimensional problems (Sun et al., 2017), therefore, it is adopted to be the base optimizer. In Fig. 10.3, an initial population will be generated using the Latin hypercube sampling technique, and evaluated using the real objective function. All initial solutions will be saved to an archive *Arc*. The following procedure will be repeated until the stopping criterion, i.e., the maximum fitness evaluations, is met. The population will be updated using the strategy of SL-PSO. Then, the way to calculate the fitness of each individual in the current population should be determined. If the amount of the data in archive *Arc* does not exceed the pre-defined threshold, all solutions in the population should be evaluated using the real objective function. Otherwise, a GP model will be trained and used for approximating the fitness value of each individual. However, some solutions are still needed to be evaluated using the real objective function, which will be selected by the multiobjective infill criterion described in detail in Chap. 10.2.2.

Algorithm 7 gives the pseudo code of the individual selection for fitness evaluation. The surrogate model will not be under fitting if there is not sufficient data. Therefore, in Algorithm 7, if not sufficient data is offered, all individuals in the current population will be evaluated using the real objective function and saved to

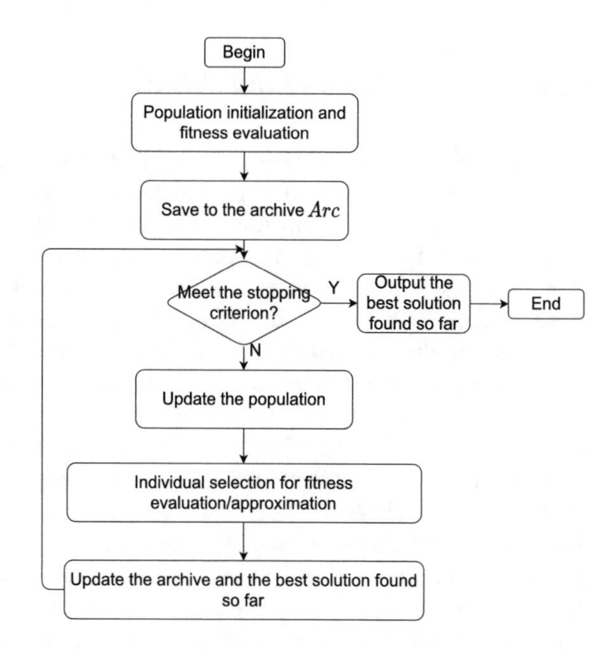

Fig. 10.3 The flowchart of MGP-SLPSO

archive *Arc*. Otherwise a GP model will be trained using the latest N_t data in archive *Arc* and used to approximate the fitness values of individuals. The multiobjective infill criterion will be employed to select a few individuals for fitness evaluation using the real objective function.As is known to all, constructing a GP model will become computationally very intensive when the number of training data increases. Furthermore, as (Wang, 2016) indicate, the difference between the estimated standard deviations (ESDs) of different solutions will vanish for high-dimensional problems, which reduces the effectiveness of existing infill criteria. Thus, the multiobjective infill criterion is proposed to address the issue of vanishing ESD information in GP-assisted evolutionary optimization of high-dimensional problems.

Algorithm 7 Pseudocode of the individual selection and exact fitness evaluation

Input: *Arc*: the archive to save solutions that have been evaluated using the real objective function;
$P(t)$: the population at current generation t;
1: **if** the amount of data in archive *Arc* does not exceed the pre-defined threshold **then**
2: Evaluate the solutions in the current population and save to the archive *Arc*;
3: **else**
4: Train a GP surrogate model using the latest N_t data in archive *Arc*;
5: Approximate the fitness value of each individual in the current population using the GP model;
6: Select individuals according to the multiobjective infill criterion for fitness evaluation using the real objective function (refer to Algorithm 8 for more detail) and save them to archive *Arc*;
7: **end if**

10.2.2 Multiobjective Infill Criterion

Different infill criteria have been proposed for GP-assisted optimization algorithms. However, most of them are only appropriate for low-dimensional problems and little research has been focused on the effectiveness of infill criteria for high-dimensional optimization problems. Figs. 10.4a and 10.4b shows the estimated standard deviations of different 100 individuals on 10- and 50-dimensional Rosenbrock functions at generations 3, 20 and 70. From Fig. 10.4a, we can see that at generation 3 for optimizing 10-dimensional Rosenbrock function, all ESD values are large but nearly the same, making it hardly possible to distinguish the degrees of the approximation uncertainty of different individuals. When the amount of the training data increases over the generation 20, the ESDs of 100 individuals become clearly distinguishable. However, at generation 70 when the optimization converges, the ESD values of different individuals become impossible distinguishable. Fig. 10.4b shows the ESDs of individuals for solving 50-dimensional Rosenbrock function at different generations. From Fig. 10.4b, we can see that the above issue becomes more serious for 50-dimensional problems. The ESD values of all individuals are difficult to be distinguished before 20-th generation, and can become distinguishable only around

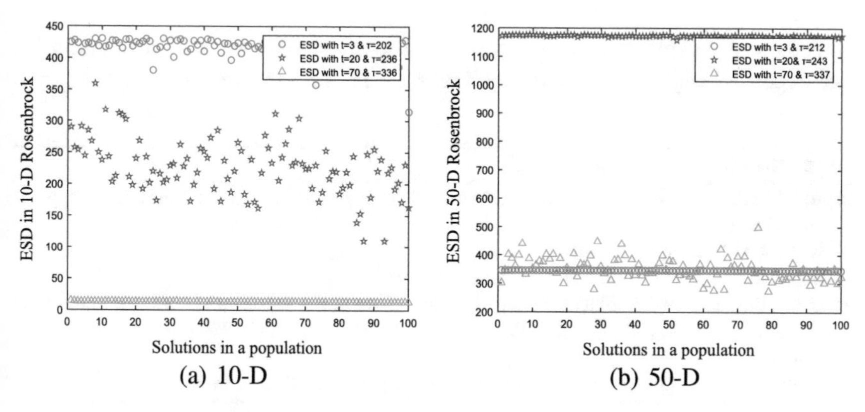

Fig. 10.4 ESDs of individuals for the Rosenbrock function

generation 70. It can be easily imagined that in a high-dimensional decision space, all individuals in the population may be equally far from the small number of training samples. Thus, the ESD values of different individuals are hardly possible to distinguishable because the ESD value of an individual is determined by the correlation between this individual and the training samples, which is eventually determined by the covariance function and the distance between the solution and the training samples.

To address the difficulties to distinguish individuals from each other, a multiobjective infill criterion is proposed, in which the approximated fitness and the approximation uncertainty are dealt as two separate objectives. Then the nondominated sorting, instead of a scalar value that linearly or nonlinearly aggregate the approximation fitness and the approximation uncertainty, is utilized to determine which individuals to be evaluated using the real objective functions. The mathematical model of the proposed multiobjective is given in the following:

$$\begin{aligned} \min_{\mathbf{x}} \quad & \mathbf{g}(\mathbf{x}) = (g_1(\mathbf{x}), g_2(\mathbf{x})) \\ \text{s.t.} \quad & \mathbf{x} \in \mathbf{S} \subset \Re^D \end{aligned} \tag{10.14}$$

where $g_1(\mathbf{x}) = \hat{f}(\mathbf{x})$ is the fitness value approximated by the GP model and and $g_2(\mathbf{x}) = \sqrt{s^2(\mathbf{x})}$ is the ESD for measuring the estimated uncertainty. S is the decision space of original expensive optimization problem.

Subsequently, the fast nondominated sorting algorithm given in (Deb et al., 2002) is employed to sort the individuals into different fronts after the fitness values of individuals are approximated using the GP surrogate model. Algorithm 8 gives lists the pseudo code of multiobjective infill criterion. The solutions on the first front will be evaluated using the real objective function to improve the exploitation of the search. Then the solutions on the last front that do not dominate any others will also be selected to be evaluated using the real objective function, which is expected to emphasize the exploration of the search. All particles that are evaluated using the real expensive objective function are then stored in the archive Arc.

Algorithm 8 Pseudo code of MIC

Input: $P(t)$: the population of current generation t with approximated fitness value and the approximated uncertainty;
M_{GP}: A GP surrogate model;

Output: individuals in the first and last non-dominated fronts.

1: Using the fast non-dominated sorting approach (Deb et al., 2002) to sort $P(t)$ into different non-dominated fronts according to the approximated fitness values and the approximated uncertainties;

2: Evaluate the individuals both on the first and the last non-dominated fronts;

To provide a clear vision to show the ability to distinguish the individuals according to the approximation fitness and approximation uncertainty, Figs. 10.5 and 10.6 give the non-dominated sorting of individuals at generation 3, 20, and 70 for the optimization of 10- and 50-dimensional Rosenbrock problem, respectively. From Figs. 10.5 and 10.6, we can see that compared the existing infill criteria, MIC is able to better distinguish the solutions under different situations, thus making it easier to select individuals for fitness evaluation using the real objective function.

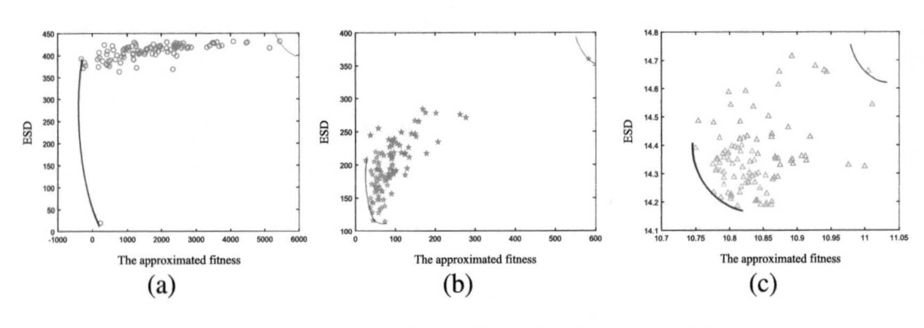

Fig. 10.5 Approximated values and ESD for 10-dimensional Rosenbrock function

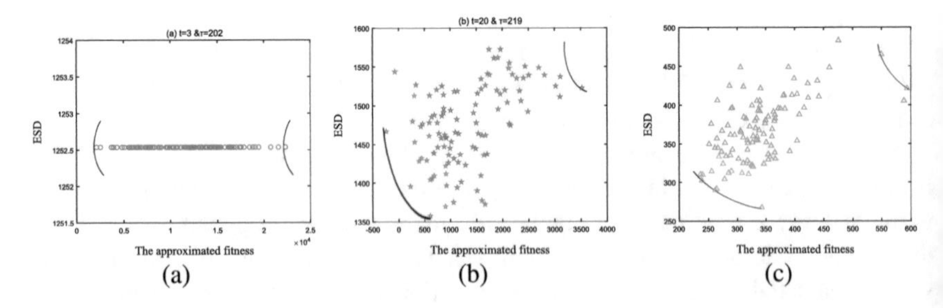

Fig. 10.6 Approximated values and ESD for 50-dimensional Rosenbrock function

10.2.3 Experimental Results and Discussions

To evaluate the effectiveness of the proposed infill criterion on high-dimensional expensive optimization problems, a set of empirical studies on six 50- and 100-dimensional benchmark problems (Liu et al., 2014a), including Ellipsoid (F1), Rosenbrock (F2), Ackley (F3), Griewank (F4), Shifted Rotated Rastrigin (F5) and Rotated hybrid Composition Function (F6), conducted by comparing with SL-PSO (without surrogates), and three variants of GP-assisted SL-PSO driven by different infill criteria, namely, GP-assisted SL-PSO based on the approximated fitness only (GP-Fit for short), the lower confidence bound (GP-LCB), and the expected improvement (GP-EI) in order to investigate whether the proposed MIC has any advantage over the scalar infill criteria. SA-COSO introduced in Chap. 10.1 is also compared. All settings of GP-Fit, GP-LCB and GP-EI are the same as MGP-SLPSO, with the infill criterion being the only difference. In GP-Fit, the particles in the current population are sorted according to their approximated fitness. The best individual is then chosen to be evaluated using the real objective function.

The parameters of the SL-PSO algorithm in MGP-SLPSO, GP-Fit, GP-LCB and GP-EI are set the same as recommended in (Cheng and Jin, 2015). The maximum number of fitness evaluations is set to 1000. The number of training data N_t in GP-Fit, GP-LCB GP-EI and MGP-SLPSO is defined to be $2N \leq N_t \leq 4N$, where N is the population size, which is dimension of the search space as $N = 100 + \lfloor \frac{n}{10} \rfloor$, $2N$ and $4N$ are the minimum and maximum number of training data, respectively. If the number of data in archive is larger than $4N$, only the most recent $4N$ data will be used to train the GP model.

Table 10.2 gives the statistical results of the compared algorithms. All algorithms perform $2N$ exact fitness evaluations to train the GP model before the optimization starts. From Table 10.2, we can find that all GP-assisted SL-PSO algorithms obtained better results than the SL-PSO without GP assistance, which confirms that the surrogate indeed assists to accelerate the convergence of the SL-PSO. Compared to GP-Fit and GP-EI, MGP-SLPSO obtained better results on all 50- and 100-dimensional benchmark problems except for F2 with 50 dimensions. Note, however, that the results of MGP-SLPSO on F2 with 100 dimension are much better than those of GP-LCB and GP-Fit. These results showed that the proposed MIC is much more effective than the scalar infill criteria for high-dimensional problems. From Table 10.2, We can also find that MGP-SLPSO outperforms SA-COSO on all six benchmark functions, which shows the good performance of MGP-SLPSO for 50- and 100-dimensional performance. However, as GP itself has serious limitations, in particular for high-dimensional problems, the MGP-SLPSO is not efficient for 200-dimensional problems.

Table 10.2 Comparative results on 50- and 100-dimensional problems, shown as mean value (standard deviation)

Prob.	Dim.	SL-PSO	SA-COSO	GP-Fit	GP-EI	GP-LCB	MGP-SLPSO
F1	50	1.78e+03 (2.74e+02)	4.93e+01 (1.60e+01)	7.71e−01 (2.33e+00)	4.01e−01 (9.80e−01)	4.34e−01 (1.00e+00)	**9.88e−16** (1.02e−15)
	100	1.17e+04 (9.60e+02)	9.29e+02 (2.36e+02)	2.72e+02 (1.21e+02)	3.91e+02 (2.15e+02)	4.03e+02 (3.06e+02)	**4.93e−05** (1.95e−04)
F2	50	2.37e+03 (4.02e+02)	2.49e+02 (5.43e+01)	1.07e+02 (3.40e+01)	1.21e+02 (3.66e+01)	**1.05e+02** (3.30e+01)	1.20e+02 (1.87e+01)
	100	9.16e+03 (1.12e+03)	2.41e+03 (7.99e+02)	1.28e+03 (3.81e+02)	1.96e+03 (6.48e+02)	1.92e+03 (4.66e+02)	**6.12e+02** (6.79e+01)
F3	50	1.78e+01 (4.26e−01)	9.54e+00 (1.21e+00)	1.10e+01 (2.53e+00)	9.95e+00 (2.03e+00)	1.04e+01 (2.74e+00)	**9.31e+00** (1.13e+00)
	100	1.90e+01 (2.43e−01)	1.59e+01 (7.44e−01)	1.60e+01 (7.74e−01)	1.70e+01 (7.62e−01)	1.72e+01 (6.25e−01)	**1.43e+01** (6.21e−01)
F4	50	2.86e+02 (4.04e+01)	5.54e+00 (1.04e+00)	7.56e−01 (3.03e−01)	7.42e−01 (3.36e−01)	6.48e−01 (4.38e−01)	**1.54e−01** (1.30e−01)
	100	8.74e+02 (8.76e+01)	6.90e+01 (1.50e+01)	1.70e+01 (1.26e+01)	3.34e+01 (1.51e+01)	2.97e+01 (1.47e+01)	**7.15e−01** (7.24e−01)
F5	50	4.09e+02 (5.20e+01)	2.14e+02 (3.33e+01)	1.03e+02 (7.90e+01)	5.62e+01 (5.38e+01)	4.67e+01 (5.52e+01)	**3.30e+01** (3.61e+01)
	100	1.52e+03 (8.53e+01)	1.34e+03 (1.13e+02)	1.70e+03 (1.44e+02)	1.78e+03 (1.66e+02)	1.93e+03 (1.96e+02)	**8.85e+02** (1.17e+03)
F6	50	1.20e+03 (2.79e+01)	1.08e+03 (3.66e+01)	1.20e+03 (5.12e+01)	1.15e+03 (3.48e+01)	1.13e+03 (3.55e+01)	**1.06e+03** (2.14e+01)
	100	1.44e+03 (2.52e+01)	1.41e+03 (3.80e+01)	1.47e+03 (4.14e+01)	1.44e+03 (2.98e+01)	1.45e+03 (2.63e+01)	**1.39e+03** (4.77e+01)

The best solutions are highlighted in bold

10.3 Multi-surrogate Multi-tasking Optimization of Expensive problems

The performance of a surrogate-assisted evolutionary algorithm relies on the contribution of the model management strategy and the efficiency of the search algorithm. The search will be effective if the optimum of surrogate model is the same to that of the real expensive problem, regardless the approximation quality. Apart from the surrogate model, an effective optimization algorithm is also essential for finding the optimal solution with a limited computational budget. Multi-tasking optimization (MTO) (Gupta et al., 2016) is proposed recently for solving multiple tasks simultaneously. In MTO, implicit knowledge transfers across different tasks via assortative mating and selective imitation are utilized. As is shown in (Gupta et al., 2016), the similarity between different tasks will substantially affect the efficiency of the optimization. Different surrogate models can be trained based on different samples to fit the original function. Thus, although the models are different, the optimal solutions of these surrogate models are expected to be the same. Therefore, in this section, we will introduce a multi-surrogate multi-tasking optimization for the computationally expensive problems. Two surrogate models, global and local, are trained for the original problems, and a multi-factorial evolutionary algorithm (MFEA) variant, called the generalized multi-factorial evolutionary algorithm (Ding et al., 2017) is adopted to optimize two surrogate models simultaneously.

10.3.1 Multi-factorial Evolutionary Algorithms

A multi-factorial optimization problem consisting of K single-objective minimization problems can be formulated as follows:

$$\{\mathbf{x}_1^*, \mathbf{x}_2^*, \ldots, \mathbf{x}_K^*\} = \{\text{argmin} f_1(\mathbf{x}_1), \text{argmin} f_2(\mathbf{x}_2), \ldots, \\ \text{argmin} f_K(\mathbf{x}_K)\} \tag{10.15}$$

where $f_k(\mathbf{x}_k)$, $k = 1, \cdots, K$ is the k-th optimization problem. $\mathbf{x}_k^* = (x_{k,1}^*, \cdots, x_{k,n_k}^*)$, $k = 1, \cdots, K$ is the optimal solution of the k-th optimization problem, n_k is the dimension of k-th optimization problem. The framework of MFEA is given in Fig. 10.7. In Fig. 10.7, one population will be initialized for simultaneously optimizing multiple tasks. Each individual in the population will be associated with one task only according to the skill factor. Note that before entering the loop, each individual will evaluate the fitness values of all tasks, and the population will sort on the fitness for each task. Then, the skill factor of each individual will be assigned according to the scalar fitness which is the reciprocal of the rank. Then the following procedures will be repeated until the computational budget is exhausted: Generate an offspring O by the assortative mating strategy, assign the skill factor to each individual by vertical cultural transmission, and update the scalar fitness for each individual in the

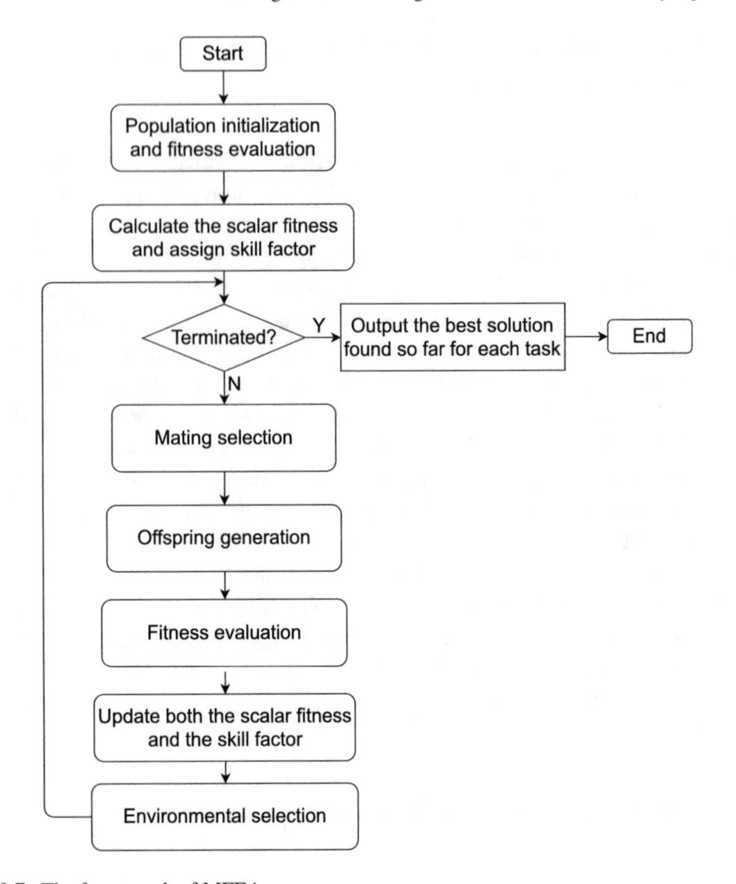

Fig. 10.7 The framework of MFEA

combined population of parent. In the multi-tasking population, the skill factor τ_i of each individual $\mathbf{x}_i$ is defined as the index of the task that individual $\mathbf{x}_i$ is associated with, i.e., $\tau_i = \text{argmin}_k\{r_k^i\}$, and the scalar fitness φ_i of individual $\mathbf{x}_i$ is defined by $\varphi_i = 1/\min\{r_i^k, k = 1, 2, \ldots, K\}$. Finally, the individuals ranked in the front after being sorted in an ascending order according to the scalar fitness will be selected to be the parent population of the next generation.

10.3.2 Main Framework

The multitasking optimization has been shown its better performance than the separate optimization of multiple tasks, especially for those problems that have high similarity. Generally, a global surrogate model is used to smooth out the local optima and assist the optimization algorithm to quickly locate the optimal solution of the problem. While a local surrogate model on a sub-region around the best position found so far is usually adopted to accurately fit the original problem so as to locate exactly at the optimal solution. Therefore, the global and local surrogate models can be seen as two tasks having high similarity on the optimal solutions. In MS-MTO, the global surrogate model is trained using all data that have been evaluated using the real objective function, which is saved in an archive *Arc*. While the local surrogate model is trained on the data that are ranked in the front after being sorted in an ascending order, i.e., from better to worse, according to their fitness values. The optimal solutions of the global and local surrogate models will be evaluated using the computationally expensive real objective function and used to update the global and local surrogate models before next round of optimization on two surrogate models. Algorithm 9 gives the pseudocode of MS-MTO. A number of solutions will be generated using the Latin hypercube sampling technique and evaluated using the real objective function. All evaluated solutions are saved to an archive *Arc* and the solution among them that has the best fitness value will be the best solution found so far. The following procedure will be repeated until the stopping criterion is met, i.e., the maximum number of allowed fitness evaluations is exhausted. Two RBF surrogate models, a global and a local model, will be trained. The global surrogate model, denoted as M_g, is trained on all data in the archive *Arc*, and the local one, denoted as M_l, is trained on N_l best data samples. Two models are treated as two tasks and the generalized multi-factorial evolutionary algorithm is used to search for the optimal solutions of these two surrogate models, which will be evaluated using the real expensive objective function and saved to the archive *Arc* for updating the global and local surrogate models in the next round.

Algorithm 9 The pseudo code of the MS-MTO algorithm

1: Sampling a number of data using LHS in the decision space, evaluate them using the real objective function and save to the archive *Arc*;
2: Determine the best position found so far;
3: **while** the stopping criterion is not met **do**
4: Train a global and a local surrogate models, respectively;
5: Search for the optimal solutions of the global and local surrogate models simultaneously using the multitasking evolutionary algorithm;
6: Evaluate the optimal solutions of global and local surrogate models using the real expensive fitness function and save to the archive *Arc*;
7: Update the best position found so far;
8: **end while**

10.3.3 Global and Local Surrogates

The surrogate models trained for a same optimization problem using different samples may significantly differ in the landscape. Therefore, their optimal solutions are

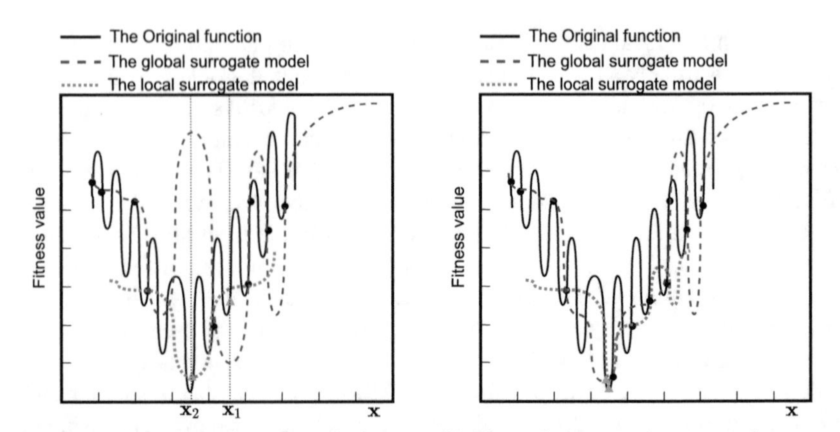

(a) The real objective function and surrogate models at t-th generation (b) The real objective function and surrogate models at $t+1$-th generation

Fig. 10.8 The global and local surrogate models at an adjacent generations. The black dots represent the samples in the archive that have been evaluated using the real objective funtion. The green triangles represent the optimal solutions found for global and local surrogate models, respectively

expected to assist finding a good solution for the expensive optimization problems in a limited computational budget. The global surrogate model M_g is used to explore the whole search space and the local surrogate model is expected to exploit the promising sub-regions around the best position found so far. Fig. 10.8 gives an illustrative example to show the contributions of the global and local surrogate models in MS-MTO. A global surrogate model M_g (denoted by the blue dashed line) is trained using all data in the archive *Arc* at the t-th generation (black dots in Fig. 10.8) and a local one M_l is trained using three data in the archive *Arc* with the best fitness values. Suppose the optimal solutions found for the global and local surrogate models are $\mathbf{x}_1$ and $\mathbf{x}_2$, respectively. Both $\mathbf{x}_1$ and $\mathbf{x}_2$ will be evaluated using the real objective function, and given in green triangles in Fig. 10.8. From Fig. 10.8a, we can see that the optimal solution of the global surrogate model is not better than that of the local surrogate model at t-th generation. However, when these two optimal solutions are added into the archive and used to update the global and local surrogate model used in the next round (given in Fig. 10.8(b)), the optimal solution of the global surrogate model is much closer to that of the real objective function, which is contributed by the optimal solution of the local surrogate model found in the previous generation. Conversely, the best solution of the global surrogate model found so far at t-th generation also contribute to the local surrogate model updating in this example, which makes the local surrogate model fit the original function more accuracy in the sub-region around the best solution found so far.

10.3.4 Multi-tasking Optimization Based on Global and Local Surrogates

The optimal solutions of the global and local surrogate models are expected to be found so as to assist finding the optimum of the real expensive optimization problem in a limited computational budget. Therefore, the methods to find the optimal solutions of both surrogate models are also very important. The multi-tasking optimization method has been shown to enhance the search performance, especially when the different tasks have similar characteristics in their fitness landscape. However, it is not able to always ensure that two surrogate models have high similarity in their fitness landscape. Thus, the generalized evolutionary multi-tasking optimization algorithm, called G-MFEA (Ding et al., 2017), is adopted as it shows better performance when the optimal solutions of different tasks are located in different region. Algorithm 10 gives the pseudo code of the multi-tasking optimization of the global and local surrogate models. In Algorithm 10, a population P will be initialized and evaluated using the global and local surrogate models, respectively. Note that a number of historical solutions in the archive Arc are also included in the initial population in order to help explore regions that are different from the region in which the current optimal solution is located. The skill factor of each individual will then be determined according to the factorial rank on each task. The following steps will be repeated before the stopping criterion, i.e., meeting the maximum number of iteration of G-MFEA. Each individual in the population will be mapped to a new location by a decision variable translation strategy, whose pseudo code is given in Algorithm 11. Then an offspring population will be generated by the assortative mating strategy. Each individual in the offspring population will also be translated using the decision variable translation strategy, and assigned a skill factor according to vertical cultural transmission. After that, each individual in the offspring population can be evaluated on the corresponding surrogate model. The next parent population will be selected in the environmental selection according to the scalar fitness of every individual in the combined population of current parent and offspring populations.

In Algorithm 11, t denotes the current generation, ϵ and θ are the threshold values to start the decision variable translation strategy and the frequency to change the translation direction, respectively. The translated direction for each task is updated every θ generations. Each individual in the combined population of the current parent and offspring population will be mapped to a new location according to the translation direction and translation distance on the task it relates to. More details on the G-MFEA can be referred to (Ding et al., 2017).

10.3.5 Experimental Results and Discussions

A comparative study on six test problems (Liu et al., 2014a), including Ellipsoid (F1), Rosenbrock (F2), Ackley (F3), Griewank (F4), Shifted Rotated Rastrigin

Algorithm 10 The pseudo code of the multi-tasking optimization of the global and local surrogate models

Input: M_g: the global surrogate model;
M_l: the local surrogate model;
$K_{\max}$: the maximum number of iteration of G-MFEA;
μ: the probability of the crossover;
σ: the probability of the mutation;
Output: The optimal solutions of two surrogate models;
Population P initialization;
Evaluate each individual with respect to M_g and M_l, and assign the skill factor τ to each individual;

Set the iteration $k = 0$;
while $k \leq K_{\max}$ **do**
Translate the decision variables of each individual in the P using the method given in Algorithm 11;
Generate an offspring population by assortative mating;
Translate the offspring population using the methods given in Algorithm 11;
Assign the skill factor τ of each individual in the offspring population by vertical cultural transmission;
Evaluate each individual in the offspring population with respect to the surrogate model its skill factor shows.
Combine the parent and offspring population, denoted as R, and update the scalar fitness ϕ of every individual in R.
Environmental selection;
end while

Algorithm 11 The decision variable translation strategy

Input: t: the current generation;
θ: the frequency to change the translation direction;
ϵ: the threshold value to start the decision variable translation strategy;
Output: A translated population
1: **if** $t > \epsilon$ **then**
2: **if** $mod(t, \theta) == 0$ **then**
3: Calculate the translated direction of each task;
4: **end if**
5: Update the position of each individual in the current population according to the translated direction of its corresponding task;
6: **end if**
7: Output the translated population;

(F5) and Rotated hybrid Composition Function (F6), with 100- and 200 dimensions are conducted to evaluate the performance of MS-MTO. The experimental results obtained by MS-MTO algorithm will be compared to some methods proposed for high-dimensional computationally expensive problems, including GORS-SLPSO (Yu et al., 2019c), SHPSO (Yu et al., 2018), SA-COSO (Sun et al., 2017), MGP-SLPSO (Tian et al., 2018a) and SAMSO (Li et al., 2021), to evaluate the performance of MS-MTO. In the experiments, the maximum number of expensive fitness evaluations is set to 1000 for fair comparisons. The size of the population is set to 100 in multi-tasking optimization. The maximum number of iterations for each running of multi-tasking optimization is set to 50. The number of data initially generated for training the global and local surrogate model is also set to $2 \times n$, where n is the dimension of the problem. The probabilities of crossover and mutation are changed from 1 to 10, the value of which is determined by $\frac{MTO_{calls}}{10} + 1$, where MTO_{calls} is the total number of calls of the multi-tasking optimization till now. Note that some history solutions in the archive *Arc* that have been evaluated using the real expensive function will also be included in the initialized population to find a better solution. The history solutions chosen to be used in the initial population include the best solutions in the archive and $25\% \times N$ solutions randomly selected from the best N data in *Arc*. All algorithms are conducted 30 independent runs for each problem.

Table 10.3 gives the statistical results obtained by the compared algorithms on high-dimensional F1-F6 problems with 50, 100 and 200 dimensions. From Table 10.3, we can see that MS-MTO obtains better results than SA-COSO on all 50-, 100-, and 200-dimensional test problems. Compared with SAMSO, MS-MTO outperforms SAMSO on 17 out of 18 test instances of F1-F6. MGP-SLPSO, GORS-SSLPSO and SHPSO did not test the performance on 200-dimensional problems, but from the results on 50- and 100-dimensional problems, we can conclude that MS-MTO is much better than MGP-SLPSO, GORS-SSLPSO and SHPSO. The MGP-SLPSO obtained better result on 50-dimensional F1 problem (Ellipsoid), which is a unimodal problem; however, its performance starts to drop as the dimension increases to 100 dimension. By contrast, MS-MTO is able to obtain much better result.

10.4 Surrogate-Assisted Large Optimization with Random Feature Selection

Surrogate-assisted evolutionary algorithms have shown better performance on high-dimensional problems which are less than 200 dimensions. However, it is not possible to be applied to solve large-scale expensive problems. In this section, a surrogate-assisted evolutionary algorithm with random feature selection, denoted as SAEA-RFS, is introduced. In SAEA-RFS, a random feature selection technique is utilized to select decision variables from the original large-scale optimization problem to form a number of sub-problems, whose dimension may differ to each other, at each generation. Then, sub-problems are sequentially optimized by searching for the optimal solutions of the surrogate models trained for the sub-problems, respectively. The best solution among all optimal solutions found for sub-problems is used to replace the decision variables of the best solution found so far, thus generating a new solution which will be evaluated using the real objective function and used to update the best solution.

Table 10.3 Comparative results on 50-, 100- and 200-dimensional problems, shown as mean value(standard deviation)

Prob.	Dim.	SA-COSO	MGP-SLPSO	GORS-SSLPSO	SHPSO	SAMSO	MS-MTO
F1	50	4.98e+01 (1.60e+01)	**9.88e−16** (1.02e-15)	4.91e+01 (1.48e+02)	6.92e+00 (2.57e+00)	5.27e−01 (2.59e−01)	1.49e−14 (5.18e−14)
	100	9.32e+02 (2.37e+02)	4.93e−05 (1.98e-04)	1.71e+02 (3.19e+02)	1.22e+02 (2.68e+01)	7.23e+01 (1.51e+01)	**1.44e−11** (2.59e−11)
	200	1.57e+04 (2.98e+03)	~	~	~	1.49e+03 (2.31e+02)	**1.27e−06** (1.29e−06)
F2	50	2.50e+02 (5.40e+01)	1.20e+02 (1.86e+01)	7.92e+01 (1.23e+02)	5.16e+01 (2.31e+00)	4.99e+01 (9.43e−01)	**4.64e+01** (1.77e−01)
	100	2.43e+03 (7.99e+02)	6.15e+02 (6.68e+01)	1.01e+02 (1.08e+01)	2.09e+02 (4.80e+01)	2.92e+02 (3.40e+01)	**9.69e+01** (1.98e−01)
	200	1.67e+04 (4.00e+03)	~	~	~	1.17e+03 (1.19e+02)	1.97e+02 (6.81e−02)
F3	50	9.57e+00 (1.22e+00)	9.34e+00 (1.12e+00)	3.71e+00 (9.44e−01)	2.57e+00 (3.45e−01)	1.62e+00 (4.34e−01)	**6.08e−06** (5.02e−06)
	100	1.59e+01 (7.45e−01)	1.43e+01 (6.21e−01)	9.04e+00 (1.79e+00)	5.63e+00 (6.88e−01)	6.02e+00 (2.94e−01)	**2.48e−05** (2.23e−05)
	200	2.08e+01 (1.24e−01)	~	~	~	1.20e+01 (3.21e−01)	**2.27e−02** (1.51e−02)
F4	50	5.56e+00 (1.04e+00)	1.12e−01 (3.78e−02)	2.08e−03 (4.17e−03)	9.35e−01 (7.45e−02)	6.89e−01 (9.85e−02)	**4.88e−09** (6.83e−09)
	100	6.91e+01 (1.49e+01)	7.15e−01 (9.31e−02)	9.24e+00 (2.76e+01)	1.12e+00 (4.50e−02)	1.06e+00 (2.27e−02)	**3.77e−06** (1.58e−05)
	200	5.81e+02 (9.04e+01)	~	~	~	1.01e+01 (1.92e+00)	**1.38e−03** (2.63e−03)
F5	50	2.15e+02 (3.22e+01)	3.31e+01 (3.61e+01)	−6.58e+01 (5.31e+01)	1.33e+02 (2.55e+01)	**−1.51e+02** (3.71e+01)	−8.07e+01 (6.30e+01)
	100	1.34e+03 (1.14e+02)	1.16e+03 (1.45e+02)	1.04e+03 (1.30e+02)	7.72e+02 (7.58e+01)	5.71e+02 (9.58e+01)	**3.79e+02** (1.19e+02)
	200	4.06e+03 (3.02e+02)	~	~	~	2.01e+03 (5.65e+01)	**1.93e+03** (1.92e+02)
F6	50	1.08e+03 (3.68e+01)	1.06e+03 (2.14e+01)	1.02e+03 (6.62e+01)	1.00e+03 (2.28e+01)	9.75e+02 (2.92e+01)	**9.10e+02** (5.17e−14)
	100	1.41e+03 (3.80e+01)	1.39e+03 (4.77e+01)	1.42e+03 (2.36e+01)	1.43e+03 (5.00e+01)	1.12e+03 (2.44e+01)	**9.10e+02** (8.30e−11)
	200	1.35e+03 (2027e+01)	~	~	~	1.22e+03 (1.20e+01)	**9.10e+02** (2.07e−09)

The best solutions are highlighted

10.4.1 Main Framework

As is well known, the larger the dimension of a problem is, the more data for training the surrogate model are required, making it impractical for assisting optimization of large-scale expensive problems. In SAEA-FES, a number of sub-problems are optimized sequentially, which is expected to reduce the search space and make it easier to build high-quality surrogates using limited amount of training data. Algorithm 12 gives the pseudo code of SAEA-FES. A number of solutions will be sampled using LHS method and evaluated using the exact expensive function, which will be saved in an archive *Arc* and used for model training. Then a population *P* for the optimization of the large-scale expensive problem is initialized. If the stopping criterion, i.e., the computational budget is exhausted, is not satisfied, the following procedure will be repeated. A temporary population, denoted as pop_t, is generated at first by copying all solutions in the population *P*. Then *K* sub-problems will be generated sequentially and optimized. The method to generate a sub-problem $sp_k, k = 1, 2, \ldots, K$ (line 6 in Algorithm 12) and optimization (lines 7–9 in Algorithm 12) will be given in detail in Sect. 10.4.2. After that, the temporary population will be updated on the corresponding decision variables of sub-problem sp_k using the current population for optimization of the sub-problem sp_k. When all sub-problems are optimized, the solutions of the temporary population pop_t are used to replace those of the population *P*. Finally, a new candidate solution will be generated by replacing the decision variables of the best position found so far using the optimal solution of the sub-problem who has the best approximated fitness, and evaluated using the real expensive large-scale optimization problems. The evaluated solution will be saved to the archive *Arc* and used to update the best position found so far for the large-scale problem again.

10.4.2 Sub-problem Formation and Optimization

A number of sub-problems will be defined and optimized sequentially in SAEA-RFS. An integer number will be generated within $[1, MD_{sp}]$ to be the dimension of a sub-problem, where MD_{sp} is the maximum dimension of the sub-problems. Then, a subset of the decision variables will be randomly selected from the original problem (line 6 in Algorithm 12). Fig. 10.9 gives an example to show how a sub-problem is formed using the random feature selection technique. In Fig. 10.9, the dimension of original problem is 10. Suppose the maximum dimension of the sub-problems is 5, in this example, the integer 3 is generated in the range $[1, 5]$. Therefore, three decision variables will be randomly selected from the original 10 variables. In this example, $\mathbf{x}_3$, $\mathbf{x}_5$ and $\mathbf{x}_9$ are selected to be the decision variables of sub-problem *sp*. A population for the optimization of the sub-problem, denoted as pop_t^{sp}, will also be generated by copying solutions on the corresponding decision variables, i.e., $\mathbf{x}_3$, $\mathbf{x}_5$ and $\mathbf{x}_9$, from the temporary population pop_t;

Algorithm 12 The pseudo code of SAEA-RFS

1: Generate a number of solutions using Latin hypercube sampling technique, evaluate these solutions and save to an archive Arc;
2: Initialize a population P in the original decision space;
3: **while** the stopping criterion is not met **do**
4: Define a temporary population $pop_t = P$;
5: **for** $k = 1$ to K **do**
6: Define a sub-problem sp_k;
7: Copy solutions from pop_t on the corresponding decision variables of the sub-problem sp_k;
8: Randomly select ND_{sp_k} solutions from Arc, and train a surrogate model M_{sp_k} for sub-problem sp_k using the selected on the corresponding variables of the sub-problem sp_k;
9: Search for the optimal solution of M_{sp_k}, denoted as $\mathbf{x}_b^{sp_k}$;
10: Replace the temporary population pop_t on the decision variables of sp_k using the current population for the optimization of sub-problem sp_k;
11: **end for**
12: $P = pop_t$;
13: Generate a new candidate solution based on the decision variables of the sub-problem that achieves the the best approximated fitness and the best solution found so far;
14: Evaluate the new candidate solution and save it in the archive Arc;
15: Update the best solution found so far;
16: **end while**
17: Output the best solution found so far;

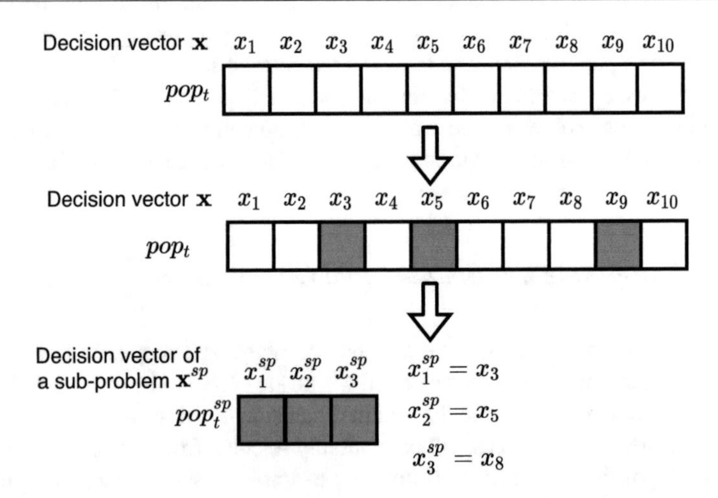

Fig. 10.9 The formation of a sub-problem

After that, a surrogate model M^{sp} will be trained using a number of data selected from the archive Arc on the corresponding decision variable. The population pop_t^{sp} will be used to search for the optimal solution of the surrogate model, which will be the best position found so far for the sub-problem. Fig. 10.10 follows the example in Fig. 10.9 to show the optimization of the sub-problem. In Fig. 10.10, ND^{sp} data will be randomly selected from the archive Arc on the corresponding x_3, x_5 and x_9

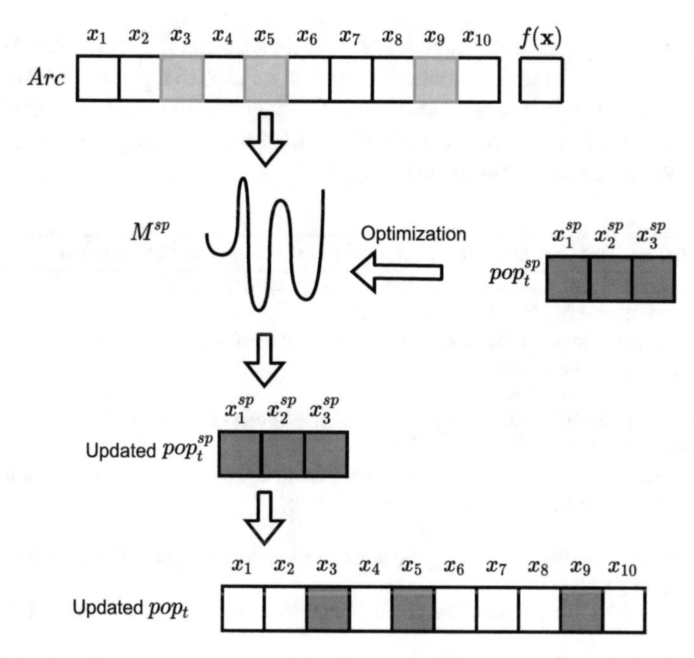

Fig. 10.10 The optimization of the sub-problem

as well as the fitness values $f(\mathbf{x})$. A surrogate model M^{sp} will be trained for the sub-problem sp using these samples. The population pop_t^{sp} is then used to search for the surrogate model M^{sp}. The updated population pop_t^{sp} will be used to replace the temporary population pop_t on the corresponding decision variables as shown in Fig. 10.10.

10.4.3 Global Best Position Update

Algorithm 13 gives the pseudo code of the method to update the global best position of the real expensive problem. After all sub-problems are optimized, the best position found far will be updated. In SAEA-RFS, a new solution will be generated by replacing the best solution found so far of the original expensive problem on the corresponding decision variables of the sub-problem who has the best approximated fitness value among all sub-problems. Suppose sub-problem sp_k has the best approximated fitness value among all optimal solutions, i.e., $k = argmin\{\hat{f}(\mathbf{x}_b^{sp_1}), \hat{f}(\mathbf{x}_b^{sp_2}), \ldots, \hat{f}(\mathbf{x}_b^{sp_K})\}$. Note that the decision variables of the sub-problem sp_k may be updated in the optimization of the following sub-problems. Therefore, the surrogate model of the sub-problem sp_k will be used to approximate the fitness value of the individual in the temporary population pop_t which is updated finally from the optimal solution of the sub-problem sp_K. Suppose individual i in pop_t is updated from the optimal solution of sp_K, and its value approximated by M^{sp_k} is denoted as $\hat{f}^{M^{sp_k}}(\mathbf{x}_i)$. Then, the solution having the smaller approximated fitness will be used to replace the best position found so far of the real expensive problems. The new solution will be evaluated using the real objective function and saved to the archive *Arc*, similarly used to update the best position found so far for the original problem.

Algorithm 13 The pseudocode of the update of the global best position

Input: The optimal solutions of all sub-problems sp_K, $k = 1, 2, \ldots, K$;
The surrogate models M^{sp_k}, $k = 1, 2, \ldots, K$;
The best position found so far *xbest* of the large-scale optimization problem;
the temporary population pop_t;
Output: The best position found so far;
1: Find the sub-problem k that has the best approximated fitness value among all optimal solutions, $k = argmin\{\hat{f}(\mathbf{x}_b^{sp_1}), \hat{f}(\mathbf{x}_b^{sp_2}), \ldots, \hat{f}(\mathbf{x}_b^{sp_K})\}$.
2: Approximate the solution i in pop_t that updated from the optimal solution of sub-problem sp_K using the surrogate model M^{sp_k};
3: $\mathbf{x}_t^{sp_k} = argmin\{\hat{f}^{sp_k}(\mathbf{x}_b^{sp_k}), \hat{f}^{sp_k}(\mathbf{x}_i^{sp_k})\}$;
4: Generate a new solution by replacing the decision variables of sp_k on the best position found so far *xbest* using $\mathbf{x}_t^{sp_k}$;
5: Evaluate the new solution using the original large-scale optimization problem and save to the archive *Arc*;
6: Update the best position found so far *xbest* if the fitness value of the new solution is better than that of *xbest*;

10.4.4 Experimental Results and Discussions

In order to evaluate the performance of SAEA-RFS, an empirical study on the CEC'2013 special session and competition on large-scale global optimization with 1000 dimensions (F1-F12, F15) and 905 dimensions (F13-F14), respectively (Li et al., 2013), is conducted. All of these benchmark problems can be classified into four categories according to their characteristics: fully-separable functions (F1-F3), partially additively separable functions with a separable subcomponent(F4-F7), partially additively separable functions with no separable subcomponents (F8-F11), overlapping functions (F12-F14) and non-separable functions (F15).

The RBF surrogate model is adopted as the surrogate model. The dimension of a sub-problem will be a random integer in the range [1, 100], i.e., the maximum dimension of each sub-problem $MaxD_{sp}$ is 100. The number of training data is set to twice of the sub-problem dimension, i.e., $ts\_num = 2 \times D_{sp_k}$. The population size *NP* is set 10, $K = 20$ sub-problems will be formed at each generation and the maximum iterations of sub-problem optimization is set to 5. The differential evolution (DE) is used as the optimizer for each sub-problem, in which DE/best/1 and binomial crossover is used as mutation and crossover strategy, respectively. The crossover rate *CR* is set to 1 and the mutation scaling factor F is set to 0.8. The algorithm will be

Table 10.4 Median and median absolute deviation of optimal fitness values obtained by different algorithms on CEC'2013 benchmark problems

Prob.		SAEA-RFS	CSO	SL-PSO	SACC-RBFN	SACC-SVR	SACC-GP	SACC-QPA
F1	Median	7.1040e+09	7.7588e+10	6.2529e+10	2.0809e+09	1.3459e+09	**9.5219e+08**	3.2809e+09
	Mad	5.5284e+08	1.6197e+09	2.3644e+09	1.1371e+08	9.4539e+07	**4.1661e+07**	2.9115e+08
F2	Median	1.9991e+04	3.9968e+04	3.8062e+04	9.5890e+03	**7.7214e+03**	9.1251e+03	1.1407e+04
	Mad	4.7842e+02	4.8994e+02	5.8755e+02	1.7731e+02	**9.1652e+01**	1.5203e+02	1.7638e+02
F3	Median	2.0923e+01	2.1627e+01	2.1632e+01	2.0591e+01	2.0580e+01	2.0576e+01	**2.0532e+01**
	Mad	1.1895e−02	4.3042e−03	5.5112e−03	1.1257e−02	9.9047e−03	1.2985e−02	**1.0999e−02**
F4	Median	**1.1983e+12**	3.2850e+12	3.8134e+12	4.8469e+12	6.7659e+12	4.8422e+12	7.9752e+12
	Mad	**2.9443e+11**	2.9211e+11	2.8020e+11	1.3494e+12	2.2941e+12	9.5760e+11	2.5130e+12
F5	Median	2.3937e+07	1.6367e+07	**1.4772e+07**	2.4079e+07	2.7173e+07	2.6763e+07	2.7873e+07
	Mad	2.3474e+06	4.6182e+05	**3.4225e+05**	2.0450e+06	2.7448e+06	3.4908e+06	1.8535e+06
F6	Median	**1.0606e+06**	1.0662e+06	1.0675e+06	1.0666e+06	1.0659e+06	1.0665e+06	1.0690e+06
	Mad	**1.7974e+03**	1.0801e+03	6.2046e+02	1.3233e+03	2.1647e+03	2.4745e+03	1.6541e+03
F7	Median	**6.9321e+09**	1.7028e+12	1.0419e+12	2.8180e+10	2.7217e+10	3.4172e+10	2.9452e+11
	Mad	**2.1430e+09**	3.0629e+11	3.3487e+11	9.1957e+09	5.8699e+09	1.7415e+10	1.4778e+11
F8	Median	**1.9640e+16**	9.8246e+16	1.1896e+17	3.1381e+17	2.7842e+17	3.0519e+17	4.4840e+17
	Mad	**6.4764e+15**	8.7887e+15	2.6829e+16	1.1836e+17	8.5000e+16	1.0684e+17	1.6490e+17
F9	Median	1.6062e+09	1.2756e+09	**1.1363e+09**	1.9539e+09	2.0153e+09	1.9795e+09	2.1442e+09
	Mad	2.3199e+08	1.5676e+07	**2.8179e+07**	1.7351e+08	2.2247e+08	2.3673e+08	2.8067e+08
F10	Median	**9.4551e+07**	9.5200e+07	9.5407e+07	9.5867e+07	9.5968e+07	9.5985e+07	9.6228e+07
	Mad	**4.2776e+05**	1.0668e+05	7.7644e+04	4.2800e+05	2.4781e+05	1.6207e+05	3.5806e+05
F11	Median	**8.8613e+11**	2.6977e+14	2.1319e+14	3.8546e+12	3.2749e+12	4.1133e+12	2.3197e+13
	Mad	**3.0049e+11**	4.6461e+13	3.2508e+13	8.1992e+11	1.4255e+12	1.6889e+12	1.0380e+13
F12	Median	4.8910e+11	2.0427e+12	1.6398e+12	2.5921e+10	**8.1716e+09**	2.8418e+10	5.9437e+10
	Mad	2.8418e+10	2.4209e+10	3.9238e+10	1.9425e+09	**7.6608e+08**	2.1324e+09	2.4218e+09
F13	Median	**1.0892e+11**	1.0776e+12	1.6420e+14	8.2671e+11	4.3888e+11	7.1633e+11	2.3902e+13
	Mad	**1.3510e+10**	1.3174e+11	3.1747e+13	1.6500e+11	1.0526e+11	1.8392e+11	1.3336e+13
F14	Median	**1.1205e+12**	7.4289e+12	3.1356e+14	4.4563e+12	2.9075e+12	4.2607e+12	2.2505e+13
	Mad	**1.1850e+11**	1.0571e+12	6.6102e+13	5.8627e+11	5.7475e+11	1.1633e+12	1.0025e+13
F15	Median	**3.1404e+08**	5.8871e+14	3.8143e+14	4.0026e+09	3.4749e+09	4.0142e+09	1.1288e+10
	Mad	**8.1643e+07**	5.7063e+13	4.3945e+13	1.6703e+09	1.4752e+09	1.5701e+09	3.8612e+09

The best median result in each row is shown in bold

run independently for 25 times on each problem, and the stopping criterion is that the maximum number of fitness evaluations $11 \times n$ is exhausted, where n is the search dimension of the problem.

Table 10.4 gives the statistical results on CEC'2013 benchmark problems. SACC-RBFN, SACC-SVR, SACC-GP and SACC-QPA are all surrogate-assisted cooperative coevolutionary (SACC) optimizer proposed by Falco et al. (Falco et al., 2019), which uniformly divide the large-scale optimization problem into a number of sub-problems and optimized assisted by the RBF, SVR, GP and QPA surrogate model, respectively. From Tabletab10:SAEA-RFS-results, we can see that SAEA-RFS can outperform SACC, CSO and SL-PSO on 9 CEC'2013 test problems. To be specific, SAEA-RFS performed better than both non-decomposition CSO and SL-PSO on 13/15 problems. Compared to the SACC framework, SAEA-RFS obtains 10/15 better results than SACC-RBFN and 11/15 better results than all others. Specially, it can also be found from Table 10.4 that the performance of SAEA-RFS becomes better when the problem is less separable.

10.5 Summary

Surrogate-assisted high-dimensional evolutionary optimization is extremely challenging due to the curse of dimensionality both in search and surrogate modeling. The exponentially large search space makes it hard to find the optimum, and the small amount of available training data seriously limit the quality of the surrogates. Therefore, it is very much unrealistic to locate the global optimum in solving high-dimensional expensive problems.

One class of techniques that is not fully discussed in the algorithms described in this chapter is dimension reduction strategies using linear and or nonlinear dimension reduction techniques such as principal component analysis and autoencoder. Dimension reduction will be particularly effective where there is a strong correlations between pairs of decision variables, which is not uncommon. Dimension reduction for surrogate-assisted medium-dimension optimization has been shown to be successful (Liu et al., 2014a), although more work is needed to examine the effectiveness of such algorithms on higher dimensional optimization problems.

References

Cheng, R., & Jin, Y. (2015). A social learning particle swarm optimization algorithm for scalable optimization. *Information Sciences*, *291*, 43–60.
Clerc, M. (1999). The swarm and the queen: towards a deterministic and adaptive particle swarm optimization. In *Proceedings of the 1999 Congress on Evolutionary Computation-CEC99 (Cat. No. 99TH8406)* (Vol. 3, pp. 1951–1957).
Deb, K., Pratap, A., Agarwal, S., & Meyarivan, T. (2002). A fast and elitist multiobjective genetic algorithm: NSGA-II. *IEEE Transactions on Evolutionary Computation*, *6*(2), 182–197.
Ding, J., Yang, C., Jin, Y., & Chai, T. (2017). Generalized multi-tasking for evolutionary optimization of expensive problems. *IEEE Transactions on Evolutionary Computation*, *23*, 44–58.
Falco, I. D., Cioppa, A. D., & Trunfio, G. A. (2019). Investigating surrogate-assisted cooperative coevolution for large-scale global optimization. *Information Sciences*, *482*, 1–26.
Gupta, A., Ong, Y.-S., & Feng, L. (2016). Multifactorial evolution: toward evolutionary multitasking. *IEEE Transactions on Evolutionary Computation*, *20*(3), 343–357.
Li, X., Tang, K., Omidvar, M. N., Yang, Z., & Qin, K. (2013). Benchmark functions for the CEC 2013 special session and competition on large-scale global optimization. *Technical report, Evolutionary Computation and Machine Learning Group*. Australia: RMIT University.
Li, F., Cai, X., Gao, L., & Shen, W. (2021). A surrogate-assisted multiswarm optimization algorithm for high-dimensional computationally expensive problems. *IEEE Transactions on Cybernetics*, *51*(3), 1390–1402.

Liu, B., Zhang, Q., & Gielen, G. G. (2014a). A Gaussian process surrogate model assisted evolutionary algorithm for medium scale expensive optimization problems. *IEEE Transactions on Evolutionary Computation, 18*(2), 180–192.

Sun, X., Gong, D., Jin, Y., & Chen, S. (2013b). A new surrogate-assisted interactive genetic algorithm with weighted semisupervised learning. *IEEE Transactions on Cybernetics, 43*(2), 685–698.

Sun, C., Jin, Y., Cheng, R., Ding, J., & Zeng, J. (2017). Surrogate-assisted cooperative swarm optimization of high-dimensional expensive problems. *IEEE Transactions on Evolutionary Computation, 21*(4), 644–660.

Sun, C., Zeng, J., Pan, J., Xue, S., & Jin, Y. (2013a). A new fitness estimation strategy for particle swarm optimization. *Information Sciences, 221*, 355–370.

Tian, J., Tan, Y., Zeng, J., Sun, C., & Jin, Y. (2018a). *Multi-objective infill criterion driven Gaussian process assisted particle swarm optimization of high-dimensional expensive problems*. Singapore: IEEE Transactions on Evolutionary Computation.

Wang, H. (2016). Uncertainty in surrogate models. In *Proceedings of the Genetic and Evolutionary Computation Conference* (p. 1279–1279). ACM.

Yu, H., Tan, Y., Sun, C., & Zeng, J. (2019c). A generation-based optimal restart strategy for surrogate-assisted social learning particle swarm optimization. *Knowledge-Based Systems, 163*, 14–25.

Yu, H., Tan, Y., Zeng, J., Sun, C., & Jin, Y. (2018). Surrogate-assisted hierarchical particle swarm optimization. *Information Sciences, 454–455*, 59–72.

Chapter 11
Offline Big or Small Data-Driven Optimization and Applications

11.1 Adaptive Clustering for Offline Big-Data Driven Optimization of Trauma Systems

The Scotland trauma system design problem (Wang et al., 2016a) assigns major trauma center (MTC), trauma unit (TU), and local emergency hospital (LEH) to existing 18 hospital centers to achieve both economic and clinical benefits (Jansen et al., 2014). However, there is no explicit expression for the objective and constraint functions. The economic and clinical benefits can be estimated by simulating the trauma system based on the historical emergency records, which makes this problem an offline data-driven optimization problem.

11.1.1 Problem Formulation

Given a configuration of the trauma system, an emergency patient can be assigned to a matched hospital center with a suitable transport according the location of the accident and the patient's injury degree (triaged to MTC or TU). Such a pre-defined allocation algorithm (Jansen et al., 2014) can work as a simulation to evaluate candidate configurations. To be specific, the allocation algorithm can output the following information of a patient: the assigned center, transport (land or air), travel time, and whether a severe patient is sent to a TU. The statistics of the allocation outputs for the historical emergency records in a long period (written as $\mathcal{D}$ and shown in Fig. 11.1) with 40,000 patients can evaluate the performance of a configuration. Therefore, the trauma system design problem can be formulated into a constrained bi-objective combinatorial optimization problem:

Y. Jin et al., *Data-Driven Evolutionary Optimization*,
Studies in Computational Intelligence 975,
https://doi.org/10.1007/978-3-030-74640-7_11

Fig. 11.1 Distribution of $\mathcal{D}$

$$
\begin{aligned}
\min\ & f_1(\mathbf{x}, \mathcal{D}), \\
& f_2(\mathbf{x}, \mathcal{D}), \\
s.t.\ & g_1(\mathbf{x}, \mathcal{D}) > V, \\
& g_2(\mathbf{x}, \mathcal{D}) < H, \\
& g_3(\mathbf{x}) > DIS,
\end{aligned}
\tag{11.1}
$$

where $f_1(\mathbf{x}, \mathcal{D})$ is the total travelling time, $f_2(\mathbf{x}, \mathcal{D})$ is the total number of severe patients who are send to a TU, $g_1(\mathbf{x}, \mathcal{D})$ is total number of patients who are sent to a MTC, $g_2(\mathbf{x}, \mathcal{D})$ is the total number of helicopter usage, and $g_3(\mathbf{x})$ is the minimal distance between any TUs. Such a problem can be solved by NSGA-II (Deb et al., 2002), but the cost of a single evaluation increases with the increasing size of $\mathcal{D}$ since the allocation algorithm is repeatedly called. Therefore, the whole optimization process needs high costs.

11.1.2 Adaptive Clustering for Offline Data-Driven Optimization

There is a dilemma between the cost and accuracy of offline data-driven optimization of trauma system design. The richer $\mathcal{D}$ is, the more accurate the estimated performance of configurations is, the higher cost the algorithm requires. One straightforward way to reduce the computing cost as well as maintain the useful information in $\mathcal{D}$ is dividing $\mathcal{D}$ into K clusters. The clustered data $\mathcal{D}_K$ with its K cluster centers and corresponding numbers of data inside clusters can be an approximation to $\mathcal{D}$. Also, $f_1(\mathbf{x}, \mathcal{D})$, $f_2(\mathbf{x}, \mathcal{D})$, $g_1(\mathbf{x}, \mathcal{D})$, and $g_2(\mathbf{x}, \mathcal{D})$ can be approximated by $f_1(\mathbf{x}, \mathcal{D}_K)$, $f_2(\mathbf{x}, \mathcal{D}_K)$, $g_1(\mathbf{x}, \mathcal{D}_K)$, and $g_2(\mathbf{x}, \mathcal{D}_K)$. Thus, one evaluation needs K calls of the allocation algorithm rather than $|\mathcal{D}|$.

It is clear that the size of K controls both the approximation accuracy and computation cost. Also, a fixed K cannot satisfy the needs of the whole optimization process. Therefore, an adaptive clustering is employed during the optimization process, where the following three questions need to be considered.

- **How to measure acceptable approximation error?** During the optimization process, the distribution of the population changes, which makes then acceptable approximation error varies. The acceptable error will not disturb the evolutionary selection. Therefore, the acceptable approximation error can be measured according to the selection operation.
- **What is the relation between the approximation error and the number of clusters K?** Assuming that the acceptable approximation error has been obtained, the number of clusters K can be determined if the relation between the approximation error ER and K is known. In fact, such relation can be regressed by different numbers of clusters and their estimated errors.
- **When to change the number of clusters K?** When the optimization on the current K cannot find better solutions (i.e. the optimization process traps in a local optimum), K should be increased.

Thus, the proposed adaptive clustering method is embedded in NSGA-II, where an improvement detection is applied in each generation. If the optimal solutions are unchanged for five generations, the cluster number adjustment is triggered.

Firstly, the acceptable approximation error is measured based on the the non-dominated sorting (Tian et al., 2017b; Wang & Yao, 2014). As Fig. 11.2 shows, predicted objective values can locate in the area within the approximation error neighborhood. If the possible location areas of both solutions are not overlapped, the comparison can be high-possibly right.

Thus, the maximum acceptable approximation error should not make solutions in the first front are ranked after the last selected front, which can be defined as follows:

$$ER^* = \frac{1}{2}\min\{f_1^k - f_1^j\}, 1 \le k \le |Front_l|, 1 \le j \le |Front_1|, \qquad (11.2)$$

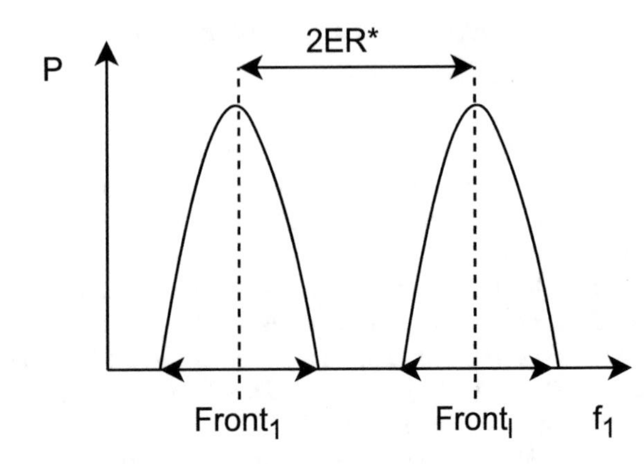

Fig. 11.2 Influence of the approximation errors on non-dominated sorting in a population

where $Front_1$ is the solution set of the first front and $Front_l$ is the solution set of the last selected front. As the optimization process proceeds, the population moves to a small region, which leads ER^* decrease.

After ER^* is obtained, the relation between ER and K can be regressed based on historical pairs (K, ER) by the following function:

$$ER = \frac{1}{\beta_1 + \beta_2 K}. \tag{11.3}$$

The new number of clusters K^* can be calculated from Eq. (11.3) by inputting ER^*. Thus, the clustering method set the number of clusters to K^* for the next generation of NSGA-II, where the dataset is changed to new clustered dataset $\mathcal{D}_{K^*}$ and the corresponding objective and constraint functions are changed.

11.1.3 Empirical Results

To study the performance and efficiency of the adaptive clustering method on offline data-driven optimization, we compare the NSGA-II variants with and without the adaptive clustering method on the Scotland dataset. Both algorithms run 20 times and stop by 1 hour. The IGD values are shown in Table 11.1 and Fig. 11.3, where the reference set is from the expensive calculation in Wang et al. (2016a). By using the Wilcoxon signed-rank test (Hollander & Wolfe, 1999), it has been found that the IGD value of NSGA-II with the adaptive clustering method is significantly better than that without the adaptive clustering method. Therefore, we can conclude that the the adaptive clustering method keeps different levels of the information of the offline data and efficiently applies to the optimization process.

Table 11.1 The IGD values of the NSGA-II variants with and without the adaptive clustering method using 1 hour runtime as the stopping criterion

	IGD
NSGA-II with adaptive clustering	**4.31e − 02 ± 1.01e − 01**
NSGA-II	9.57e−02±5.08e−02

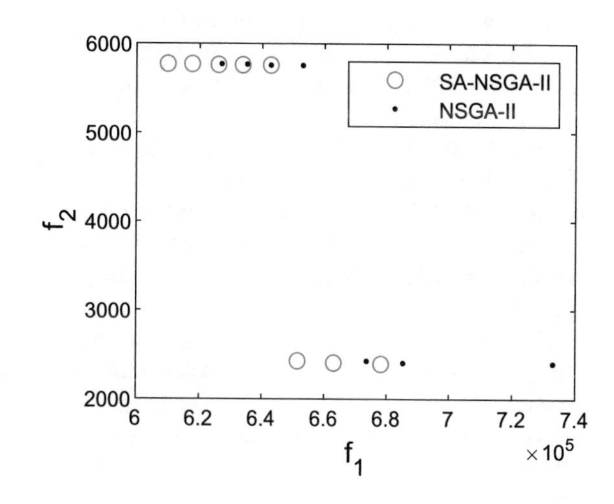

Fig. 11.3 Non-dominated solution sets of the SA-NSGA-II and NSGA-II obtained by stopping the algorithms after 1 h

11.1.4 Discussions

In this work, the challenges inherent in designing a trauma systems is addressed, which is a real-world off-line big data-driven evolutionary optimization problems. Although surrogate models are employed to reduce the heavy computation burden from the big data, it employs a specific model using domain knowledge rather than a generic model. It is a new idea to to group the data into a number of clusters for surrogate models to reduce the cost.

To balance the computation cost and the optimal solution accuracy, the surrogate management scheme in this section is to adaptively adjust the proper number of clustering according to the optimization state. Such an idea can be applied to many big data-driven optimization problems and multi-fidelity optimization problems in deep learning task.

In addition, if we view the data with the configuration as input and the objective and constrains as label, the trauma system design problem can be viewed as a small data-driven optimization problem, because the computation expense is high. In Wang and Jin (2018), random forest is employed to approximate those expensive function evaluation for this the trauma system design problem.

11.2 Small Data-Driven Multi-objective Magnesium Furnace Optimization

11.2.1 Model Management Based on a Global Surrogate

Two main challenges in offline data-driven evolutionary optimization are that no new data can be collected to further improve the surrogate and that the quality of the optimized solutions cannot be verified before they are implemented in the real-world. Nevertheless, it is usually helpful to introduce some model management mechanisms to guide the search rather than building a single surrogate offline and then performing the optimization on the surrogate. This is particularly true when the the amount of data is limited while the search space is relatively large, in which case, the surrogate may contain several false optimums. Thus, realizing a proper form of model management in offline small data driven optimization is equally important as in online data driven optimization.

Here we are going to present a model management method that relies on a coarse surrogate model, which is supposed to capture the global landscape of the original problem (Guo et al., 2016). In addition to this global surrogate, a fine surrogate model is employed to learn the local features to help more efficiently find the optimum. To prevent the evolutionary search from being seriously misled by the fine model, the coarse model is then used as the real objective function, from which new data can be sampled for managing the fine surrogate.

A diagram of a Gaussian process assisted NSGA-II with the help of a low-order polynomial model is shown in Fig. 11.4, which is called NSGA-II_GP. The main structure of the algorithm is very similar to that of the surrogate-assisted dominance based MOEAs, except that m (m is the number of objectives) second-order polynomial surrogates (the coarse models) are built before the optimization starts. In the optimization, m Gaussian process models are built based on a set of selected training data using the fuzzy C-mean clustering method (FCM), which is detailed in Sect. 7.3.3. Then, the NSGA-II is run to optimize the expected improvement of all objectives. Thus, this can be seen as an evolutionary multi-objective Bayesian optimization algorithm. After NSGA-II runs a pre-defined number of generations, a set of non-dominated solutions can be obtained. These solutions are then clustered using the k-means clustering algorithm in the search space and the one closest to the cluster center will be chosen to be evaluated using the coarse polynomial surrogates. Note that redundant samples are removed before clustering. From the above, we can see that the coarse surrogates plays the role of real function evaluations in this algorithm, therefore, the evaluated solutions are added into the training data for updating the Gaussian process models. This process continues until a termination condition is satisfied.

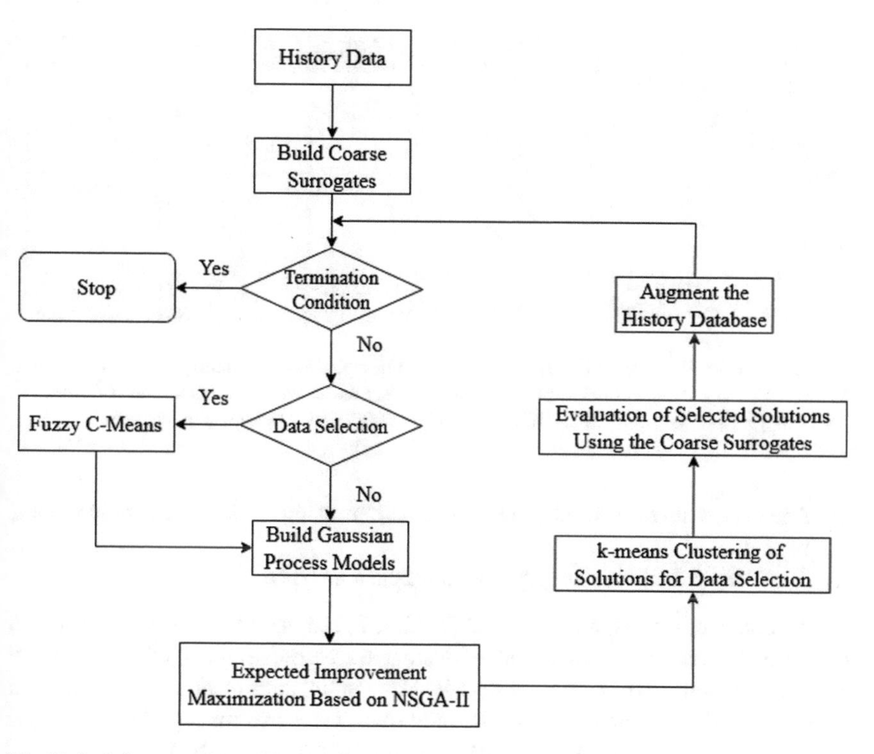

Fig. 11.4 A diagram of the offline Gaussian process assisted NSGA-II using low-order polynomials as for model management

11.2.2 *Empirical Verification on Benchmark Problems*

Before NSGA-II_GP is applied to solve real-world problems, it is first verified on benchmark problems. To this end, the performance of NSGA-II_GP is compared with that of NSGA-II without the assistance of surrogates, and ParEGO discussed in Sect. 7.2. The hypervolume (HV) is adopted for comparing the three algorithms. In calculating the hypervolume, the reference point is calculated by $b_j = \max_j + \delta(\max_j - \min_j)$, $j = 1, ..., m$, where $\delta = 0.01$, $\max_j$ and $\min_j$ are the maximization and minimization of the functions in all results to be assessed to make the results comparable. HV values will be normalized by dividing the them by $\prod_{j=1}^{m} b_j$. Other experimental settings are as follows:

1. In each run, $11n-1$ data points are generated, where n is the dimension of decision space. They represent the offline data, and no noise is added.
2. In NSGA-II_GP, the number of training data is limited to 80 for 5D or 6D test problems.
3. In NSGA-II_GP and ParEGO, the population size and the maximum number of generations are both set to 50. The population size of NSGA-II is also set to 50.

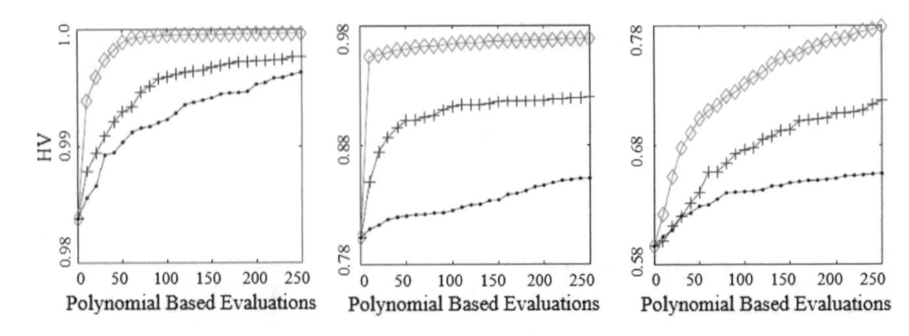

Fig. 11.5 The results obtained by NSGA-II_GP on three benchmark functions. From left to right: 5D DTLZ1, 5D ZDT3, and 6D WFG2. The diamonds, crosses and dots denote the HV values of NSGA-II_GP, ParEGO and NSGA-II, respectively, averaged over ten independent runs

4. The maximum number of function evaluations using the low-order polynomials to set to 250.
5. Ten independent runs are performed for each instance.

The comparative results on the 5D DTLZ1, ZDT3 and 6D WFG2 are shown in Fig. 11.5. The results were selected to illustrate the cases NSGA-II_GP significantly outperforms both ParEGO and NSGA-II. More simulation results can be found in Guo et al. (2016), which indicate that NSGA-II_GP performs significantly better than the two compared algorithms on 21 out of 27 test instances. Note that the results on the test problems are verified using the real objectives functions. Thus, these results empirically confirm that it is helpful to use coarse, low-complexity surrogates for model management in offline data-driven optimization where verification of the surrogates using real objective functions is not allowed.

11.2.3 Optimization of Fused Magnesium Furnaces

Magnesia is a class of purified MgO crystals that are widely used in chemical and other process industries. Typically, submerged arc furnaces are used that heats the caustic calcined magnesia to above 2800 degrees Celsius to produce final magnesia product after natural cooling and crystallization of melt (Yang & Chai, 2016) and a systems with two furnaces are shown in Fig. 11.6.

It takes about ten hours for the melting process and consumes a huge amount of electricity. Control of the magnesium manufacturing process is typically based on human heuristics in setting the target value of electricity consumption for a ton of magnesia (ECT), then the operational control and optimization system decides the positive-negative rotating of motors to adjust the height of the electrodes on the basis of real current, ECT, material ingredient and granularity. The main objectives to optimize the magnesium production process is maximize the total output and high-quality rate, and minimize the electricity consumption by tuning the ECT setpoint

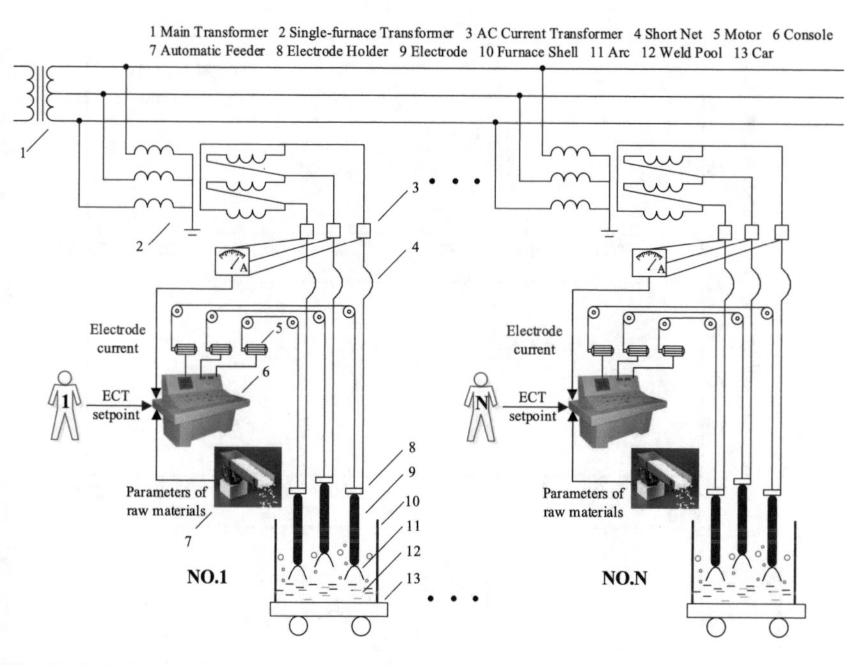

Fig. 11.6 A diagram illustrating two fused furnaces for producing magnesia

of the furnaces. Usually, it is very hard to build a mathematical model describing the relationship between ECT and performance indices and consequently, optimization of the process can largely rely on the history data collected from the production process. Since performing experiments is extremely expensive and conducting numerical simulations are less likely, performance optimization of fused magnesia furnaces is a typical off-line data-driven optimization problem.

Denote the ECT set point and performance indices of the i-th furnace as r_i, y_i, q_i, and e_i, and assume there are N furnaces in total, the magnesia furnace optimization problem can be formulated by:

$$\begin{aligned}
&\min -Y, \min -Q, \min E \\
&s.t. \\
&Y = \sum_{i=1}^{N} y_i,\ Q = \frac{\sum_{i=1}^{N} y_i \times q_i}{\sum_{i=1}^{N} y_i},\ E = \sum_{i=1}^{N} e_i, \\
&y_i = h_{1,i}(r_i),\ q_i = h_{2,i}(r_i),\ e_i = h_{3,i}(r_i), \\
&r_{i,min} \le r_i \le r_{i,max},\ i = 1, 2, ..., N,
\end{aligned} \tag{11.4}$$

where E is the total energy consumption, Y is the total product, and Q is the high-quality ratio. $h_{1,i}$, $h_{2,i}$ and $h_{3,i}$ denote the relationship between ECT and the performance indices of the i-th furnace.

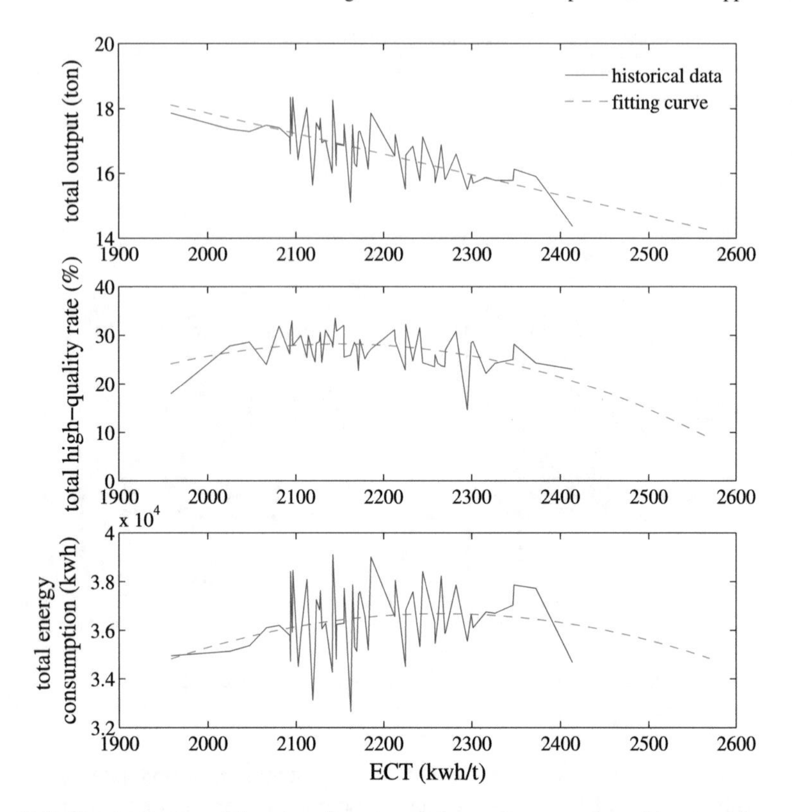

Fig. 11.7 The data collected from one furnace and the results approximated by the polynomial model

The data is collected from a factory in Liaoning Province, China, in which five furnaces are connected to one transformer, that is, $N = 5$ in Eq. (11.4). Data collection for this optimization problem is highly time-consuming, since only one group of performance indices can be measured within 24 hours. In the present example, 60 pairs of performance indices for the five furnaces, including output, high-quality rate, energy consumption, and ECT are collected. Figure 11.7 plots the data from one furnace, which is very noisy.

The coarse surrogate model adopted here to fit these data is a second-order polynomials, and the resulting results of the model are denoted by the dashed lines in Fig. 11.7. Although the data collected from the five furnaces are supposed to be the same, they differ a lot due to user interventions and other environmental disturbances. Therefore, the parameters of the low-order models are averaged over the five furnaces and consequently, the optimization problem in Eq. (11.4) can be re-written as follows:

$$
\begin{aligned}
&\min -Y, \min -Q, \min E \\
&s.t. \\
&Y = \sum_{i=1}^{5} y_i,\ Q = \frac{\sum_{i=1}^{5} y_i \times q_i}{\sum_{i=1}^{5} y_i},\ E = \sum_{i=1}^{5} e_i, \\
&y_i = -5942.1 r_i + 29521, \\
&q_i = -40.7548 r_i^2 + 170.1776 r_i - 148.1144, \\
&e_i = -17353 r_i^2 + 78201 r_i - 51871, \\
&1958.5 \le r_i \le 2569.5, i = 1, 2, ..., 5.
\end{aligned}
\tag{11.5}
$$

The same parameter settings used for the benchmark problems are adopted for offline operation optimization of the furnaces. Simulation settings are the same with experiments on the test problems. Fig. 11.8 plots the obtained optimal solutions (denoted by circles) together with the history data denoted by an '$*$', and the '$\times$' indicates the ideal optimal solution. It should be noted that the obtained solutions have not been verified real production data.

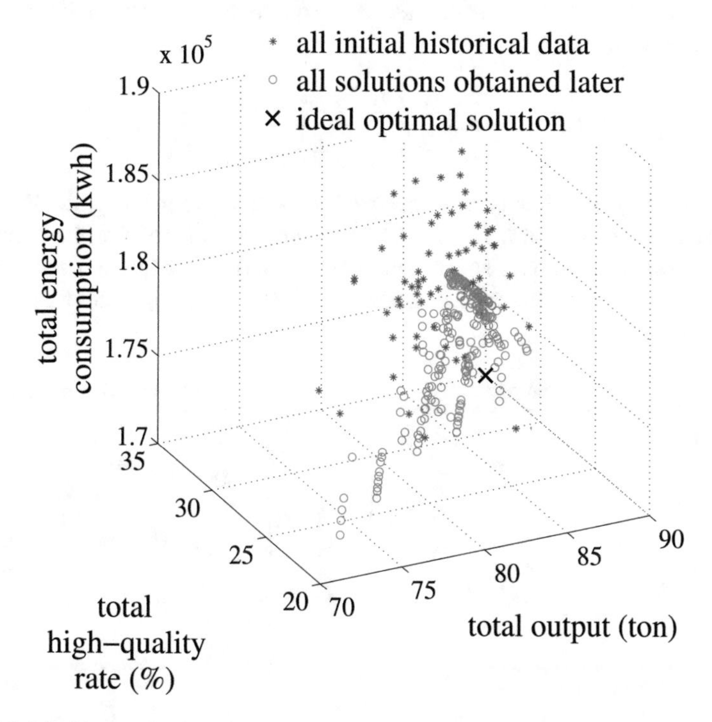

Fig. 11.8 The ideal optimal solution denoted by '$\times$', the collected data denoted by '*', and the obtained solutions denoted by circles

11.3 Selective Ensemble for Offline Airfoil Optimization

In practice, some airfoil optimization problems are based on computational fluid dynamic (CFD) simulations for function evaluation. Since CFD simulations are computational expensive, those problems are small data-driven optimization problems. In some particular cases, CFD simulations cannot provide any interface for optimization algorithms, which can be modelled as offline data-driven optimization problems.

We optimize the RAE2822 airfoil, a test case used in the Group for Aeronautical Research and Technology in Europe (GARTEUR) AG52 project (Andres et al., 2017). The CFD simulations are performed using the VGK simulator (Ashill et al., 1987; Freestone, 2004) to generate the offline data.

11.3.1 Problem Formulation

The airfoil design problem has 14 decision variables to control the shape of this RAE2822 airfoil as shown in Fig. 11.9. The objective function in Eq. 11.6 is to minimize the average drag over lift ratio in two different operation conditions, which is calculated using CFD simulations.

$$f_{Airfoil} = \min \frac{1}{2}\left(\frac{D_1}{L_1}\Big/\frac{D_1^b}{L_1^b} + \frac{D_2}{L_2}\Big/\frac{D_2^b}{L_2^b}\right), \tag{11.6}$$

where two design conditions are considered, D_i and L_i are the drag and lift coefficients in design condition i, D_i^b and L_i^b are the drag and lift coefficients of baseline design in design condition i. The objective value is normalized by the baseline design Fig. 11.9. 70 random geometries are generated with their $f_{Airfoil}$ values for the offline

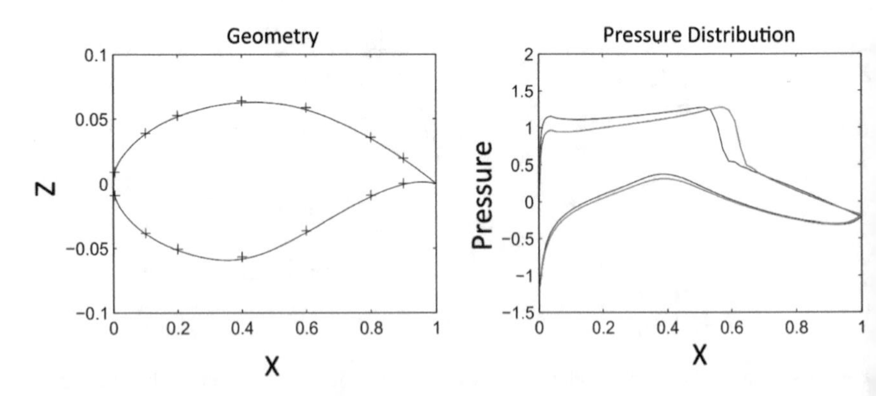

Fig. 11.9 The baseline design of RAE2822 airfoil

data $\mathcal{D}$. During the optimization process, there is no new data that can be sampled using the VGK simulation.

11.3.2 *Selective Ensemble for Offline Data-Driven Optimization*

To solve this offline data-driven airfoil optimization problem, we use the following DDEA-SE (Wang et al., 2018e), which is an offline data-driven EA using selective surrogate ensemble (Zhou, 2012).

Ensemble learning employs a number of base models and combine them to create a strong model. Ensemble models have advantages over single learners on accuracy and robustness. Both the computation cost and accuracy increase with the number of base models in ensemble models, but the improvement of accuracy is not significant when the number of base models is too large. Therefore, it is not economical to combine too many base models. Selective ensemble learning (Zhou, 2012) is an effective way to address that issue on the computation cost, where a subset of base learners is selected for the ensemble model.

As shown in Fig. 11.10, DDEA-SE employs the common process of EAs for the optimization search, except the evaluation step. All the candidate solutions are evaluated using the surrogate ensemble, which is combined by the selected models from a model pool. Considering the computation cost of building the model pool, we train T RBFN models based on T subsets of $\mathcal{D}$, which are obtained using bootstrap

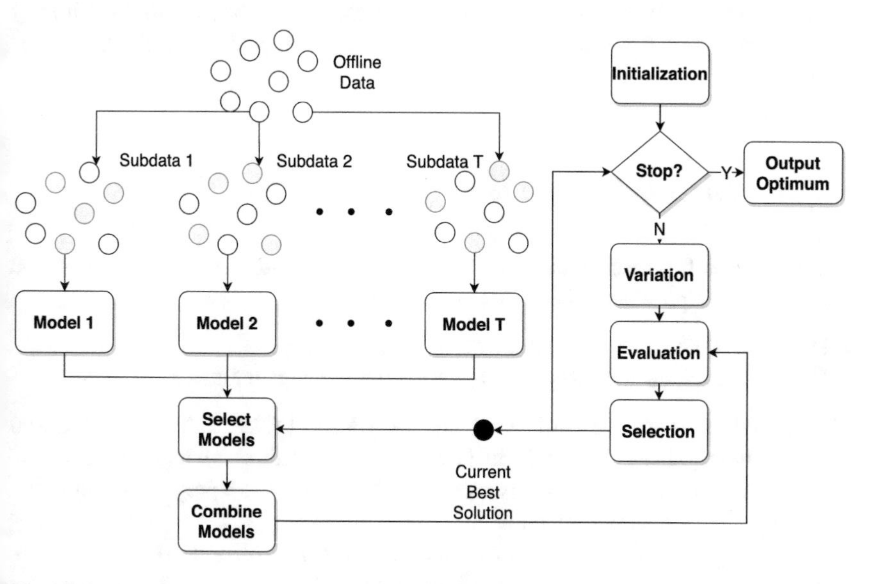

Fig. 11.10 A diagram of selective ensemble in DDEA-SE

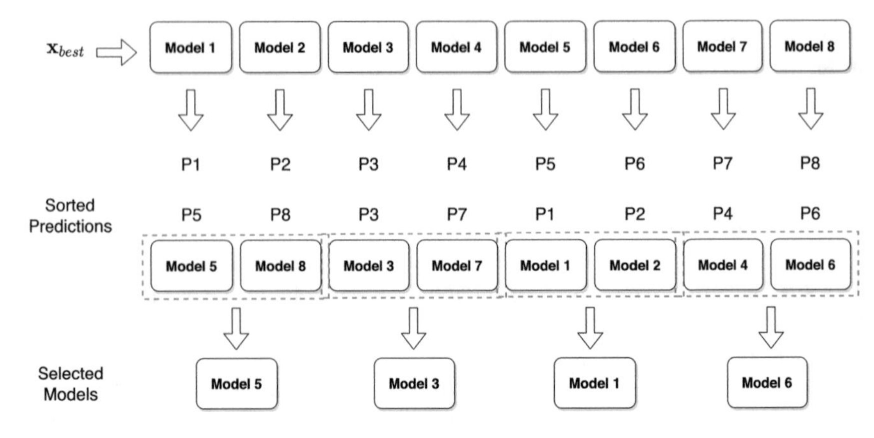

Fig. 11.11 An example of model selection in DDEA-SE

sampling. To be specific, each data point has a 50% probability to be selected in a dataset.

After the model pool is built, the evolutionary optimization process begins. In the each generation, DDEA-SE selects Q RBFN models to combine the ensemble model as the approximated fitness function, i.e. the average of those Q predicted fitness is the approximated fitness. When the stopping criterion is met, DDEA-SE outputs the predicted optimum.

In each generation, DDEA-SE combines Q diverse RBFN models in the local area around the current best solution $\mathbf{x}_{best}$. Taking Fig. 11.11 as an example, T RBFN models are sorted based on their prediction on $\mathbf{x}_{best}$, then they are divided into Q groups. Only one RBFN model is randomly selected in each group for the model combination.

11.3.3 Comparative Results

To study the performance of DDEA-SE on the offline data-driven airfoil optimization problem, we employ two other DDEA-SE variants:

- DDEA-E: DDEA-SE without the model selection step,
- DDEA-RBFN: an EA using one RBFN model which is built from $\mathcal{D}$.

The special parameters T and Q for DDEA-SE and DDEA-E are 2000 and 100 correspondingly. The common EA in the compared algorithms is a real-coded genetic algorithm with the simulated binary crossover (SBX) ($\eta = 15$), polynomial mutation ($\eta = 15$), and tournament selection.

Before applying DDEA-SE to the offline data-driven airfoil optimization problem, DDEA-SE, DDEA-E, and DDEA-RBF are compared on the Ellipsoid and Rastrigin test problems are shown in Table 11.2. From these results, we can see that DDEA-SE

Table 11.2 Optimal solutions obtained by DDEA-SE, DDEA-E and DDEA-RBFN, where I# means the instance number of offline data

P	*n*	I#	DDEA-SE	DDEA-E	DDEA-RBFN
Ellipsoid	10	1	**1.0± 0.1**	1.7±0.7	3.2±2.0
		2	**0.6± 0.1**	1.2±0.6	2.7±1.6
		3	**1.5± 0.1**	2.0±0.9	5.6±2.7
	30	1	**4.2± 0.6**	5.4±1.1	15.8±5.5
		2	**2.8± 0.2**	5.5±1.6	12.4±4.1
		3	**4.3± 0.4**	7.0±1.7	16.0±4.9
	50	1	**11.6± 2.0**	18.5±3.5	54.2±22.9
		2	**14.3± 2.7**	20.4±2.9	52.1±20.7
		3	**12.1± 2.3**	18.6±4.2	65.4±23.4
	100	1	**317.2± 74.4**	371.0±89.2	2186.0±1665.8
		2	**330.8± 48.8**	489.0±189.5	2593.2±897.5
		3	**294.9± 36.3**	364.0±66.5	1245.5±776.5
Rastrigin	10	1	**34.0± 4.6**	76.6±11.7	80.1±21.5
		2	**52.4± 4.6**	93.3±16.7	76.7±33.6
		3	**57.1± 1.8**	109.2±14.0	90.8±26.2
	30	1	**116.8± 7.2**	208.3±29.2	286.8±39.2
		2	**90.5± 4.5**	122.4±22.3	191.5±37.8
		3	**100.8± 5.0**	134.5±22.4	238.7±44.2
	50	1	**189.5± 16.4**	233.3±41.0	408.4±74.2
		2	**158.6± 16.0**	233.7±32.8	421.4±41.0
		3	**180.0± 18.1**	263.9±40.8	441.0±48.2
	100	1	**833.8± 70.2**	891.8±103.3	1053.3±57.3
		2	**848.3± 82.7**	949.4±75.2	1068.7±96.8
		3	**762.2± 99.7**	860.6±94.9	1003.1±74.0
Average rank			1.0	2.1	2.9
Adjusted *p*-value			NA	**0.0002**	**0.0000**

The results are shown in the form of mean ± standard deviation. The results are analyzed by the Friedman test with the Bergmann-Hommel post-hoc test (DDEA-SE is the control method and the significance level is 0.05). The best fitness values among all the compared algorithms for each problem are highlighted

performs the best, followed by DDEA-E, and DDEA-RBF performs the worst. In other words, surrogate ensemble improves the performance of the offline data-driven EAs and selective surrogate ensemble can further enhance the performance.

Then, those three compared algorithms are applied to the RAE2822 airfoil test case. After 100 generations, the average obtained fitness values are shown in Table 11.3. DDEA-SE is the best algorithm, and DDEA-RBFN is the worst, which is same as the results in Table 11.2.

Table 11.3 Exact fitness values obtained by DDEA-SE, DDEA-E, and DDEA-RBFN on the RAE2822 airfoil test case

DDEA-SE	**0.8470 ± 0.0079**
DDEA-E	0.9473±0.0358
DDEA-RBFN	3.4194±10.4958

The best results are highlighted

From the results in Table 11.3, we can conclude:

- A surrogate ensemble is more robust than a single model, as both DDEA-E and DDEA-SE are better than DDEA-RBFN.
- A selective surrogate ensemble can describe more local features than a surrogate ensemble. That is the reason why DDEA-SE outperforms DDEA-E.

11.4 Knowledge Transfer in Offline Data-Driven Beneficiation Process Optimization

11.4.1 Introduction

Off-line data-driven optimization is challenging since the surrogate cannot be updated during the optimization and the obtained optimal solutions cannot be verified. In the previous two sections, we introduced two algorithms, NSGA-II_GP and DDEA-SE, one uses a low-order polynomials as the real objective functions for surrogate management, and the other builds a large number of base learners off-line and then adaptively selects a subset of these base learners to ensure that the surrogate is able to best approximate the local fitness landscape. Here, we aim to go a step further to address two main challenges in off-line data-driven multi-objective optimization where very limited historical data are available:

- Design surrogate models that are able to correctly guide the evolutionary search using the limited historical data only. While NSGA-II_GP uses one single coarse surrogate together with an adaptive Gaussian process and DDEA-SE employs an adaptive ensemble, it is of interest to explore the use of an adaptive coarse surrogate together with an adaptive fine surrogate.
- Select the promising solutions to be implemented when the optimization is completed, where the optimized solutions have never been validated using the real objective functions during the search. It should be noted that the "optimal" solutions achieved at the end of the optimization may not be really optimal due to the approximation errors introduced by the surrogates.

In the next, we present a multi-surrogate approach (Yang et al., 2020b) with knowledge being transferred between the surrogates so that the surrogates are able to reliably guide the search. Furthermore, a method for choosing final solutions guided by a set of reference vectors is suggested to improve the reliability of the solutions.

11.4.2 Knowledge Transfer by Multi-surrogate Optimization

One main challenge in surrogate-assisted evolutionary optimization is the curse of dimensionality, since the higher the search dimension is, the more data is needed to build a reasonably good surrogate, and the larger the search space. Several algorithms have been proposed for online data-driven optimization, as presented in Sect. 10.1. One idea is to reduce the dimension of the optimization problem by randomly sampling a subset of the decision variables as used in Sect. 10.4. Here, we introduce a slightly different idea for handling offline data-driven optimization of high-dimensional problems. In each round of optimization, a low-dimensional coarse surrogate is built by randomly subsampling the decision space, which will be treated as a low-fidelity formulation of the original problem. This low-dimensional low-fidelity surrogate serves two purposes. First, evolutionary search is performed to acquire knowledge about the problem to be optimized. Second, the knowledge acquired in the coarse search will be transferred to a fine search process based on a surrogate model that approximates the original problem. Knowledge transfer is achieved blending the coarse surrogate built in a subspace of the original problem with and the fine surrogate that is built in the original decision space. Note that in each round of the search, the subset of decision variables for building the coarse surrogate is randomly sampled anew without replacement, and all optimal solutions found at the end of the fine search are stored in an archive for final solution selection.

The overall framework of the algorithm, called MS-RV (standing for multi-surrogate assisted optimization and reference vector assisted solution selection), is given in Fig. 11.12. In both coarse and fine search, any popular crossover and mutation such as simulated binary crossover and mutation can be used for generating offspring population, and the elitist non-dominated sorting can be used in the coarse search, i.e., search based on the low-dimensional coarse surrogate. The reproduction process of fine search is slightly different, and the pseudo code for fine search is presented in Algorithm 1.

Algorithm 1 Fine Search Process

1: **Input:** PF_{iter}: Current population of the fine search; P_{TL}: population of last generation of coarse search; $Dind$: Indices of the selected decision variables in coarse search; N: Population size; DB: Database to store all solutions
2: **Output:** $PF_{\text{iter}+1}$: Population of fine search for the next iteration;
3: Set $PL = PF_{\text{iter}}$;
4: **for** i=1:N **do**
5: Set $PL(i,Dind)$=$P_{TL}(i,:)$;
6: **end for**
7: $P = PL$ U PF_{iter};
8: P = randomly shuffled P;
9: Q=crossover + mutation(P);
10: $P = Q$ U PF_{iter};
11: PF_{iter+1} = environmental selection(P);
12: Add $PF_{\text{iter}+1}$ to DB;

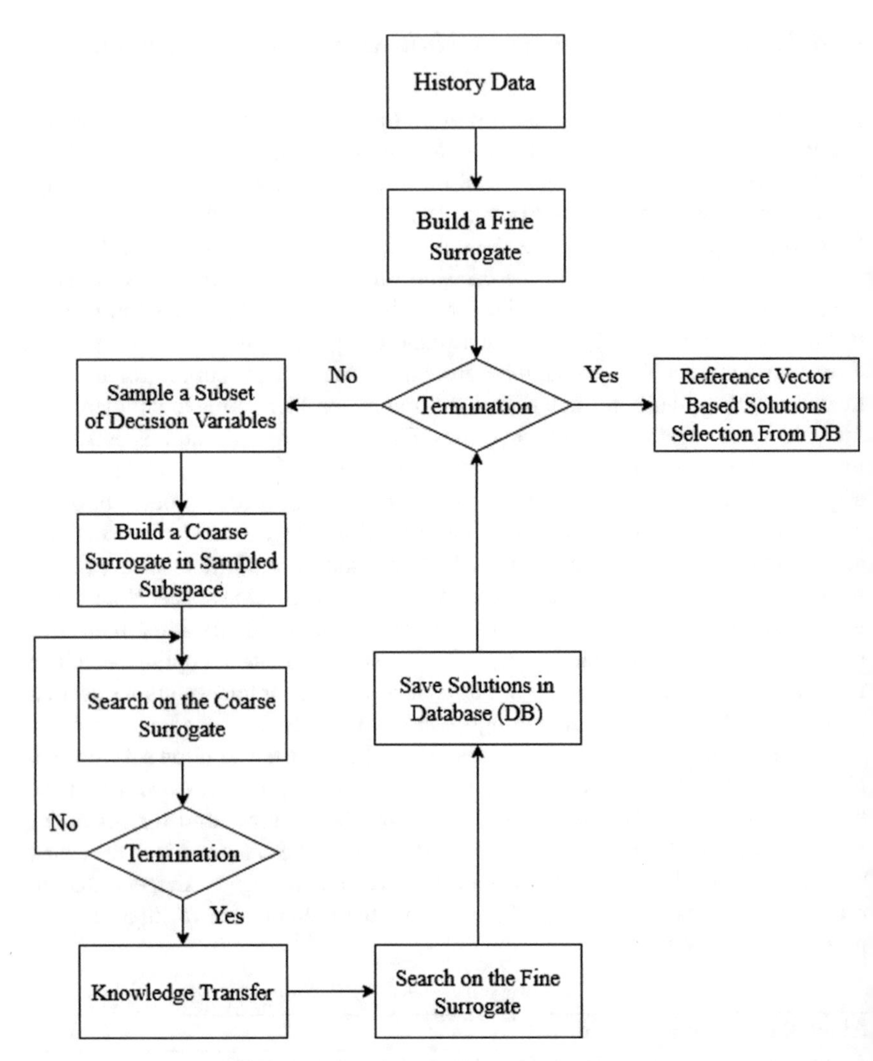

Fig. 11.12 A diagram of the multi-surrogate assisted offline data-driven optimization and reference vector guided final solution selection

11.4.3 Reference Vector Based Final Solution Selection

Once the offline search is complete, it appears reasonable to choose the non-dominated solutions in the final population and select some of them for final implementation. However, none of these optimal solutions have been validated by the real objective functions, meaning that these optimal solutions might not be optimal indeed.

To ensure that the highly promising, we present below an idea of choosing the final solutions from all solutions in the database (DB) based on a set of reference vectors as in RVEA described in Sect. 8.2.1. Without particular user preferences, the reference vectors are evenly distributed and the number of reference vectors is a user specified parameter. It is also possible to generate the reference vectors based on the user preferences. Like in RVEA, all objective values are normalized between [0, 1] before they are clustered.

Once the solutions in DB are clustered, we then check if each cluster contains a sufficient number of potentially good solutions. In case there are too few solutions in a cluster, i.e., smaller than the predefined threshold, solutions in the neighboring reference vectors will be added to the cluster. If there are too many solutions in a cluster, we can remove some solutions from the cluster and do not use them for averaging. Typically, for a minimization problem, solutions that are far from the origin in the objective space will be removed. Finally, all solutions in each group are averaged in the decision space, which is the final solution for that group.

11.4.4 Optimization of Beneficiation Process

To verify the benefits of using multi-surrogate based search and reference vector guided final solution selection, empirical studies on a few suites of benchmark problems are reported in Yang et al. (2020b). In the implementation, NSGA-II is used as the basic search algorithm. The distribution indexes of both crossover and mutation in NSGA-II are both set to 20. The crossover and mutation probability are $pc = 1.0$ and $pm = 1/n$, respectively, where n is the number of decision variables of the original optimization problems. Latin hypercube sampling method is used to generate $10n$ historical data. The population size is set to 50, and a maximum number of 40 generations is run for the fine search, each containing 15 generations of coarse search. In the decision space subsampling, 30% of the original decision variables are sampled. In the simulations, 100 and 150 final solutions are chosen for bi- and three-objective optimization. The required minimum number of solutions in each group is 20. All results are averaged over 20 runs.

To illustrate the performance of MS-RV, the HV value of the solutions it obtained over the fine search generations, together with a variant that uses a fine surrogate based search only (i.e., removing the coarse search from the loop), are presented in Fig. 11.13, from which a clear advantage of having multi-surrogate search and knowledge transfer is shown. Note that the HV values are calculated based on the fine surrogate and the HV values validated using the real objective functions are also shown. This is possible for benchmark problems, however, the real objective functions are for validation only and they should not be used in the optimization for offline optimization.

Operational indices optimization of the beneficiation process (Fig. 11.14) is a typical off-line data-driven problem since no precise mathematical equations of the objective functions can be given owing to the complex physical and chemical reac-

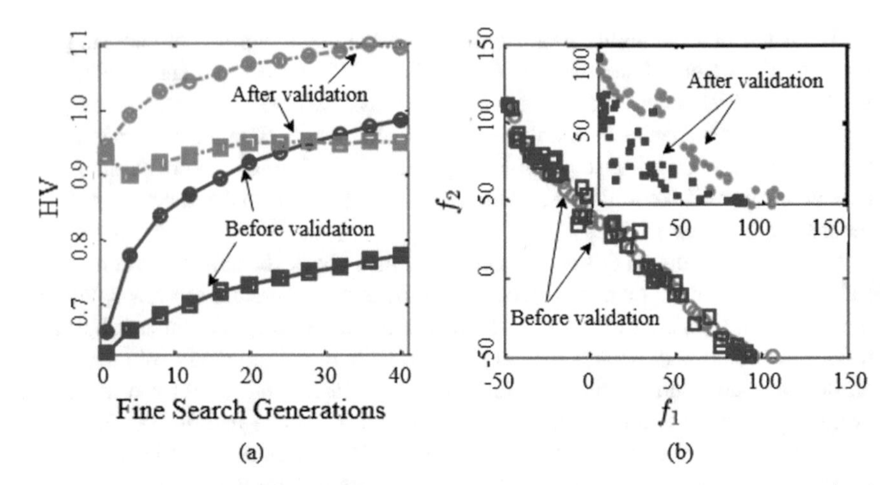

Fig. 11.13 Comparative results on 30D DTLZ1. a. HV values over generations of MS-RV (circles connected by solid line), and the variant without coarse search (squares connected by solid line). The HV values of the solutions obtained by MS-RV and its variant validated by the real objective functions are denoted by circles and squares connected by dashed lines, respectively. b. The obtained non-dominated solutions before and after validation using the real objective functions

tions in the process. In addition, only a small amount of historical data can be collected. The main objectives the optimization are to optimize the concentrate grade (G), concentrate yield (Y), and reduce the energy consumption (E). The following 15 operational conditions are considered as the decison variables, including particle sizes of the raw ore entered in LMPL and HMPL (pl, ph), the grades of the raw ore entered in LMPL and HMPL (gl, gh), the capacity and run time of the shaft furnace roasting (sc, st), the grade of waste ore (gw) and feed ore of grindings in LMPL and HMPL (gfl, gfh), the capacity of grindings in LMPL and HMPL (gcl, gch) and runtime of grindings in LMPL and HMPL (gtl, gth), and the grade of tailings from LMPL and HMPL (tl, th). The objectives are defined in the following form:

$$\begin{aligned} &\min \{-G, -Y, E\} \\ &s.t.\ \ G = \Phi_1(\mathbf{x}), \\ &\qquad Y = \Phi_2(\mathbf{x}), \\ &\qquad E = sc + 0.3st + gcl + gch + gtl + gth \end{aligned} \tag{11.7}$$

where $\Phi_1(\cdot)$ and $\Phi_2(\cdot)$ is the correlation between the objectives and the decision variables in operational indices optimization, which is however, theoretically unknown.

Unfortunately, the obtained solutions cannot be validated using real objective functions. Here, MS-RV is again compared with its variant without the support of the coarse search, and additionally with a multiform optimization algorithm (Yang et al., 2019), which is the state-of-art method for solving the operational indices optimization problem. Altogether 150 pairs of historical data are collected. In the experiments, 30 independent runs are performed for each algorithm under comparison, and 40 generations of fine search are run.

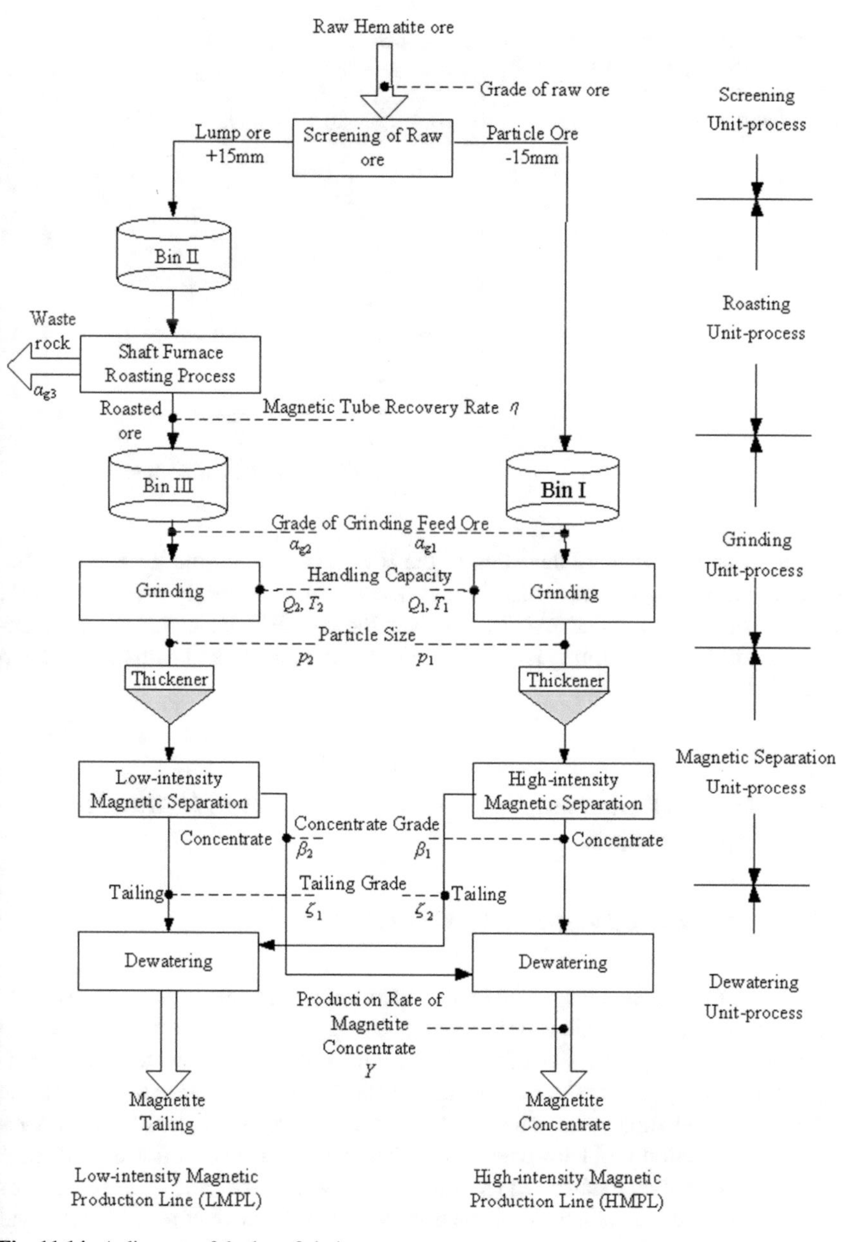

Fig. 11.14 A diagram of the beneficiation process

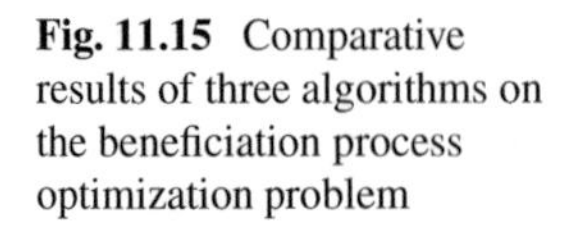

Fig. 11.15 Comparative results of three algorithms on the beneficiation process optimization problem

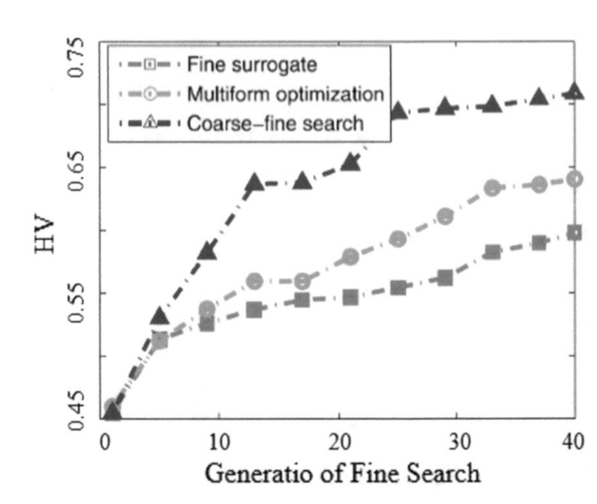

The mean and standard deviation of the HV values of the solution sets over the generations obtained by three algorithms under comparison are shown in Fig. 11.15. These results indicate that MS-RV achieves the best HV values among the compared algorithms, confirming its efficient search ability compared with the other two algorithms.

11.5 Transfer Learning for Offline Data-Driven Dynamic Optimization

11.5.1 Dynamic Data-Driven Optimization

Many complex systems may operate in dynamic environments where their parameters in the objective functions or constraints change over time. For example for the operational index optimization of the beneficiation process we discussed in the previous section, the type of raw ore or capacity of equipment may change over time since the demand and supply may change requirements (Ding et al., 2012). Such data-driven optimization problems whose objective and / or constraint functions change over time are offline data-driven dynamic optimization problems. A large body of research on dynamic optimization has been reported (Yazdani et al., 2021a, b), and most of the research therein assume that the objective and constraint functions are analytically known. Most recently, a data-driven evolutionary algorithm for dealing with online data-driven optimization problems in dynamic environments is reported (Luo et al., 2018c), where surrogates are re-built when new data is available. The surrogate assisted evolutionary algorithm is equipped with a memory scheme reusing previous optimal solutions so that a fast tracking of the dynamic moving optimum can be obtained.

The mechanisms for reusing knowledge developed in dynamic evolutionary optimization algorithms mainly reply on evolutionary techniques, such as initialization based on predicted optimums, use of memory and multi-population. However, in data-driven optimization, machine learning based surrogates are always needed, and it is easily conceivable that knowledge transfer can be achieved in training the surrogates. In this section, we introduce a new ensemble learning technique developed in incremental learning for training surrogates. In this approach, each base learner learns the objective functions in a selected number of previous environments. Then, a multi-tasking evolutionary, known as multifactorial evolutionary algorithm (MFEA) (Gupta et al., 2016), also referring to Sect. 10.3.1, is adopted that optimizes the multiple selected problems the same time. This way, the knowledge about the optimums of the previous problems can be transferred to the present problem, thereby enhancing the performance of tracking the dynamic optimums (Yang et al., 2020a). Because the obtained optimal solutions cannot be verified before implementation in offline data-driven optimization, a support vector domain description (SVDD) (Tax & Duin, 1999) is made use of to select the solution for implementation from the obtained solutions.

11.5.2 Data Stream Ensemble for Incremental Learning

Incremental learning is developed to capture the changing nature of the data distribution in machine learning. Among many methods, data stream ensemble (DSE) learning (Gomes et al., 2017) is a popular and effective approach to learning in non-stationary environments. DSE maintains a pool of weak base learners and changes the structure of the ensemble (Gomes et al., 2017). Specifically, we adopt the accuracy updated ensemble algorithm (AUE2) (Brzezinski & Stefanowski, 2013) for building a surrogate model in dynamic environments. The structure of AUE2 is illustrated in Fig. 11.16. AUE2 is composed of K base learners $S_k, k = 1, 2, ..., K$ together with a new base learner for the t-th environment S_t. In response to a new environment, AUE2 creates a new base learner S_t and updates $S_k, k = 1, 2, ..., K$ on the newly collected data DB_t. The structure of AUE2 also changes either adding new base learners or removing some base learners trained in the past when the size is too big. Finally, the weights of of the base learners are also updated to give a larger weight to the members that are likely to be the most accurate in the current environment based on their accuracy on the current data data chunk DB_t.

Given a training data B_t in the new environment, all K base learners $S_k, k = 1, 2, ..., K$ are tested on the current data batch B_t. Thus, the root mean square error (RMSE) of the base learners are calculated as follows:

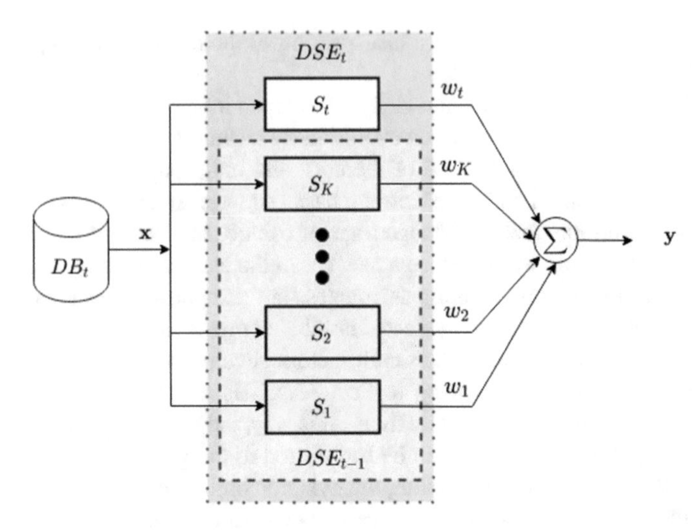

Fig. 11.16 An illustrative example of a data stream ensemble (DSE)

$$E_k = \sqrt{\frac{1}{|DB_t|} \sum_{(\mathbf{x}_j, y_j^d) \in DB_t} \left(y_j - y_j^d\right)^2}, \ k = 1, 2, ..., K \tag{11.8}$$

where $(\mathbf{x}_j, y_j^d)$ is the j-th data sample in DB_t, y_j is the prediction of S_k on $\mathbf{x}_j$. $|DB_t|$ denotes the number of data pairs in DB_t.

The new base learner S_t is trained on DB_t, and its RMSE, denoted as E_t, is calculated using leave-one-out cross-validation. After the errors of all base learners are calculated, the weights of $S_k, k = 1, 2, ..., K$ are updated by:

$$w_k = \frac{1}{E_t + E_k}, \ k = 1, \ldots, K \tag{11.9}$$

$$w_t = \frac{1}{E_t}. \tag{11.10}$$

If the number of the existing base learners K is less than the predefined maximum number K_{max}, S_t is directly added to the ensemble. Otherwise, K base learners with the largest RMSE will be replaced by S_t. K_{max} is recommended to be set to 10.

In addition to weight adaptation, AUE2 also updates all previous base learner using a combination of its own training data and part of DB_t. For example, the k-th base learner S_k is updated using a combination of DB_k and DB_t. Meanwhile, the training data of S_k is replaced with the union of $DB_k \bigcup DB_t$.

The final output of DSE_t is given by:

$$y_t = \sum_{k=1}^{K} w_k S_k + w_t S_t \tag{11.11}$$

11.5.3 Ensemble Based Transfer Optimization

The offline data-driven optimization algorithm presented here is called DSE-assisted MFEA with SVDD enhanced solution selection, DSE_MFS for short. In the multi-tasking optimization, each base learner is seen as a task, and therefore there are $K + 1$ tasks in total, which will be optimized using the MFEA.

The surrogate for assisting the optimization of the t-th environment is DSE_t, consisting of S_k, $k = 1, 2, ..., K$ and S_t. Note that DSE_1 is equal to S_1 when $t = 1$. Different to the original MFEA which which all tasks are considered to be equally important, DSE_MFS assigns a larger number of individuals to the current optimization problem than to the previous ones. For a given population size $|P|$, $\left\lceil \frac{|P|}{2} \right\rceil$ individuals are assigned to the current task, and $\left\lceil \frac{|P|}{2K} \right\rceil$ to individuals related to each previous task, where K is the number of base learners (previous tasks). The population is randomly initialized and each of its individual is assigned to a task by means of a skill factor τ. An individual p_i is assigned to the k-th base learner if its skill factor $\tau_i = k$, $k = 1, \ldots, K$, otherwise to DSE surrogate if $\tau_i = K + 1$.

In MFEA, knowledge transfer between the tasks is realized with the help of crossover and mutation. For a pair of parents p_i, p_j, whose skill factors are τ_i, τ_j, two offspring are created with crossover and mutation if $p_i$and p_j are assigned to the same task, and the offspring will be assigned to the same task. If $p_i$and p_j are from different tasks, then a randomly number smaller than a predefined random mating probability (rmp) will be generated. DSE_MFS adaptively calculates the probability for each task to keep the number of individuals assigned to it constant. For example, if the number of individuals for tasks τ_i and τ_i are N_i and N_j, respectively, then the probability of assigning an offspring to $\tau_i(\tau_{o_i})$ is $\frac{N_i}{N_i+N_j}$ and that to $\tau_j(\tau_{o_j})$ is $\frac{N_i}{N_i+N_j}$. At a probability of rmp, one of the parents is mutated to create an offspring, to which the same task is as its parent is assigned.

By simultaneously evolving the tasks in the previous environments (represented by the base learners) to the current task, knowledge between these tasks can be transferred and the search efficiency on the current task can be enhanced.

11.5.4 Support Vector Domain Description for Final Solution Selection

As discussed in the previous sections, the optimal solutions obtained at the end offline data-driven optimization may not be optimal and it remains challenging to select promising solutions for implementation. Different from the reference vector based solution selection, we adopt here the support vector domain description (SVDD) technique to enhance the reliability of solutions selected for final implementation, since it is found to be able to describe the underlying pattern of the given training data and identify whether a test data is an outlier.

Given N pairs of data $x = (\mathbf{x_1}, \ldots, \mathbf{x_N})$, SVDD projects them to a feature space with the help of a kernel function, in which a hypersphere containing the maximum number of the samples with a minimum hypervolume is found. However, a few very remote data sets will be left outside the hypersphere. The hypersphere is defined by the center μ and a radius R by minimizing:

$$F(R, \mu, \xi_i) = R^2 + \eta \sum_i \xi_i \tag{11.12}$$

$$s.t. \ \|\mathbf{x_i} - \mu\| \leq R^2 + \xi_i, \tag{11.13}$$

$$\xi_i \geq 0 \tag{11.14}$$

where η is a constant between [0, 1], balancing a trade-off between the volume of the hypersphere and the number of data pairs that are outside the hypersphere.

Th above optimization problem can be solved by constructing the Lagrangian:

$$\begin{aligned} L &= \sum_i \alpha_i \left(\mathbf{x_i} \cdot \mathbf{x_i}\right) - \sum_{i,j} \alpha_i \alpha_j \left(\mathbf{x_i} \cdot \mathbf{x_j}\right) \\ s.t.\ &0 \leq \alpha_i \leq \eta, \ \textstyle\sum_i \alpha_i = 1, \ \mu = \sum_i \alpha_i \mathbf{x}_i. \end{aligned} \tag{11.15}$$

For a given new sample $\mathbf{x}'$, SVDD can then be used to check if it is part of the training data. To this end, the similarity between $\mathbf{x}'$ to the training data is assessed according to its distance to the center of the hypersphere using the following equation:

$$d_{\mathbf{x}'} = \left(\mathbf{x}' \cdot \mathbf{x}'\right) - 2 \sum_i \alpha_i \left(\mathbf{x}' \cdot \mathbf{x_i}\right) + \sum_{i,j} \alpha_i \alpha_j \left(\mathbf{x_i} \cdot \mathbf{x_j}\right), \tag{11.16}$$

Finally, we have

$$s_{\mathbf{x}'} = \operatorname{sgn}(d_{\mathbf{x}'} - R) \tag{11.17}$$

Then the new sample $\mathbf{x}'$ is accepted by SVDD if $s_{\mathbf{x}'} = -1$; otherwise the sample is considered as an outlier. Thus, an SVDD can be trained using all offline data, and then all obtained optimal solutions are verified using the trained SVDD to exclude all solutions that are considered to be an outlier.

The overall framework of DSE_MFS is provided in Fig. 11.17.

11.5.5 Empirical Results

To demonstrate the performance of DSE_MFS, we compare it with three offline data-driven optimization algorithms, two for stationary data-driven optimization, CALSAPSO (Wang et al., 2017a) and DDEA-SE (Wang et al., 2018e), one for offline data-driven dynamic optimization, SAEF (Luo et al., 2018c), and GPMEM (Liu et al., 2014a), which is for online data-driven optimization. The comparative

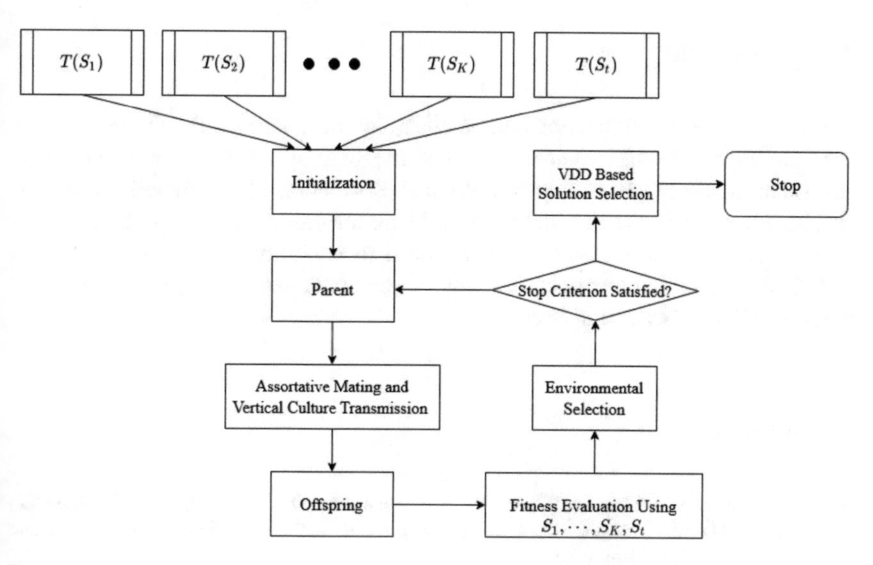

Fig. 11.17 A diagram of the overall DSE_MFS algorithm

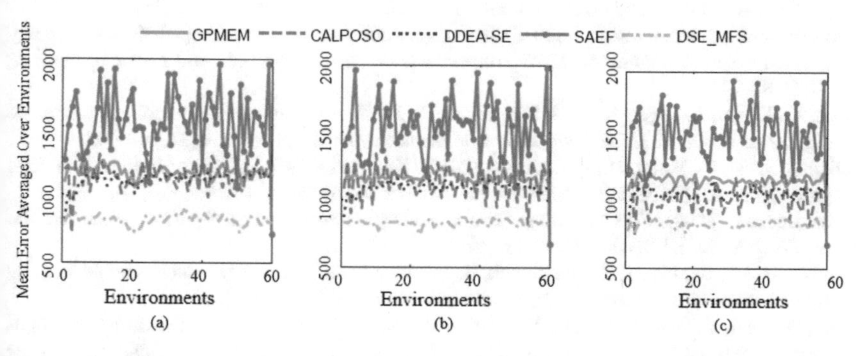

Fig. 11.18 Comparative results obtained by five data-driven algorithm in terms of the mean tracking within averaged over 60 environments. From left to right: The available data is $5n$, $10n$ and $15n$

study is conducted on the 15D $F6$ (Li et al., 2008), which is a composition of Sphere, Rastrigin, Weierstrass, Griewank and Ackley, and the amount of available data is set to $5n$, $10n$ and $15n$, respectively. In the experiments, the total number of environments is set to 60 and each environment lasts for 20 generations. The probability for mating rmp is set to 0.3 and the percentage of the selected best solutions for training the SVDD is set to $sp = 0.25$.

The results are given in Fig. 11.18. From these results, we can conclude that DSE_MFS performs the best among the five compared algorithms in tracking the moving optimum in terms of mean tracking error averaged over the environments.

11.6 Summary

Offline data driven optimization is of practical significance, but difficult to ensure that the found optimal solutions are indeed optimal, or at least better than (or dominate in case of multi-objective optimization) the best solutions. Nevertheless, the results that have been achieved on benchmark problems are promising. Indeed, the selective ensemble method in Sect. 11.3 has also been successfully applied to evolutionary search of deep neural architectures, which is computationally very intensive. More detailed will be provided in Sect. 12.3.

References

Andres, E., Gonzalez, D., Martin, M., Iuliano, E., Cinquegrana, D., Carrier, G., et al. (2017). GARTEUR AD/AG-52: Surrogate-based global optimization methods in preliminary aerodynamic design. In *EUROGEN* (pp. 1–6).

Ashill, P., Wood, R., and Weeks, D. (1987). An improved, semi-inverse version of the viscous garabedian and korn method (VGK). *RAE TR*, 87002.

Brzezinski, D., & Stefanowski, J. (2013). Reacting to different types of concept drift: The accuracy updated ensemble algorithm. *IEEE Transactions on Neural Networks and Learning Systems, 25*(1), 81–94.

Deb, K., Pratap, A., Agarwal, S., & Meyarivan, T. (2002). A fast and elitist multiobjective genetic algorithm: NSGA-II. *IEEE Transactions on Evolutionary Computation, 6*(2), 182–197.

Ding, J., Chai, T., Wang, H., & Chen, X. (2012). Knowledge-based global operation of mineral processing under uncertainty. *IEEE Transactions on Industrial Informatics, 8*(4), 849–859.

Freestone, M. (2004). *VGK method for two-dimensional aerofoil sections part 1: principles and results* (p. 96028). ESDU: Technical report.

Gomes, H. M., Barddal, J. P., Enembreck, F., & Bifet, A. (2017). A survey on ensemble learning for data stream classification. *ACM Computing Surveys (CSUR), 50*(2), 23.

Guo, D., Chai, T., Ding, J., and Jin, Y. (2016). Small data driven evolutionary multi-objective optimization of fused magnesium furnaces. In *IEEE Symposium Series on Computational Intelligence* (pp 1–8). IEEE, Athens, Greece

Gupta, A., Ong, Y.-S., & Feng, L. (2016). Multifactorial evolution: toward evolutionary multitasking. *IEEE Transactions on Evolutionary Computation, 20*(3), 343–357.

Hollander, M., & Wolfe, D. (1999). *Nonparametric statistical methods*. Hoboken: Wiley-Interscience.

Jansen, J. O., & Campbell, M. K. (2014). The GEOS study: Designing a geospatially optimised trauma system for scotland. *The Surgeon, 12*(2), 61–63.

Jansen, J. O., Morrison, J. J., Wang, H., Lawrenson, R., Egan, G., He, S., & Campbell, M. K. (2014). Optimizing trauma system design: The GEOS (geospatial evaluation of systems of trauma care) approach. *Journal of Trauma and Acute Care Surgery, 76*(4), 1035–1040.

Li, C., Yang, S., Nguyen, T., Yu, E. L., Yao, X., Jin, Y., Beyer, H., and Suganthan, P. (2008). Benchmark generator for cec 2009 competition on dynamic optimization. Technical report.

Liu, B., Zhang, Q., & Gielen, G. G. (2014a). A Gaussian process surrogate model assisted evolutionary algorithm for medium scale expensive optimization problems. *IEEE Transactions on Evolutionary Computation, 18*(2), 180–192.

Luo, W., Yi, R., Yang, B., & Xu, P. (2018c). Surrogate-assisted evolutionary framework for data-driven dynamic optimization. *IEEE Transactions on Emerging Topics in Computational Intelligence, 3*(2), 137–150.

Tax, D. M., & Duin, R. P. (1999). Support vector domain description. *Pattern Recognition Letters, 20*(11–13), 1191–1199.

Tian, Y., Wang, H., Zhang, X., & Jin, Y. (2017). Effectiveness and efficiency of non-dominated sorting for evolutionary multi-and many-objective optimization. *Complex & Intelligent Systems, 3*(4), 247–263.

Wang, H., & Jin, Y. (2018). A random forest-assisted evolutionary algorithm for data-driven constrained multiobjective combinatorial optimization of trauma systems. *IEEE Transactions on Cybernetics, 50*(2), 536–549.

Wang, H., Jin, Y., & Doherty, J. (2017). Committee-based active learning for surrogate-assisted particle swarm optimization of expensive problems. *IEEE Transactions on Cybernetics, 47*(9), 2664–2677.

Wang, H., Jin, Y., & Janson, J. O. (2016). Data-driven surrogate-assisted multi-objective evolutionary optimization of a trauma system. *IEEE Transactions on Evolutionary Computation, 20*(6), 939–952.

Wang, H., Jin, Y., Sun, C., & Doherty, J. (2018). Offline data-driven evolutionary optimization using selective surrogate ensembles. *IEEE Transactions on Evolutionary Computation, 23*(2), 203–216.

Wang, H., & Yao, X. (2014). Corner sort for Pareto-based many-objective optimization. *IEEE Transactions on Cybernetics, 44*(1), 92–102.

Yang, C., Ding, J., Jin, Y., and Chai, T. (2020a). A data stream ensemble assisted multifactorial evolutionary algorithm for offline data-driven dynamic optimization. In *IEEE Transactions on Cybernetics*. Submitted.

Yang, J. and Chai, T. (2016). Data-driven demand forecasting method for fused magnesium furnaces. In *12th World Congress on Intelligent Control and Automation*. IEEE.

Yang, C., Ding, J., Jin, Y., & Chai, T. (2020). Off-line data-driven multi-objective optimization: Knowledge transfer between surrogates and generation of final solutions. *IEEE Transactions on Evolutionary Computation, 24*(3), 409–423.

Yang, C., Ding, J., Jin, Y., Wang, C., & Chai, T. (2019). Multitasking multiobjective evolutionary operational indices optimization of beneficiation processes. *IEEE Transactions on Automation Science and Engineering, 16*(3), 1046–1057.

Yazdani, D., Cheng, R., Yazdani, D., Branke, J., Jin, Y., & Yao, X. (2021a). A survey of evolutionary continuous dynamic optimization over two decades: Part A. *IEEE Transactions on Evolutionary Computation.*

Yazdani, D., Cheng, R., Yazdani, D., Branke, J., Jin, Y., & Yao, X. (2021b). A survey of evolutionary continuous dynamic optimization over two decades: Part B. *IEEE Transactions on Evolutionary Computation.*

Zhou, Z.-H. (2012). *Ensemble Methods: Foundations and Algorithms*. CRC Press.

Chapter 12
Surrogate-Assisted Evolutionary Neural Architecture Search

12.1 Challenges in Neural Architecture Search

In recent years, deep neural networks have witnessed vast successful real-world applications, in particular in computer vision and natural language processing. Such success can be partly attributed to the rapid development of data collection storage techniques and computation hardware, in particular computationally powerful graphics processing units.

Building a deep neural network usually includes two steps: designing the neural architecture, choosing a learning algorithm and then training the weights and all other parameters of the chosen architecture. Among them, the architecture is key to the performance of the network. However, the architecture design requires strong expertise in both machine learning and the problem domain, which makes design of deep neural networks unfriendly to beginners or practitioners. Therefore, neural architecture search (NAS), as an automated approach to design deep neural networks, has attracted much attention from both academia and industry (Elsken et al., 2019).

In fact, the NAS can be formulated as an optimization problem to find one or multiple neural architectures to achieve the best performance as described below:

$$\min f(A, \mathcal{D}), \tag{12.1}$$

where A is the network architecture, $\mathcal{D}$ is the training data for this learning task using the network, and $f(A, \mathcal{D})$ is a quantitative but black-box metric for the architecture. Although the expression in Eq. (12.1) is very simple, the optimization problem can be large-scale, multi-objective, bi-level, or highly constrained, and computationally expensive, heavily depending on the learning task.

Therefore, the neural architecture search process can be viewed as an optimization process as shown in Fig. 12.1. Firstly, the architecture of a neural network needs to be properly represented, so that a search algorithm can be applied to the defined search space. Candidate architectures are quantitatively assessed based on their per-

Y. Jin et al., *Data-Driven Evolutionary Optimization*,
Studies in Computational Intelligence 975,
https://doi.org/10.1007/978-3-030-74640-7_12

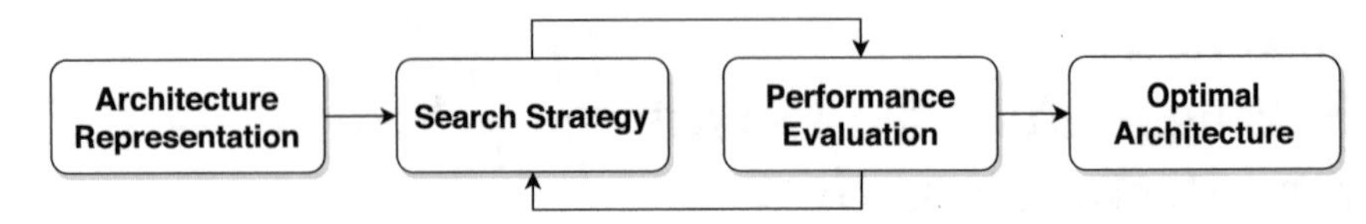

Fig. 12.1 A flowchart of neural architecture search

formance, which guides the search process. After a number of iterations, an optimal architecture is found. Thus, from the optimization point of view, the parameters (coding) that represent the neural architecture are the decision variables, the learning performance, as well as the network complexity, robustness, memory consumption, and interpretability (mostly a subset of them) are the objectives. Neural architecture search is usually an unconstrained optimization problem.

In the following, we will discuss the challenges with respect to architecture representation, search strategies, and performance evaluations.

12.1.1 Architecture Representation

How neural architecture is encoded or represented determines the search space of NAS, thus heavily affects the search efficiency and the resulting performance of NAS. Without considering the limitation of the computation cost, the search space should be as large as possible, so that the NAS algorithm can have the possibility to find new and good architecture. However, a large search space reduces the seach efficiency of NAS algorithms when the computation budget is limited. Therefore, how to design a proper search space to balance the desired performance and the required computational resources. As already discussed in Sect. 1.2.1, there is a trade-off between compactness and flexibility, causality and locality, and robustness and evolvability (Jin & Sendhoff, 2009).

To design a compact and efficient search space, the most straightforward way is to learn the features from existing deep neural networks, where much human experience is required.

For the architecture of convolutional neural networks, the simplest and most natural representation is a chain of layers. As shown in Fig. 12.2a, the chain structure is a neural network with k hidden layers (Baker et al., 2017), where the input of L_i is the output of L_{i-1}. Usually, a fixed k is set to limit the size of search space. For each layer, its type, e.g., pooling or convolution, and the corresponding hyperparameters need to be determined.

Among recent popular hand-crafted convolutional neural networks, many new design elements (like skip connections) have been proposed and shown to perform well on many tasks. To further increase the global representation ability of the chain structure, skip connections are added to Fig. 12.2b, which can produce more complex networks than the purely chain structure.

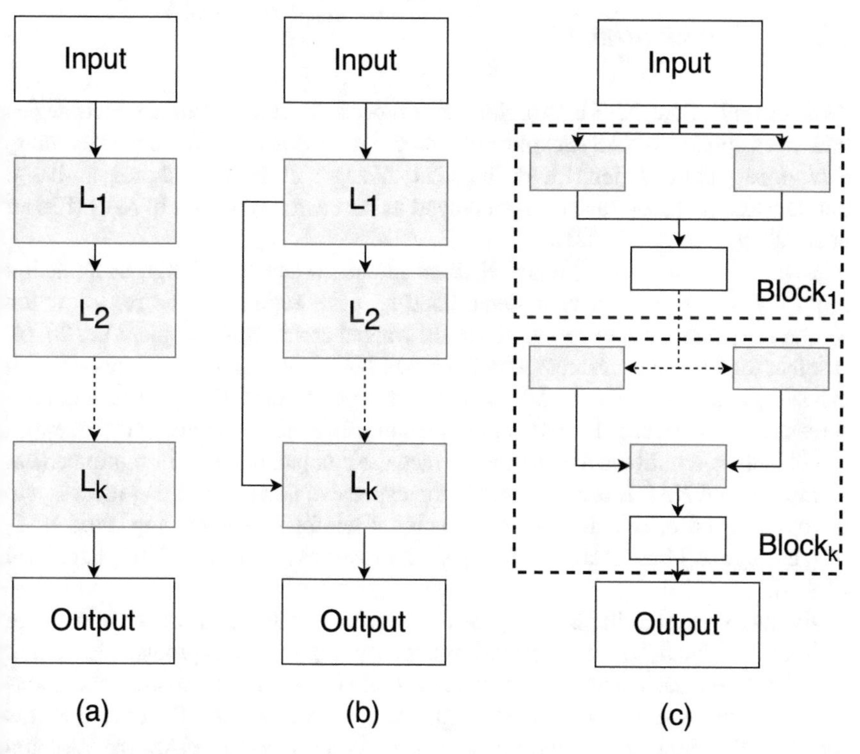

Fig. 12.2 Different neural architecture representations: **a** chain structure, **b** chain structure with skips, and **c** block structure

Although the two chain structure variants have a good global representation ability, their degree of freedom is very large, which makes the search space large. Thus, it will be hard for the search strategy to efficiently find the optimal architecture. To enhance the search ability, different cells, blocks, or segments of networks are extracted from the existing successful networks, then a combination of those blocks can be used to represent new architectures (Zoph et al., 2018), as shown in Fig.12.2c.

Compared with the chain structure, the block structure has following advantages:

- The search space is significantly reduced.
- Block-based architectures are easy to be transferred to other learning tasks.
- Repeating blocks has been proven as a useful design principle, which makes block structure robust.

However, the block structure requires *a priori* knowledge for block design.

Thus, the search space determines the effectiveness and efficiency of the architecture search. So far, most existing work on NAS are based on the block structure, which has been shown effective (Pham et al., 2018). Therefore, one challenge for NAS is that how to break the bottleneck of expertise guided search within an acceptable computational cost.

12.1.2 Search Strategies

As discussed before, NAS is an optimization process. Thus, different search strategies have been applied to NAS to explore the search space. So far, reinforcement learning, evolutionary computation (Liu & Jin, 2021; Zhang et al. 2020), and gradient-based optimization methods have been employed as the search strategies in NAS (Elsken et al., 2019; Zhu et al., 2021).

Reinforcement learning based NAS considers architecture design as an action sequence, where each layer is determined by each action and the reward is the estimated quantitative performance for the trained architecture (Zoph & Le, 2016). Another kind of reinforcement learning based NAS learns an optimal policy which can sample actions to sequentially generate the architecture. Thus, the state in reinforcement learning based NAS contains the sampled actions sampled, and the reward can be obtained until the final action. In fact, the computation cost of reinforcement learning-based NAS is too high due to the expensive network training process and large action space. To reduce the complexity of reinforcement learning based NAS, weight sharing (Pham et al., 2018) and block structures (Zoph et al., 2018) have been applied.

Evolutionary algorithms (EAs), as a powerful global search methodology, are well suited or NAS, which is a black-box expensive optimization problem. EA-based NAS firstly encodes neural architectures and employs genetic variation operations to explore the search space, which is guided by a performance the related fitness function (Angeline et al., 1994). Due to the characteristics of NAS, the variation operators (e.g., crossover and mutation) and the fitness functions are needed to be re-designed. The variation operators in EA-based NAS should be tailored for architecture representation (encoding). Also, the fitness function in EA-based NAS is flexible. Multiple performance indicators have been simultaneously considered in multi-objective NAS (Liu & Jin, 2021; Lu et al. 2019; Zhu & Jin, 2020). Similarly to reinforcement learning based NAS, EA-based NAS suffers from the high computation cost from the expensive fitness evaluations.

One hardness for both reinforcement learning and EA-based NAS is their discrete search space, which makes the search process less efficient. Therefore, continuous relaxation is employed in (Liu et al., 2018b) to convert the architecture search space into a continuous space, so that the gradient-based method can be used in NAS. In addition, the layer hyperparameters have been optimized using the gradient-based methods (Shin et al., 2018).

Although those different search strategies have been shown to have good ability to find good architectures on different tasks, much work has been conducted to quantitatively compare their performance on various tasks.

12.1.3 *Performance Evaluation*

Existing NAS algorithms guide their search process according to various performance indicators. The most straightforward way is to use the accuracy on training or validation datasets. However, such an accuracy can be obtained until the training process is complete, which makes it computationally highly intensive to evaluate a single candidate architecture (Real et al., 2017). The high computational cost of evaluation has become one main challenge to NAS.

To reduce the high computation complexity of architecture evaluations, different methods have been applied to NAS.

- **Low fidelity estimation**: A certain number of estimation methods in a low fidelity level are used to replace an accurate performance assessment method with an approximate one. As the evaluation process is an iterative network training process based on the dataset, low fidelity estimation methods can reduce the training iterations (Zela et al., 2018), data set (size and resolution) (Klein et al., 2017; Chrabaszcz et al., 2017), or lower the complexity of the architecture blocks (Real et al., 2019).
- **Multi-fidelity estimation**: There is a trade-off between the accuracy and computational cost of low fidelity estimation. In some cases of NAS, the fidelity level can be controlled, for example, the network training process can stop at different numbers of iterations as different fidelity levels. During the process of NAS, multiple fidelity levels can be employed according to the optimization need, which might lead to both effective and efficient performance.
- **Curve prediction**: During the process of NAS, expensive evaluations cost on a large number of poor-performing architectures. In fact, it is possible to early terminate their training processes by learning curve extrapolation (Domhan et al., 2015). Also, curve features can be learned to speed up the performance prediction of architectures (Liu et al., 2018a).
- **Weight inheritance and sharing**: An alternative way to save cost is to inherit weights from parent networks (Wei et al., 2016) or hypernetworks (Zhang et al., 2018). Thus, repeated training processes in NAS can be reduced to some extent. Such estimation methods are based on one hypothesis that the trained over-parameterized network can compare architectures quality.
- **Performance predictors (surrogate models)**: The expensive performance evaluations can be predicted by cheap data-driven surrogate models (Sun et al., 2019c). Those models are trained from the data of architecture samples and their performance values.

12.2 Bayesian Optimization for Neural Architecture Search

As discussed in the previous section, NAS can be viewed as a black-box optimization problem as described in Eq. (12.1), whose main challenge is the high computational

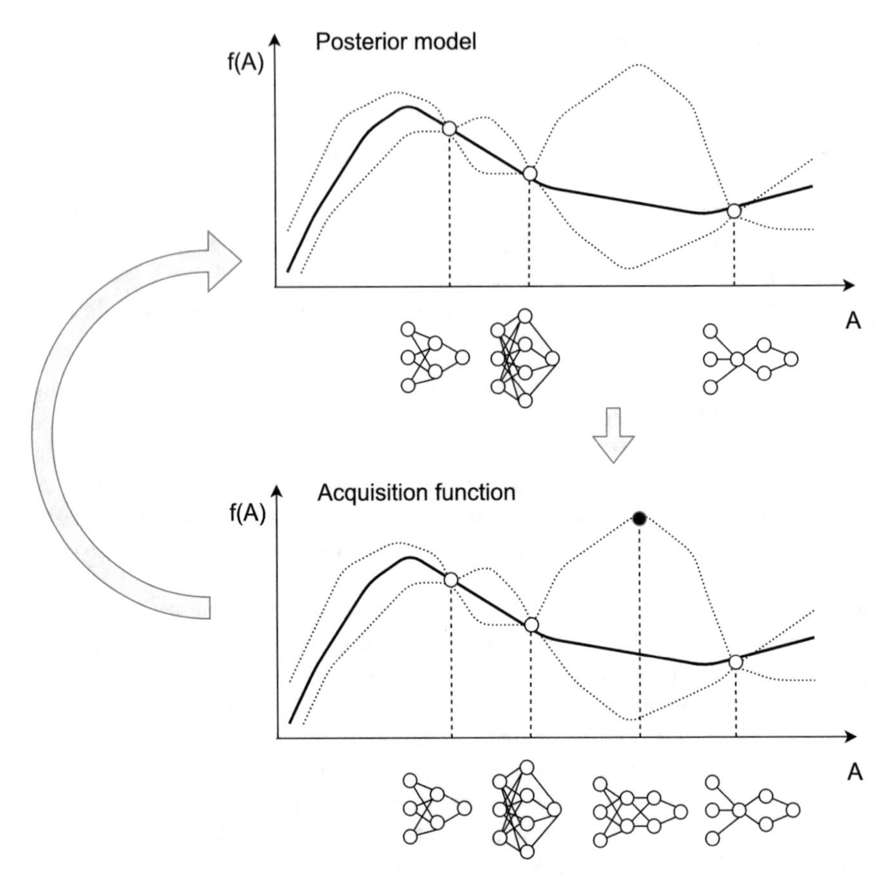

Fig. 12.3 Main steps of BO for NAS

cost. As a query-efficient optimization tool for expensive problems, BO (Shahriari et al., 2015) has been used to speed up NAS (Bergstra et al., 2013), where a probabilistic model is employed as the performance predictor. BO for NAS follows the main steps of BO in Fig. 12.3:

- **Initial architecture sampling**: A number of random architectures are chosen to evaluate their performance by training each network.
- **Build probabilistic model**: One probabilistic model is built to approximate the exact objective function in Eq. (12.1).
- **Sequential sampling**: An acquisition function (Mockus et al., 1978) is adopted to sample new architecture to update the probabilistic model.

More details of BO can be found in Sect. 5.4.

The most popular surrogate used in Bayesian optimization for NAS is the GP model (Williams & Rasmussen, 2006), apart from several ensemble models (Suk et al., 2017) to a replace the mean prediction and variance. As discussed in Sect. 4.2.5,

GP models predicts the mean fitness and the variance based on covariance functions or kernel functions, which measure the similarity of two architectures in the search space. However, the search space is discrete, which leads the difficulty of choosing proper Kernel functions. Therefore, existing BO methods for continuous problems cannot be directly used for NAS. To address the above issue, Bayesian NAS needs to carefully design its encoding method and kernel functions.

12.2.1 Architecture Encoding

In fact, network architectures are connections of layers or blocks, which are directed acyclic graphs (DAGs) with a number of nodes and edges. Therefore, the encoding of architectures can refer to that of DAG.

Adjacency matrix, as a matrix to represent whether two nodes are connected or not, is a classical encoding method for DAGs, which is applicable to network architectures. Thus, the architecture encoding using adjacency matrix can be one-hot or categorical with indices of nodes. Due to the arbitrarily assigned indices, it is possible that one architecture can have different codes.

An alternative encoding method is to find all the paths from the input node to the output node of the architecture, then those paths are encoded to represented the network. As adjacency matrix, the architecture encoding using path can one-hot or categorical with indices of nodes, which results in the only representation for one architecture. However, multiple architectures can be represented into one same code.

In addition to the discrete encoding methods, autoencoder is employed to convert the discrete space into a continuous space (Luo et al., 2018b), then the model is trained in the transferred space.

12.2.2 Kernel Functions

The kernel function measures the similarity of two solutions, which play an important role of building the probabilistic model. Unfortunately, in NAS, the architectures are presented in discrete codes, which makes many popular kernels such as Gaussian kernels fails. To address this issue, two types of kernel functions or distances are employed: genotypic and phenotypic kernels.

- Genotypic kernels are based solely on the architecture representation. Since most Bayesian NAS adopts the DAG-based encoding method, graph kernels are usually used (Kriege et al., 2020), with Weisfeiler-Lehman Kernel (Ru et al., 2020) for instance.
- Phenotypic kernels does not utilize any information of architecture representation. The phenotypic kernel functions measure the distances between architectures by behaviors, which are the outputs for a certain sampled inputs (Stork et al., 2019).

The performance of existing kernels is limited. Therefore, recently, multiple kernel learning (Gönen & Alpaydın, 2011) has been employed to combine multiple kernels to improve the performance.

12.2.3 *Discussions*

Although BO can speed up NAS, there are many open issues to be addressed in the future.

- **Approximation error**: Models in Bayesian NAS cannot avoid approximation error, which might mislead the search. How to decrease the approximation error in the local promising area is still challenging.
- **Kernel function**: The search space of NAS is discrete, whose landscape is different from continuous space. The distances of between architectures are hard to measure, which affects the accuracy of the models in BO. Therefore, the kernel function in the models needs special consideration for NAS.
- **Search space mapping**: To avoid the above hardness of the discrete search space, the original space can be mapped into a continuous space. However, whether the mapping brings more benefits than loss to the optimization remains elusive.

12.3 Random Forest Assisted Neural Architecture Search

As a deep neural network, convolutional neural networks (CNNs) have found many successful applications in image processing. As shown in Fig. 12.4, one CNN consists of an input layer, a number of hidden layers and an output layer. The hidden layers include layers that perform convolutions (convolutional layers) and pooling (pool layers), which are followed by fully connected layers and normalization layers, also referring to Sect. 4.5.1. Convolutional layers convolve the input and pass its result to the next layer, while pooling layers reduce the dimensions of the data. With these hidden layers, individual neurons respond to only a restricted region of the input image, which is inspired by the connectivity pattern between neurons in human and animal visual systems. However, the performance of a CNN highly depends on its architecture. Searching a well-performing CNN architecture is a typical NAS problem.

As mentioned in the previous section, Bayesian NAS needs to deal with the discrete search space using proper kernel functions or mapping methods, which might result in accuracy loss of the performance predictor. To alleviate such issues, surrogate models which are good at modeling discrete variables, such as decision tree-based models can be employed as the performance predictor in NAS. In this section, a random forest (RF) assisted NAS (termed AE-CNN+E2EPP) (Sun et al., 2019a) is tested on the CNN object classification problems.

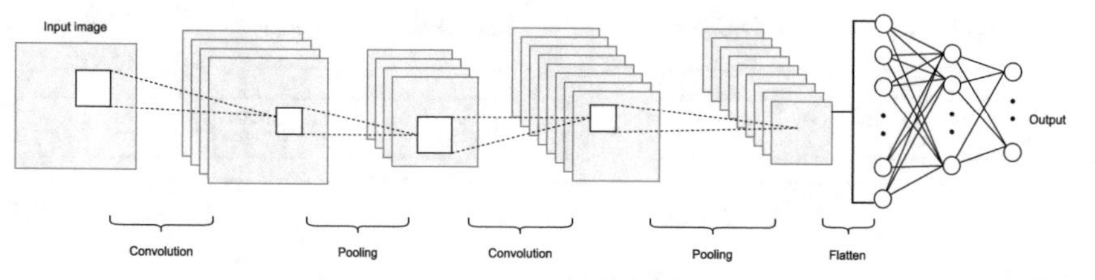

Fig. 12.4 An illustrative example of CNNs

12.3.1 Block-Based Architecture Representation

ResNet (He et al., 2016) and DenseNet (Huang et al., 2017a) are two popular CNNs to address the vanishing gradient issue (Sun et al., 2019b). The basic structures of ResNet and DenseNet are shown in Fig. 12.5. The ResNet block has a chain of convolutional layers with a shortcut connection from the input to the output, while each layer of the DenseNet block is connected to all feedforward layers. Due to the good performance of both structures, AE-CNN+E2EPP employs the ResNet and DenseNet blocks as its basic convolutional blocks.

There are three types of blocks in the block-based encoding of AE-CNN+E2EPP, which are widely used in existing CNNs.

- **DenseNet blocks** are composed of multiple DenseNet units.
- **ResNet blocks** are composed of multiple ResNet units.
- **Pooling blocks** consist of only one pooling layer, either MAX or MEAN.

Thus, DenseNet and ResNet blocks need three variable to present their structures: $type$, out, and $amount$, which are the type of blocks, the number of outputs, and the number of units. Pooling blocks need two variable to present their structures: $type$, and $position$, which are type of pooling and their location in the CNN architecture. Therefore, the length of the code is $3n_b + 2n_p$, where n_b is the maximal number of

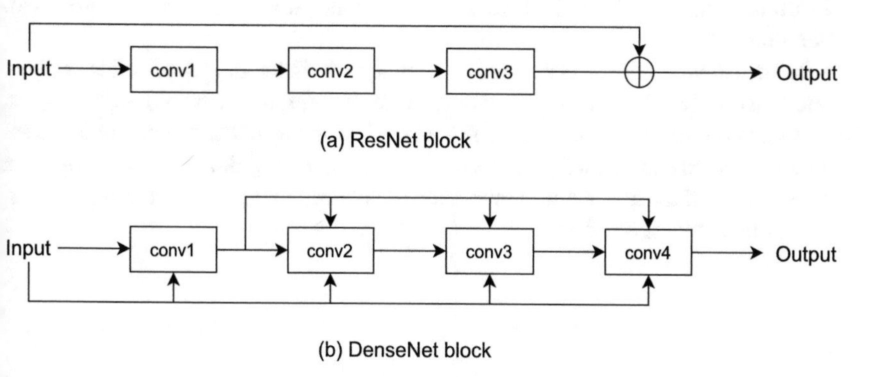

Fig. 12.5 Examples of ResNet and DenseNet blocks

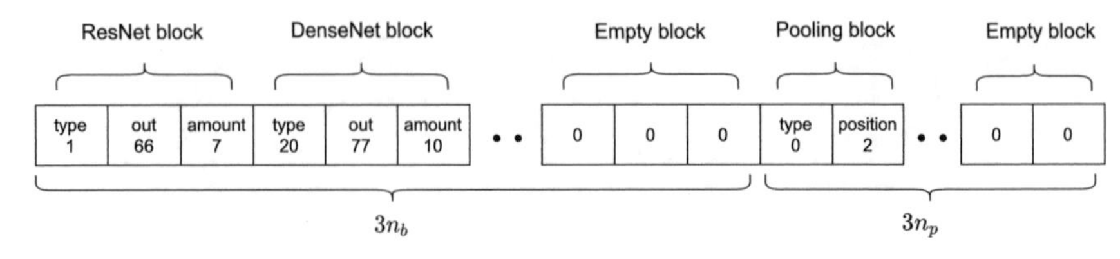

Fig. 12.6 An encoding example in AE-CNN+E2EPP

DenseNet and ResNet blocks and n_p is the maximal number of Pooling blocks. The code is divided into two parts with $3n_b$ bits and $2n_p$ bits, the first $3n_b$ bits are ResNet and DenseNet blocks and the following $2n_p$ bits are the Pooling blocks. To represent variable-length architecture, a number of empty blocks is added to the code.

To encode ResNet and DenseNet block, '1' represents the ResNet block type, and the increasing factor k represents the DenseNet block type. In AE-CNN+E2EPP, three types of DenseNet blocks are employed, i.e., $k = 12, 20, 40$. The maximum numbers of units are different: 10 for the ResNet block and the DenseNet blocks ($k = 12, 20$), and 5 for the DenseNet block ($k = 40$). The range of *out* is (Bottou, 2012) and that of *amount* is (Elsken et al., 2019; Angeline et al., 1994) Since there are two types, i.e., MAX and MEAN pooling, AE-CNN+E2EPP use 0 and 1 to represent those two types of pooling blocks, respectively. Parameter *position* means the index of the ResNet or DenseNet block that is before the pooling block, which makes its range as $[1, n_b]$. In addition, (0,0,0) or (0,0) means an empty block. Figure 12.6 is an example code of an architecture.

12.3.2 Offline Data Generation

In this section, we use AE-CNN+E2EPP to search the architecture for an object classification task. The optimization objective is set as the training accuracy. Therefore, the offline data $\mathcal{D}$ is a pair of encoded neural architecture and its classification error after weight training.

Before running AE-CNN+E2EPP, d samples within the ranges of codes are generated randomly. Then, the d chromosome are decoded into d neural architectures. Each sampled neural architecture is trained by the training dataset for the object classification task to learn its weights, and its corresponding training error is its objective value. When all those d neural architectures have completed their training process, the offline data $\mathcal{D}$ is obtained.

12.3.3 Random Forest Construction

Since a discrete encoding method is adopted, AE-CNN+E2EPP employs the random forest (RF) (Liaw & Wiener, 2002) as the surrogate model to predict the training error of the candidate architectures.

As an ensemble learning method, an RF model contains Q classification and regression trees (CARTs) (Steinberg & Colla, 2009), which divides the decision space into rectangle regions with leaf nodes, and the output of each leaf node is the average output of the samples in each divided region. Growing the pool of CARTs needs to repeat the following steps for Q times.

1. Randomly choose 50% decision variables in $\mathcal{D}$ as the training data for this CART.
2. The CART is initialized with one root node.
3. Collect the training data whose inputs are dropped in the range of the current node.
4. If the splitting criterion, i.e., the mean squared error decrease of this node after splitting is larger than a threshold, is met, the node is split, otherwise this node become a leaf node.
5. Visit the next node by the breadth-first search until there is no non-leaf node or the maximum depth is reached.

Thus, those Q different CARTs are trained for the RF model. The combination of CARTs in the RF model is the average of all the outputs of the trained CARTs.

12.3.4 Search Methodology

AE-CNN+E2EPP is a variant of DDEA-SE (Wang et al., 2018e) (introduced in Sect. 11.3), where the RBFN models are replaced by a RF model for the NAS task. The original DDEA-SE employs a large number of RBFN models for the surrogate ensemble, which is appropriate for continuous optimization problems. Since the RF model is ensemble of decision trees (CARTs), it is naturally a model pool for the discrete search space, which is the reason why the RF model is used in AE-CNN+E2EPP.

With the above offline data $\mathcal{D}$, modified DDEA-SE using RF is applied to search the optimal architecture. The main steps are shown below:

1. **Initialization**: A random population of N architectures is generated and encoded. A RF model with 2000 CARTs is built based on $\mathcal{D}$.
2. **Genetic variation**: Offspring is generated using crossover and mutation.
3. **Model selection**: Selected 100 diverse CARTs based on the prediction of the current best architecture.
4. **Performance Prediction**: Both parent and offspring populations are predicted by the ensemble of the selected 100 CARTs.

Table 12.1 Parameter settings for CNN training

Parameter	Values
Batch size	128
Weight decay	5000
Training epochs	350
Learning rate	0.01 (1, 151–249 epochs), 0.1 (2–150 epochs), 0.001 (251–350 epochs)
Maximal number of units for DenseNet blocks	10
Maximal number of units for ResNet blocks	10
Maximal n_p	4
Maximal n_b	8

5. **Environmental selection**: Select the best N architectures as the next parent population.
6. **Stopping criterion**: If the stopping criterion is met, output the predicted optimum, otherwise go the "Genetic variation" step.

12.3.5 Experimental Results

To test the performance, we choose CIFAR10 and CIFAR100 datasets, which are two popular open datasets [1] for object recognition. The CIFAR10 dataset has 60000 32×32 colour images in 10 classes (i.e. airplanes, cars, birds, cats, deer, dogs, frogs, horses, ships, and trucks), while the CIFAR100 dataset has 60000 32×32 colour images in 100 classes.

In the experiment, only 123 random architectures are sampled as the offline data $\mathcal{D}$ and stochastic gradient descent (Bottou, 2012) is employed to train those architecture for obtaining the classification errors. The detailed setting for the architecture training is shown in Table 12.1.

[1] https://www.cs.toronto.edu/~kriz/cifar.html.

After 100 generations of optimization on a population with 100 individuals, AE-CNN+E2EPP achieves 94.70 and 77.98% accuracy on CIFAR10 and CIFAR100 datasets, which is similar to existing NAS algorithms. However, AE-CNN+E2EPP only takes 8.5 GPU days, which is much shorter than most existing NAS algorithms. It is clear that AE-CNN+E2EPP is an effective and efficient algorithm.

12.4 Summary

This chapter discusses the challenges of NAS with respect to the definition of the search space, selection of search strategies, and the high computational cost for performance evaluations. Focusing on reducing the computational cost of NAs, Bayesian optimization, an efficient optimization technique for expensive black-box optimization, is introduced. Finally, a random forest assisted evolutionary NAS is presented and empirically verified on CIFAR10 and CIFAR100 datasets. This shall offer inspirations for using data-driven evolutionary optimization to improve the search efficiency of evolutionary NAS.

References

Angeline, P. J., Saunders, G. M., & Pollack, J. B. (1994). An evolutionary algorithm that constructs recurrent neural networks. *IEEE Transactions on Neural Networks, 5*(1), 54–65.

Baker, B., Gupta, O., Naik, N., & Raskar, R. (2017). Designing neural network architectures using reinforcement learning. *International Conference on Learning Representations.*

Bergstra, J., Yamins, D., & Cox, D. (2013). Making a science of model search: Hyperparameter optimization in hundreds of dimensions for vision architectures. *International Conference on Machine Learning* (pp. 115–123).

Bottou, L. (2012). Stochastic gradient descent tricks. In *Neural networks: Tricks of the trade* (pp. 421–436). Berlin: Springer.

Chrabaszcz, P., Loshchilov, I., & Hutter, F. (2017). A downsampled variant of imagenet as an alternative to the cifar datasets. *arXiv preprint*arXiv:1707.08819.

Domhan, T., Springenberg, J. T., & Hutter, F. (2015). Speeding up automatic hyperparameter optimization of deep neural networks by extrapolation of learning curves. *Twenty-Fourth International Joint Conference on Artificial Intelligence.*

Elsken, T., Metzen, J. H., & Hutter, F. (2019). Neural architecture search: A survey. *Journal of Machine Learning Research, 20,* 1–21.

Gönen, M., & Alpaydın, E. (2011). Multiple kernel learning algorithms. *The Journal of Machine Learning Research, 12,* 2211–2268.

He, K., Zhang, X., Ren, S., & Sun, J. (2016). Deep residual learning for image recognition. *Proceedings of the IEEE Conference on Computer Vision and Pattern Recognition* (pp. 770–778).

Huang, G., Liu, Z., Van Der Maaten, L., & Weinberger, K. Q. (2017a). Densely connected convolutional networks. *Proceedings of the IEEE Conference on Computer Vision and Pattern Recognition* (pp. 4700–4708).

Jin, Y., & Sendhoff, B. (2009). A systems approach to evolutionary multiobjective structural optimization and beyond. *IEEE Computational Intelligence Magazine, 4*(3), 62–76.

Klein, A., Falkner, S., Bartels, S., Hennig, P., & Hutter, F. (2017). Fast Bayesian optimization of machine learning hyperparameters on large datasets. *Artificial Intelligence and Statistics*, 528–536.

Kriege, N. M., Johansson, F. D., & Morris, C. (2020). A survey on graph kernels. *Applied Network Science, 5*(1), 1–42.

Liaw, A., & Wiener, M. (2002). Classification and regression by random forest. *R News, 2*(3), 18–22.

Liu, J., & Jin, Y. (2021). Multi-objective search of robust neural architectures against multiple types of adversarial attacks. *Neurocomputing*.

Liu, H., Simonyan, K., & Yang, Y. (2018b). Darts: Differentiable architecture search. In *International Conference on Learning Representations*.

Liu, C., Zoph, B., Neumann, M., Shlens, J., Hua, W., Li, L.-J., et al. (2018a). Progressive neural architecture search. *Proceedings of the European conference on computer vision (ECCV)* (pp. 19–34).

Lu, Z., Whalen, I., Dhebar, Y., Deb, K., Goodman, E., Banzhaf, W., & Boddeti, V. N. (2019). Multi-criterion evolutionary design of deep convolutional neural networks. *arXiv preprint*arXiv:1912.01369.

Luo, R., Tian, F., Qin, T., Chen, E., & Liu, T.-Y. (2018b). Neural architecture optimization. *Advances in Neural Information Processing Systems*, 7816–7827.

Mockus, J., Tiesis, V., & Zilinskas, A. (1978). The application of Bayesian methods for seeking the extremum. *Towards Global Optimization, 2*(2), 117–129.

Pham, H., Guan, M., Zoph, B., Le, Q., & Dean, J. (2018). Efficient neural architecture search via parameters sharing. *International Conference on Machine Learning* (pp. 4095–4104).

Real, E., Moore, S., Selle, A., Saxena, S., Suematsu, Y. L., Tan, J., et al. (2017). Large-scale evolution of image classifiers. *International Conference on Machine Learning* (pp. 2902–2911).

Real, E., Aggarwal, A., Huang, Y., & Le, Q. V. (2019). Regularized evolution for image classifier architecture search. *Proceedings of the AAAI Conference on Artificial Intelligence, 33,* 4780–4789.

Ru, B., Wan, X., Dong, X., & Osborne, M. (2020). Neural architecture search using Bayesian optimisation with Weisfeiler-Lehman kernel. *arXiv preprint*arXiv:2006.07556.

Shahriari, B., Swersky, K., Wang, Z., Adams, R. P., & De Freitas, N. (2015). Taking the human out of the loop: A review of bayesian optimization. *Proceedings of the IEEE, 104*(1), 148–175.

Shin, R., Packer, C., & Song, D. (2018). Differentiable neural network architecture search.

Steinberg, D., & Colla, P. (2009). CART: classification and regression trees. *The Top Ten Algorithms in Data Mining, 9,* 179.

Stork, J., Zaefferer, M., & Bartz-Beielstein, T. (2019). Improving neuroevolution efficiency by surrogate model-based optimization with phenotypic distance kernels. In *International Conference on the Applications of Evolutionary Computation (Part of EvoStar)* (pp. 504–519). Berlin: Springer.

Suk, H.-I., Lee, S.-W., Shen, D., Initiative, A. D. N., et al. (2017). Deep ensemble learning of sparse regression models for brain disease diagnosis. *Medical Image Analysis, 37,* 101–113.

Sun, Y., Wang, H., Xue, B., Jin, Y., Yen, G. G., & Zhang, M. (2019a). Surrogate-assisted evolutionary deep learning using an end-to-end random forest-based performance predictor. *IEEE Transactions on Evolutionary Computation, 24*(2), 350–364.

Sun, Y., Xue, B., Zhang, M., & Yen, G. G. (2019b). Completely automated cnn architecture design based on blocks. *IEEE Transactions on Neural Networks and Learning Systems, 31*(4), 1242–1254.

Sun, Y., Xue, B., Zhang, M., & Yen, G. G. (2019c). Evolving deep convolutional neural networks for image classification. *IEEE Transactions on Evolutionary Computation, 24*(2), 394–407.

Wang, H., Jin, Y., Sun, C., & Doherty, J. (2018e). Offline data-driven evolutionary optimization using selective surrogate ensembles. *IEEE Transactions on Evolutionary Computation, 23*(2), 203–216.

Wei, T., Wang, C., Rui, Y., & Chen, C. W. (2016). Network morphism. *International Conference on Machine Learning* (pp.564–572).

Williams, C. K., & Rasmussen, C. E. (2006). *Gaussian processes for machine learning* (Vol. 2). MA: MIT press Cambridge.

Zela, A., Klein, A., Falkner, S., & Hutter, F. (2018). Towards automated deep learning: Efficient joint neural architecture and hyperparameter search. arXiv preprint arXiv:1807.06906.

Zhang, H., Jin, Y., Cheng, R., & Hao, K. (2020). Efficient evolutionary search of attention convolutional networks via sampled training and node inheritance. *IEEE Transactions on Evolutionary Computation*.

Zhang, C., Ren, M., & Urtasun, R. (2018). Graph hypernetworks for neural architecture search. *International Conference on Learning Representations*. <pagination /> <pagination />

Zhu, H., Zhang, H., and Jin, Y. (2021). From federated learning to federated neural architecture search: A survey. *Complex & Intelligent Systems*.

Zhu, H., & Jin, Y. (2020). Multi-objective evolutionary federated learning. *IEEE Transactions on Neural Networks and Learning Systems, 31*(4), 1310–1322.

Zoph, B., & Le, Q. V. (2016). Neural architecture search with reinforcement learning. *arXiv preprint*arXiv:1611.01578.

Zoph, B., Vasudevan, V., Shlens, J., & Le, Q. V. (2018). Learning transferable architectures for scalable image recognition. *Proceedings of the IEEE Conference on Computer Vision and Pattern Recognition* (pp. 8697–8710).

Index

Y. Jin et al., *Data-Driven Evolutionary Optimization*,
Studies in Computational Intelligence 975,
https://doi.org/10.1007/978-3-030-74640-7

D

E

F

G

H

Printed in the United States
by Baker & Taylor Publisher Services